# Disassembly Modeling for Assembly, Maintenance, Reuse, and Recycling

## The St. Lucie Press Series on Resource Management

### Titles in the Series

**Applying Manufacturing Execution Systems**
*by Michael McClellan*

**Back to Basics: Your Guide to Manufacturing Excellence**
*by Steven A. Melnyk and R.T. "Chris" Christensen*

**Basics of Supply Chain Management**
*by Lawrence D. Fredendall and Ed Hill*

**Collaborative Manufacturing: Using Real-Time Information to Support the Supply Chain**
*by Michael McClellan*

**Enterprise Resources Planning and Beyond: Integrating Your Entire Organization**
*by Gary A. Langenwalter*

**ERP: Tools, Techniques, and Applications for Integrating the Supply Chain**
*by Carol A. Ptak with Eli Schragenheim*

**Handbook of Supply Chain Management**
*by Jim Ayers*

**Integral Logistics Management: Planning and Control of Comprehensive Supply Chains, Second Edition**
*by Paul Schönsleben*

**Integrated Learning for ERP Success: A Learning Requirements Planning Approach**
*by Karl M. Kapp, with William F. Latham and Hester N. Ford-Latham*

**Introduction to e-Supply Chain Management: Engaging Technology to Build Market-Winning Business Partnerships**
*by David C. Ross*

**Inventory Classification Innovation: Paving the Way for Electronic Commerce and Vendor Managed Inventory**
*by Russell G. Broeckelmann*

**Lean Manufacturing: Tools, Techniques, and How to Use Them**
*by William M. Feld*

**Lean Performance ERP Project Management: Implementing the Virtual Supply Chain**
*by Brian J. Carroll*

**Macrologistics Management: A Catalyst for Organizational Change**
*by Martin Stein and Frank Voehl*

**Restructuring the Manufacturing Process: Applying the Matrix Method**
*by Gideon Halevi*

**Supply Chain Management: The Basics and Beyond**
*by William C. Copacino*

**The Supply Chain Manager's Problem-Solver: Maximizing the Value of Collaboration and Technology**
*by Charles C. Poirier*

**Supply Chain Networks and Business Process Orientation: Advanced Strategies and Best Practices**
*by Kevin P. McCormack and William C. Johnson*

# Disassembly Modeling for Assembly, Maintenance, Reuse, and Recycling

A. J. D. (Fred) Lambert
Surendra M. Gupta

CRC PRESS
Boca Raton London New York Washington, D.C.

**Library of Congress Cataloging-in-Publication Data**

Lambert, A. J. D.
Disassembly modeling for assembly, maintenance reuse, and recycling/A.J.D. (Fred) Lambert, Surendra M. Gupta.
p. cm. (The St. Lucie Press Series on Resource Management)
Includes bibliographical references and index.
ISBN 1-57444-334-8 (alk. paper)
1. Assembly-line methods. 2. Production planning. I. Gupta, Surendra M. II. Title.

TS178.4.L35 2005
670.42—dc22 2004054451


International Standard Book Number 1-57444-334-8
Library of Congress Card Number 2004054451
Printed in the United States of America 1 2 3 4 5 6 7 8 9 0
Printed on acid-free paper

*Dedicated to Our Families*

---

*Teuntje, Henrike, Florian, and Frans*

*-AJDL*

*Sharda, Monica, Neil, and my parents Mam and Sarla*

*-SMG*

# Preface

Human beings have been using a variety of products for a long time. In the beginning, they made simple products such as utensils, jewelry, and ordinary tools. As time passed, the products became more sophisticated. Initially, the substance used principally consisted of organic materials. For example, early models of looms were made entirely of wood. Their construction and repair depended on the craftsmanship and skills of a few talented people. The discovery and availability of numerous other materials, chiefly metals, enhanced the variety and complexity of products. Craftsmanship was replaced by industrial production. New and intricate functionalities became available. As of late, information technology has led to the present day situation where millions of complex products are available to billions of consumers with an assortment of new products, materials, and techniques introduced daily. Today, with the advancement of technology, sophisticated products are assembled in a variety of ways. Similarly, when a product needs repairs or reaches its end-of-life, it might require efficient disassembly to replace/retrieve parts or materials. Because of the complex nature of assembly and disassembly, the need for a scientific approach to both processes is obvious.

Evidently, designing disassembly processes with minimum damage and cost has always played a crucial role in maintenance and repair. Rationalization of the production challenged the design of products and production processes that has ultimately resulted in concepts such as design for assembly and concurrent engineering. Scientific methods for assembly optimization that have been stimulated with the advent of computers and robots have played a key role for decades. Surprisingly, a breakthrough in this domain came by considering the assembly process as a reverse of disassembly, which is valid under some specific assumptions.

It is clear that scientific consideration of assembly and disassembly processes is needed when operating in harsh (or hazardous) and inaccessible environments, such as in nuclear reactors and in space. In addition, the need for planning under time pressure and the desire for profit maximization further highlight the importance of scientific methods. All these aspects compel the deployment of optimization methods that are based on mathematical modeling of products and processes. Various methods ranging from network theory to mathematical programming have become available in this domain. Numerous applications of network theory that have already been introduced in activities such as task planning and assembly line balancing, can also be adapted and applied in assembly and disassembly studies. Mathematical programming has also been used for selecting the best sequence of (dis)assembly operations from a large number of possible sequences that are usually viable in such problems.

In more recent times, additional challenges in disassembly theory started to emerge due to the growing need for end-of-life processing. This subject became increasingly important during the last decade of the 20th century. The systematic approach of disassembly processes was fueled by both the desire for precious and valuable materials recovery and the challenge for environmentally benign processing of postconsumed products. This resulted in the discovery of new types of problems that had to be solved, such as those related to the optimum disassembly level and the optimum clustering of waste streams.

Although much work has been done in recent years, a systematic and integrated approach of the various aspects of disassembly theory has never resulted in a coherent body of knowledge. Indeed, several review papers on disassembly have appeared recently. However, they merely make the reader aware of their existence rather than attempting to integrate various approaches of disassembly theory.

This book is written with the intention to fill this gap with a presentation of a coherent and comprehensive discussion of disassembly planning theories and methodologies. Firmly based on hundreds of papers that have been published that reflect various approaches to disassembly theory, the authors present an overview of the state-of-the art in this field, frequently adding new materials that enhance the transparency of the methodology that is discussed. Although the subject that is dealt with seems restricted, completeness cannot be pursued to its full extent, for open ends are present both in the chain approach and in the aggregation level. This implies that the theory on a low level of aggregation, which corresponds to a purely physical and mechanical approach, is not addressed. Theory on a high level of aggregation, which embraces scheduling and reverse logistics issues, is also not included to its full extent. The product life cycle is discussed within the framework of industrial ecology, but phases other than the disassembly phase are not fully expounded upon.

From a methodological point of view, there are three main approaches to disassembly planning; exact methods, heuristics, and metaheuristics. From these, the emphasis will be on the exact methods, which are widely applicable in various disassembly planning problems. Heuristics and metaheuristics are discussed without the details of the multiple methods that have been reported thus far.

For the same reason, the modeling theory is confined to various network approaches and related matrix representations. This book does not deal with a detailed discussion of the different models that are based on Petri nets, for these are essentially analogous to the network approach. It does, however, engage the three main applications of disassembly theory, which include assembly optimization, maintenance and repair, and end-of-life processing.

In the last decade, much optimism existed with regard to the possibility of robotically operated disassembly lines for end-of-life products. This initial optimism, however, has since been mitigated to some extent, mainly because uncertainties in supply and quality of those products put to rigorous challenges to robotic operation. Even so, the development of disassembly theory has made significant progress since then, which has been fruitful in all the phases of complex product life cycle including aspects of reverse logistics.

The optimization of disassembly processes remains an important and promising field of research. Integrated design, resulting in design for life cycle, might eventually result in a renaissance of the robotic disassembly line. This is desirable from both the environmental and the working condition points of view as high labor costs are presently impeding in exploiting the full benefits of end-of-life disassembly. The authors express their hope that this book will contribute to the establishment of environmentally sound and economically feasible product design and end-of-life processing.

# About the Authors

**Dr. ir. A.J.D. (Fred) Lambert** is an Assistant Professor of Industrial Ecology in the Department of Technology Management at the University of Technology at Eindhoven, The Netherlands. He received his B.S. in Electrical Engineering from the Technical College at Vlissingen, his M. Sc. in Technical Physics and his Ph.D. in Theoretical Plasma Physics from the University of Technology at Eindhoven, all in The Netherlands. In addition to the University of Technology at Eindhoven, Dr. Lambert has participated in research projects at the Philips Company, the University of Greifswald (Germany), the University of Trieste (Italy), the Technical University at Lausanne (Switzerland), and the FOM Institute for Plasma Physics at Rijnhuizen (The Netherlands). Dr. Lambert has published papers on many different topics, including nuclear fusion, nonequilibrium thermodynamics, MHD power generation, energy systems modeling, process integration, materials flow modeling, and more recently on disassembly sequencing. He has published more than 30 research papers in various scientific journals and has made contributions to numerous books, conference proceedings, and professional papers. Dr. Lambert teaches undergraduate classes on energy efficiency, managerial aspects of reuse and recycling, and supervises group projects on industrial ecology related topics. He coaches graduate students on such topics as energy and waste management in industry, and sustainable energy resources. His recent activities can be viewed at <http://www.tm.tue.nl/vakgr/aw/> (technology studies). Dr. Lambert can be reached via e-mail at <A.J.D.Lambert@tm.tue.nl>.

**Dr. Surendra M. Gupta, P.E.** is a Professor of Mechanical and Industrial Engineering and Director of the Laboratory for Responsible Manufacturing at Northeastern University in Boston, Massachusetts. He received his BE in Electronics Engineering from Birla Institute of Technology and Science, Pilani, India; MBA from Bryant College, Smithfield, Rhode Island; and MSIE and Ph.D. in Industrial Engineering from Purdue University, West Lafayette, Indiana. Dr. Gupta's research interests are in the areas of production/manufacturing systems and operations research. He is mostly interested in environmentally conscious manufacturing, electronics manufacturing, MRP, JIT, and queueing theory. He has authored and coauthored more than 275 technical papers published in prestigious journals, books, and international conference proceedings. He has traveled widely and presented his work at international conferences in Africa, Asia, Australia, Europe, North America, and South America. Dr. Gupta has been elected to the membership of several honor societies and is listed in various Who's Who publications. He is a registered professional engineer in the State of Massachusetts and a member of ASEE, DSI, IIE, INFORMS, and POMS. Dr. Gupta is a recipient of the Outstanding Research Award and the Outstanding Industrial Engineering Professor Award (in recognition of Teaching Excellence) from Northeastern University. His recent activities can be viewed at <http://www.coe.neu.edu/~smgupta/> and he can be reached via e-mail at <gupta@neu.edu>.

# Acknowledgments

We have worked in the field of disassembly for more than a decade. However, to the best of our knowledge, this is the first book written solely on disassembly. As such, it is somewhat bewildering to write the first book in any area, because it is difficult to know how one should harness and organize some important topics without ignoring some others; there are no precedents available that could serve as guides. We, therefore, started pondering with ideas of how to explain to the new generation of students, researchers, and practitioners the general concepts of disassembly. The obvious starting point was to explore the history and reason for disassembly. However, as we started to dig deeper, we realized that the field of disassembly has latently grown quite vast without anyone taking the initiative to organize it. This book is an attempt to start this process.

This book would not have been possible without the encouragement of many people. We would like to thank hundreds of researchers, colleagues, and graduate students whose works we have read and benefited from and whom we have met and interacted with, at conferences around the world. Therefore, the knowledge in this book is a collective effort of many people working in this field. We especially thank the people of the joint European REVLOG (Reverse Logistics) project, and in particular Dr. ir. S.D.P. Flapper, whose fellowship and enthusiasm substantially influenced this book. In addition, we thank our current students who continually help us propel toward future discoveries.

Finally, we are indebted to our families, to whom this book is lovingly dedicated, for constantly giving us their unconditional support and "quiet time," which made the challenging task of writing this book much more pleasurable.

**A. J. D. (Fred) Lambert,** Eindhoven, The Netherlands
**Surendra M. Gupta,** Boston, Massachusetts, U.S.A.

# Table of Contents

## Part 3
## Disassembly Planning ........................................ 339

# *Part I*

## *Disassembly Practice*

# 1 Introduction

This chapter provides an overview of how disassembly theory developed and continues to develop. To that end, a historical perspective of assembly and disassembly is presented in section 1.1. Section 1.2 highlights different levels of aggregation in assembly and disassembly. Section 1.3 introduces various methodologies that are frequently used for modeling and solving disassembly problems and articulates the choices that have been made in developing this book. Section 1.4 introduces the basic terminology used in the book in an attempt to unify the field as different authors use different aliases for comparable concepts. In section 1.5, the two basic approaches for product representation that appear in this book are discussed. The chapter concludes with a brief outline of subsequent chapters of the book (section 1.6).

Every chapter cites an extensive list of references in the literature. By no means do the authors claim these lists to be exhaustive, the lists nevertheless embrace a substantial portion of the knowledge that has been generated by various authors during the past two decades. Apart from that, some material presented in the book is new and has never been published before.

## 1.1 ASSEMBLY AND DISASSEMBLY HISTORY

### 1.1.1 General

The history of disassembly cannot be considered in isolation without also considering the history of assembly, for both obviously represent two characteristics of the same product and the solutions to both problems have been influenced by comparable techniques.

Disassembly is virtually as old as mankind and thus is even older than assembly! The oldest example of disassembly comes from the retrieval (or disassembly) of various parts of animals by human beings for meat. The industrial counterpart of this was the establishment of large-scale slaughterhouses during the later part of the 19th century, which at that time were known as disassembly lines. Obviously, the experience gained from these disassembly lines played an important role in the introduction of automobile assembly lines a few decades later. The assembly lines facilitate the production of complex products in an efficient manner. In assembly, components and modules are united together to obtain a desired product. Craftsmen originally executed this process. However, division of labor and associated cooperation of various specialized craftsmen led to the production of large-scale complex products, such as ships. A further advancement in this evolution was the assembly of products with components that were produced separately and in advance by specialized suppliers. An appealing example of this is the equipment used in ships. The first well-documented assembly line was introduced in 1438 where Venetian galleys were outfitted using the *division of labor* principle (Wild, 1972), a principle that was also described by Adam Smith in 1746 as one of the factors that led to a successful economy at the time.

Although organizational features were of great importance, the achievement of technical innovations was also indispensable in the development of assembly and disassembly processes. One of these innovations was the *interchangeability* of components. Successful application of this concept is sometimes attributed to the French general, Jean-Baptiste De Gribeauval whose work enabled the interchangeability of components for guns in the second half of the 18th century, which was obviously very useful in military practice. This had a major impact on the outcome of the U.S. Civil War! The need for interchangeable components stimulated the development of standardization and specialized machinery. The Englishman, Sir Joseph Whitworth, who first proposed this idea, established the standards for screw threads in the 1800s. His work is regarded as the cornerstone of the development of modern day industrial standards, which currently number around 800,000.

The concept of interchangeability provided the foundation of mass production that enabled the advent of novel products and industries during the latter part of the 19th century. Examples of such products included sewing machines (Singer), agricultural machines (McCormick), bicycles, and typewriters. This trend led to the production of numerous complex products for both households and industries. Disassembly of these products was mainly practiced for maintaining and repairing the products. End-of-life products were either discarded or were recycled as scrap metal.

As mentioned before, disassembly *as a production process* was first implemented in slaughterhouses. *The Encyclopedia Americana* (1965) describes this as follows:

> *The meat-packing business is unusual in that its basic operating principle is not one of assembling but of disassembling: the meat animal is split up into many separate cuts and such various byproducts as hides, hair, bones, and glandular derivatives.*

It is in this type of industry that the first *moving* (dis-) assembly line appeared. Commenting on the slaughterhouses of Cincinnati and Chicago that existed around 1900, the *Encyclopedia Britannica* (1994) remarks that "mechanized disassembly procedures and conveyor procedures were being incorporated into regular operations." It must be stressed that the slaughterhouse line is far removed from the ideal disassembly line, as the interchangeability of components is virtually absent here. This fact frustrated the introduction of further mechanization and robotization in slaughterhouses in later times, although such a development was advocated because of hygienic purposes. However, the use of moving conveyors had an unmistakable impact on the evolution that was taking place in the assembly industry.

The beginning of the 20th century was marked by two major developments; *Taylorism* and *Fordism*, which substantially modified the practice of assembly. Taylorism was based on "The Principles of Scientific Management," written by F.W. Taylor in 1911, which emphasized work measurement and *task allocation*. Essential to this philosophy was the opinion that breaking the tasks into smaller subtasks was beneficial to production efficiency. The unpopular icon of the man-with-the-stopwatch and the movie parody titled "Modern Times" by Charlie Chaplin, were motivated by this idea. It is evident that such practice only works if the tasks are repetitive to a large extent. The basic philosophy behind Taylorism has since been deemphasized because of its dehumanizing effects that are incompatible with the evolved societal norms. The principles of work measurement and task allocation, however, have been proven useful, albeit in slightly different contexts, such as product and process design, and modern developments such as mechanization and robotization.

Fordism (Batchelor, 1994) is synonymous with the moving assembly line, and addresses both technical and organizational issues. It was aimed at boosting the production of relatively complex (about 5000 components) and expensive products, such as automobiles, to a mass production level. The well-known model-T Ford of 1908 was no longer custom-made but was produced in large quantities on moving assembly lines starting in 1913 by, of course, sacrificing variety.

From then on, the development of complex products intensified, including the introduction of electric and electronic products. This was accompanied by the application of new materials, such as thermoset plastics on the basis of phenolic acid, and the massive application of nonferrous metals, such as copper, for electrical applications. Enhanced interchangeability enabled the introduction of mechanization and automation of assembly processes, although human labor remained indispensable in a multitude of tasks.

The rapid development of information technology in the second half of the 20th century had a major impact on various aspects of products and processes:

1. *In the production process*, where computerized numerical control (CNC) machines were introduced in the manufacturing of mechanical components. Robots assisted both in manufacturing and in the execution of assembly tasks.
2. *In the design process*, where computer aided design (CAD) became common, which resulted in *virtual prototypes* that could be tested and modified at the conceptual phase.
3. *In the information exchange process*, where enhanced communication between the different partners in the chain became possible, such as between suppliers, original equipment manufacturers (OEMs), and customers. This boosted the introduction of management techniques such as just-in-time (JIT) production and lean production.
4. *In the products itself*, as new types of products emerged that focused on information processing for both entertainment and professional purposes. Traditional products, such as mechanical typewriters, became obsolete and were replaced by electronic products. Other mechanical and electromechanical products became available that provided with additional functionality, which required an increasing amount of electric and electronic hardware in them.

CNC machines gradually entered the industry in the late 1960s followed in the 1970s by the computer aided design/computer aided manufacturing

(CAD/CAM) concept. Robotic assembly was also introduced to the industry in the course of the same decade.

The availability of new types of high-grade plastics, the so-called engineering plastics, made it possible to replace the traditional steel components with plastic components. This not only reduced the mass of the product, but also enabled new types of fasteners (such as snap fitting) and the integration of functions. Although positive in many ways, the introduction of plastics also had a drawback, as the easily recyclable ferrous metals no longer dominated the products. This had an impact on end-of-life processing, as discarded products could no longer be considered as scrap metal (Isaacs and Gupta, 1997).

Another relevant trend was the tendency toward *miniaturization*. Radio lamps became electron tubes, which in turn were replaced by transistors and later microchips. Surface mounted devices (SMDs) replaced passive components such as capacitors and resistors, if they could not be integrated in the microchips. Bulky chassis made of steel sheets evolved into printed wiring boards (PWBs) that became increasingly sophisticated. Specialized PWBs were introduced, with integrated optical and mechanical components, such as lasers, mirrors, and the coils of electric motors. In modern products such as notebooks, PWBs are distributed through virtually the whole product, and electronic components are even mounted on foils. Cathode ray tubes (CRTs), which were extensively introduced since the middle of the 20th century, provided for bulky products, which contained a considerable amount of glass, copper, and lead. These are now gradually being replaced by liquid crystal display (LCD) and plasma screens, which represent far less mass and volume. Miniaturization also requires considerably reduced power, which in turn reduces the power supply circuitry of the product. A modern phone, for example, is no longer a massive product that consists of many discrete components, but is rather a sandwich structure that is composed of some metal sheets and plastic foils, and one or more PWBs. The principal value of such products is in embedded software, and a substantial share of its mass is in a piece of metal that adds some weight to the device for stability. The ultimate consequence of miniaturization is *dematerialization*, which puts the service that is provided by the product in the foreground.

Evidently, miniaturization is counteracting materials use and, consequently, tends to reduce the waste problem. However, it also stimulates the proliferation of small electronic devices in virtually every domain of life, including ancillary functions in already existing products, such as cars. Recycling of miniaturized products, which are considered heterogeneous, is not easy and often economically and environmentally detrimental. Thus the reduction of the amount of post consumer waste is accompanied by its increased complexity.

Although there has always been some connection between product design and the design of the assembly process, this concept was explicitly emphasized by the Japanese, German, and Danish authors in the early 1980s. Originally, they called it: *Assembly oriented design.* Presently, known as *design for assembly* (DfA), this methodology was proposed and extensively treated by Boothroyd et al. (1982) and Dewhurst (1993).

The concept of *concurrent engineering* (also known as *simultaneous engineering*) emerged around 1987. This philosophy was aimed at shortening the development time of complex products, not only for weapon systems, but also for civilian industries such as automotive and electronics. Concurrent engineering was considered a means for obtaining competitive advantage by reducing costs and quickly introducing novel products to the market. It involved the simultaneous execution of tasks in a conceptual product life cycle (for example, the simultaneous development of the design of a product and its production). This required communication between the product designers and the production engineers.

A further concept that appeared relevant to assembly is *artificial intelligence*, with roots that go back to the 1960s. Although the initial optimism on literally creating artificial intelligence has long since faded, it did result in the developments of heuristic and metaheuristic methodologies, which were embedded in expert systems. *Expert systems* were intended to contain and apply professional knowledge in a systematic way that became popular in the 1980s and 1990s. Their applicability, however, is usually confined to restricted domains. An interesting aspect of these systems is that they include abstract models of the real systems that are intended to support the decision making process of, say, a production system. These *decision support systems* became customary in the late 1970s.

Besides the developments in the technology and information domain, the logistics and management systems have also advanced. *Material requirements planning* (MRP) emerged around 1975, the JIT concept became popular in the 1980s and *lean production* philosophy came into use in the late 1980s. These developments had an impact on the way products were designed. For instance, the order-driven production in which customers could individually select the product configuration became an expectation of consumers. Another consequence was the increased involvement of suppliers in product design, for lean production resulted in the outsourcing of all activities that were not considered core activities. Recent developments in this field are boosted by the adaptation of information exchange via the Internet, such as in e-business.

The "systematic approach" portion of the history of disassembly is relatively short, which spans no more than the past 25 years. As was mentioned previously, *disassembly as a production process* was performed in the slaughterhouses more than a century ago. However, the first scientific paper titled *disassembly line balancing* that was inspired by slaughterhouse lines was published relatively recently (Donnan and Makan, 1983). The balancing of a slaughterhouse line, for example, involved splitting of the workstation for veterinary control into two parallel-operated sections, so as to balance the longer duration of the control operation with the shorter task times of the cutting operations. This publication, of course, predated by far, any publication that is related to the disassembly of the end-of-life complex products.

However, disassembly for the purposes of *maintenance and repair* has been studied for a while. Many papers focused on disassembly for maintenance, particularly if these had to be performed in hostile environments, such as in nuclear power plants, fusion reactors, and space stations. In these cases, the design had to allow for automated or remote disassembly and reassembly. One of the challenges in repairing a product involves replacing a damaged component by disassembling the minimum number of "good components" in order to minimize the repair time and maximize the ease of replacement. This involves designing products specifically to facilitate such operations and disassembly planning for a predetermined product structure.

A major incentive for studying disassembly processes in a systematic way came from the success of *assembly planning*. As has been pointed out previously, assembly has played an important role in industry for a long time. Assembly-planning issues have been studied extensively, particularly in the area of *assembly line balancing*. Some studies in this area date back to the early 1960s (Kilbridge and Wester, 1961; Prenting and Battaglin, 1964). The principal purpose of assembly line balancing is to distribute the various assembly tasks over all the workstations such that the idle times are minimized, and the throughput from the assembly line is maximized.

In assembly process, sequencing is one of its fundamental characteristics. The sequencing of assembly operations has traditionally been performed on an adhoc basis. An example of an early paper on *assembly sequencing*, which is considered part of the balancing problem, is due to Tabucanon and Somnasang (1981). The authors compared two algorithms: a heuristic one and one that generates all possible sequences and selects the optimum sequence. Both were incorporated in decision support systems for determining the optimum sequence of assembly operations. Each method has its characteristic disadvantage: the heuristic one usually returned a suboptimal sequence while the exact one required an exponentially increasing time to obtain the solution as the number of components in the product increased. An exact algorithm for generating all the possible assembly sequences did not exist. One of the obstacles was the appearance of deadlocks while attempting to generate those sequences. It was then speculated that in such cases, it might be beneficial to study the problem in the reverse direction, i.e., study disassembly.

Bourjault made the first such attempt and formulated the problem using a systematic approach in his doctoral thesis (Bourjault, 1984). Dornan (1987) reported one of the world's first automated disassembly planners, in the form of an expert system. Based on Bourjault's work, the algorithm for finding the complete set of (dis-) assembly sequences was developed by Homem de Mello and Sanderson (1990 and 1991). The algorithm included a set of constraints, which were governed by precedence relationships that accounted for obstruction of some disassembly operations due to the presence of some components. De Fazio, Whitney, and Baldwin emphasized the importance of the precedence relationships (De Fazio and Whitney, 1987; Baldwin et al., 1991).

The above-mentioned work established a firm basis for the systematic analysis of disassembly processes, and opened the way for the rapid development of disassembly theory, which is reflected in many papers that are discussed in this book.

### 1.1.2 End-of-Life Disassembly

Disassembly as a process on its own started to gain momentum during the 1990s when the number and variety of discarded complex products increased rapidly. Growing environmental consciousness paved the way for the introduction of take-back systems for discarded electric and electronic products. In various countries, a fee had to be paid in advance when the consumer bought such products. Once the product is discarded, the fee is used for a free take-back and subsequent environmentally conscious processing of the product. Such processing usually includes disassembly, shredding, and materials separation and retrieval. This was considered environmentally beneficial as it saved virgin resources, materials and energy as well as avoided environmental contamination by reducing the amount of final waste sent to landfill.

In the 1970s, component reuse and materials recycling were essentially limited to discarded cars, for these were originally rich in ferrous metals constituting about three fourths of the car by weight. Usually, processing of discarded cars started with the disassembly of some components that could be purchased as spare parts. Bisides this, shredding of cars and subsequent separation of steel and some nonferrous materials turned out to be profitable. Steel scrap was useful in the iron and steel industry as secondary material. What remained was the automotive shredder residue (ASR), which consisted of a mix of light materials such as glass, rubber, and plastics. This substance was usually heavily contaminated with working fluids. ASR was originally landfilled. As the amount of steel content in cars gradually decreased and the environmental regulations tightened, measures were taken for simultaneously counteracting the contamination and reducing the amount of ASR. This necessitated the draining of the working fluids from the car, the removal of batteries, and some of the nonmetallic components such as tires, windows, seats, and bumpers before further processing. In European Communities (EC) countries, this practice has become a standard regulation via the Directive 2000/53/EC of the European Parliament on end-of-life vehicles (EC, 2000).

Traditionally electric appliances also contained a large amount of metal, but it frequently also had some hazardous substances such as asbestos in heating equipments and chlorofluorocarbons (CFCs) in cooling equipments. This prompted the governments to ban the use of such substances in new products as well as to issue directives for processing the end-of-life products making sure to remove substances such as CFC from the product before further processing or destruction.

End-of-life processing of electronic equipment was mainly motivated by the existence of precious metals in them (such as silver, gold, and palladium). Precious metals were present in the circuit boards of such equipments as mainframe computers and telecommunications products. However, because of the changing composition of electronic scrap, there has been a gradual decrease in the amount of valuable materials retrieved from such products, which has made the recycling process somewhat less lucrative from a purely economic point of view.

With the increasing number of discarded products, the urgency for environmentally driven end-of-life processing of electronic equipment has been widely recognized. Just as in the case of cars, selective disassembly prior to shredding and separation, for obtaining satisfactory homogeneous materials and avoiding huge amounts of shredder residue, also makes sense here. In electronic products, the metal contents are much smaller than in cars or electrical equipment. The mass of the individual components is relatively small, which pushes down the economic viability of disassembly. In addition, the variety of electronic products is large and their designs are constantly changing.

With the need to perform disassembly operation on even larger numbers of discarded electronic products in the future, robot-operated disassembly lines have recently been proposed. This approach, however, has met with high skepticism from researchers because of the presence of high degree of uncertainties in the end-of-life products such as incomplete information on the product structure, unknown conditions of components and fastener, and variability in supply and product type. Unfortunately, the alternative, which is manual labor, is too expensive, although this is partly caused by inappropriate costing methods that are based on direct labor. Since a much lower added value process (as compared to assembly) characterizes disassembly operations, a serious bottleneck occurs if material recycling has to be justified as economically viable.

Disassembly theory is not confined to studying the disassembly process of a given product, but it also addresses the design issues. *Design for disassembly* (DfD), as an extension of design for assembly, has been popular since the early 1990s. One of the first authors who formulated this paradigm was Dewhorst (1993). Almost simultaneously, the problem of robotic disassembly aimed at recycling was studied by authors from Germany, Japan and later by U.S. researchers (Feldmann and Hopperdietzel, 1993; Seliger et al., 1993; Knapp and Jansen, 1993). The idea of disassembly factories with a high degree of automation was launched. Since then, much research has been devoted to this topic.

Simultaneously considering in the design all phases of the product life, including production, consumption, maintenance, and the end-of-life, is called *design for life cycle*. This implies that, even in the conceptual phase of a product, the design of appropriate production systems and dismantling systems should be considered, because these systems depend strongly on the design of the product itself and vice versa.

All these developments and applications have further stimulated and enticed researchers to continue studying the process of disassembly.

## 1.2 LEVELS OF AGGREGATION

Complex problems, such as the ones that are dealt with in decision support systems, are usually tackled using a hierarchical approach. In assembly robot planning, a four-level hierarchy is frequently applied, the levels of which are:

1. Servo level
2. Robot level
3. Object level
4. Task level

As can be observed, the planning becomes increasingly complex as we go through the steps. Consequently, a corresponding decrease in the level of details takes place in modeling.

Comparable approaches have been implemented for the analysis of assembly tasks aimed at computerized process planning. A useful division in aggregation levels has been presented by Heemskerk (1990), who suggests the following hierarchy:

1. Primitive level
2. Component level
3. Product level
4. Batch level

A substantially modified hierarchy, which covers a broad range of disassembly problems, is presented below:

1. Physical level
2. Surface level
3. Component level
4. Modular level
5. Product level
6. Batch level

The *physical level* deals with the physical properties of components, with an emphasis on forces and deformability. Technical constraints and stability are some of the issues addressed at this level.

The *surface level* deals with the aspects of individual components, such as free and mating surfaces, and is applied if detailed analyses of a disassembly operation are required. In many analyses, a rigid body approximation is assumed. Accessibility and movability analysis take place at this level.

The *component level* deals with the movement of components in the course of disassembly operations and their possible interaction with other components. It deals with topics such as geometric and topological constraints and, consequently, precedence relationships.

The *modular level* considers functional subsystems of a product. It is in-between the component and the product level. Modularity analysis is done at this level.

The *product level* is applied if the analysis of a product as a whole is required, for studying the relationships between the disassembly operations and the sequence of those operations. It includes the establishment of disassembly precedence graphs, AND/OR graphs, and other representations of the possible sequences of disassembly operations. It also embraces the selection of appropriate sequences.

The *batch level* is used if the processing of multiple products has to be considered. This level involves demand-to-order problems and scheduling issues.

This book discusses all the levels, with an exception of the physical level, which is principally the domain of mechanical engineering.

A distinction can also be made between the product and the system approaches, also known as the *product-oriented* and *process-oriented* approaches.

The product-oriented approach focuses on how components interact while the process-oriented approach deals with how workstations and tasks interact. Both approaches are tightly intertwined. This is schematically depicted in Figure 1.1.

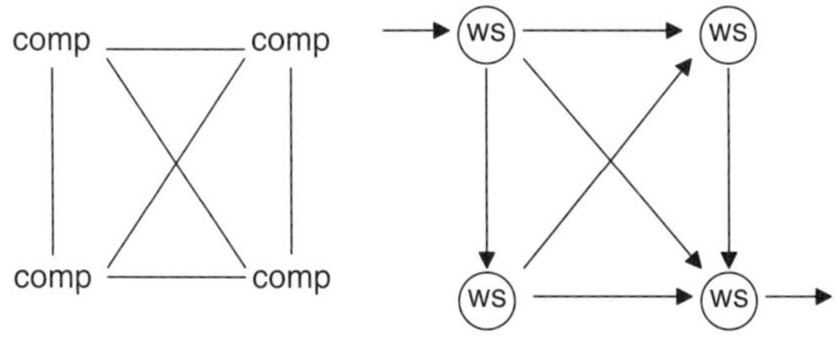


**FIGURE 1.1** (a) Product- and (b) process-oriented approaches.

Figure 1.1a illustrates the basic representation of the product-oriented approach, via a connection diagram. The nodes are components and the arcs represent connections between them and thus are undirected. Figure 1.1b shows the basic representation of the process-oriented approach, in which the nodes are workstations and the directed arcs are product flows that move between these stations.

A related distinction, which is commonly used, is due to Fox and Kempf (1985) who discern assembly planning and scheduling. According to the authors, the modeling and control of assembly processes is carried out in three steps:

1. A detailed analysis on a technical basis.
2. The design of a network of related assembly tasks, which defines the assembly process and is called *assembly planning.*
3. The proper operation of the assembly process, which is part of the complete logistic process in the production unit. This is called *assembly scheduling.*

The distinction between *disassembly sequencing* and *disassembly planning* also need to be made. Disassembly sequencing addresses the question, "How to disassemble?" while disassembly planning delineates "How much to disassemble?" The domains of disassembly sequencing and planning both have given rise to a considerable number of papers, which cover the various topics of those domains. This book covers both sequencing and planning issues (with more emphasis on sequencing) and some excursions to the interface between the sequencing and planning domains.

## 1.3 DISASSEMBLY SEQUENCING AND OPTIMIZATION

The identification of the optimum sequence for disassembly operations is one of the important objectives of disassembly planning. The following three types of methods can be used to accomplish this:

1. *Heuristic methods.* These methods make use of sets of rules, usually compiled in an algorithmic form, which provide a systematic way for finding "good" solutions. Heuristic methods are quick. A principal disadvantage is that there may exist unexpected classes of products for which the heuristic may not work properly.
2. *Metaheuristic methods* embrace a class of modeling techniques and search methods that typically do not go straight to an optimum, but rather make use of fuzzy properties and converge to some outcome in an unpredictable way. These methods account for the vagueness in figures for parameters such as cost, which are decisive in selecting an optimum solution. There are several metaheuristic methods, such as fuzzy logic, genetic algorithms, ant colonies, simulated annealing, tabu search, Bayesian networks, neural networks, distributed agents, etc. (Rekiek et al., 2002). The disadvantages of these methods include: (i) the need for specialized software; (ii) relatively long CPU times; (iii) generation of a suboptimum solution without any knowledge of its distance from optimum; (iv) restricted gain of insight

in the problem. However, the advantage is that they can deal with a lot of complicated problems that cannot be solved using exact methods. Metaheuristic methods can also be applied if the model size becomes too large for being solved via exact methods in a reasonable CPU time. Metaheuristic methods can be applied such that the algorithm stops after a predefined number of iterations, or if some increment drops below a predefined value. By this, a reasonable solution is obtained that may be indicative for the optimum solution.

3. *Exact methods* are based on mathematical programming. They involve: (i) search algorithms that visit all possible solutions in a systematic way, and (ii) search algorithms that more quickly converge to the optimum solution without visiting all the possible solutions. The latter category frequently makes use of a simplex or a branch and bound algorithm. A simplex algorithm is central to many *linear programming* (LP) solvers. Branch and bound algorithms are systematic search algorithms that effectuate the exclusion of certain parts of the solution space (branches in a solution tree) from further search. These algorithms are applied in integer programming problems. Modified versions are applied in *0-1 (binary) integer linear programming problems*. Binary programming methods in particular are frequently encountered in optimizing disassembly sequences. The principal advantages of this approach are convergency, transparency, and flexibility. Drawbacks are in the restricted applicability of the method. If nonlinear problems are encountered, which require *nonlinear programming* (NLP) procedures, further drawbacks are encountered. These are related to initialization, scaling, the possibility of arriving at locally optimum solutions, and decreased flexibility. If models are used that need a CPU time for their exact solution that exponentially increases with the size of the model (typically, the number of components in a product), which reflects the problem of NP-completeness, the exact method can be applied to models of moderate size for the evaluation of heuristic and metaheuristic models.

This book mainly deals with exact methods because these are promising and have been proven to work efficiently in disassembly optimization. The claim of "exactness" should however be qualified, as fuzziness is hidden in the assumptions and simplifications that often have to be made in modeling. Additional fuzziness is interjected in the assessment of the model parameters, such as the cost of the disassembly operations and the revenues of the components. Apart from this, exact methods are neither easily expandable to detailed analysis, nor can they include full recovery chains. Nevertheless, some major advantages of the exact method are apparent:

1. The method offers a clear insight in the problem.
2. Modeling is relatively simple.
3. The problem's structure can be easily modified.
4. Sensitivity analysis with respect to the model parameters is easy.

5. Calculations can be carried out with standard software.
6. The required CPU time may remain manageable, for increasingly large problems.
7. Adaptive planning is possible.

The use of exact methods has been counteracted for a long time because of the fact that the size of the problem increases exponentially, i.e., by $O(e^N)$, if $N$ is the number of components in the product to be disassembled, which is mathematically expressed by asserting that the problem is *NP-complete.* The same is true for the number of possible disassembly sequences. Although mitigated by the constraints that are imposed by the particular structure of the product, NP-completeness is persistent. This is only true, however, if the complete search space has to be investigated. However, the application of LP methods tend to avoid this problem.

The search for an optimum solution is preceded by the conversion of a real world problem to a linear or binary linear programming model. This includes a thorough analysis of the product structure, which is usually accomplished by reducing the data on the product to only the essential one. Appropriate modeling requires knowledge of several mathematical techniques including network and graph theory, set theory, combinatorial analysis, Boolean logic, and discrete mathematics.

Several authors have made use of *Petri nets* to solve disassembly problems (Moore and Gupta, 1996). Petri nets, which are a flexible modeling tool for simulating various processes, have been used since the 1980s for the modeling of both assembly and disassembly processes. One of the first applications of Petri nets in this field involved the control of robotic work cells (Yuditskii, 1983). Although Petri nets are characterized by time discreteness, they are quite applicable to the continuous approach of disassembly processes, and their study has even contributed considerably to the development of exact methods in disassembly sequencing following this approach. Assembly Petri nets were first proposed by Bourjault et al. (1987). Solving assembly sequencing problems by the formulation of assembly Petri nets is due to Kanehara et al. (1993). The applications of Petri nets to end-of-life disassembly problems are due to Moore et al. (1998, 2001); Zussman et al. (1998); Erdös and Xirouchakis (1999), and Tiwari et al. (2002). We will not cover Petri nets modeling in this book because equivalent alternative methods are available for disassembly sequencing.

## 1.4 BASIC TERMINOLOGY

This section presents some of the terminology and basic concepts used in this book, which is categorized into three parts.

### 1.4.1 The Product

The disassembly process starts with the *product*. The product represents a particular *functionality* (ability to provide services). The product consists of a number of discrete parts, which are called the *components*. A component cannot be further disassembled and is sometimes called an *atomic part* for this reason.

Components can be grouped in subassemblies. A *subassembly* is a connected set of components. A *subset* is also a set of components of the product that is not necessarily connected. Yet another way of grouping of the components is in *modules*, which are functional units, composed of components.

If components are physically linked, such a link is called a *connection*. The terms *liaison* and *joint* are also used by several authors. If the components are nearly in touch with each other, this can be considered a *virtual* connection in some cases. Connections restrict the freedom of motion of the components involved. This can be established in different ways, the most uncomplicated way of which is *mating*. In many cases, specialized components, or parts of components, called *fasteners*, are used for connections. Fasteners can be discrete components such as screws, or nondiscrete material objects such as snap fits. If the fastener is a component, but not considered as such in modeling, it is considered a *quasi-component*.

### 1.4.2 The Process

*Disassembly operations* accomplish the basic transformations in the disassembly process. These can be defined according to two different approaches:

1. The *disestablishment* of connections (other than mating surfaces)
2. The *detachment* of components or subassemblies

These steps can be generalized by defining the disassembly operation as the removal of quasi-components, components, or subassemblies.

Disassembly operations are subdivided into *disassembly tasks*, which are unit operations on a lower level of aggregation. These involve *preparatory tasks* such as the establishment of a fixture, changing tools, repositioning the product, etc. Apart from this, there are *proper disassembly task*s, which include:

1. Getting access to a component/subassembly
2. Moving a component
3. Removing a component
4. Collecting a component

The spatial trajectory of the component that is detached together with its orientation, is called the *disassembly path*. If the disassembly path is such that *collisions* with remaining components inevitably occur, the disassembly path is said to be *obstructed*.

Basic objects that enable disassembly operations are *fixtures* (jigs, clamps, templates, etc.), *tools*, which include *grippers,* screwdrivers, etc., and *bins* for the collection of components. Tools are for exerting forces on components.

At this stage, it is convenient to make some subtle distinctions in disassembly operations.

*Nondestructive disassembly operations* are disassembly operations that do not impair the components. These can be further classified into:

1. *Reversible operations* are nondestructive actions that can be accomplished with relative ease, such as screwing and unscrewing.

2. *Semireversible operations* are also nondestructive operations that can be accomplished but with slightly more difficulty. For example, snap fits are easy to assemble but a bit more laborious to disassemble.

*Semidestructive operations* are irreversible because they usually destroy the fasteners by cutting, folding, or breaking operations. There is usually no damage or limited damage to the components. Although not strictly nondestructive, these types of actions frequently occur in disassembly and can be incorporated in the extended concept of disassembly that is used in this book.

With these concepts established, the disassembly process can now be defined as follows:

*Disassembly* is a process in which a product is separated into its components and/or subassemblies by nondestructive or semidestructive operations. Operations that cause substantial destruction are not part of this. In the component approach, the binary disassembly unit operation is the separation of a parent subassembly in two child subassemblies. The *disassembly process* is a sequence of disassembly operations. It starts with the *initial state*, which is the complete product, and the *final state*, which is either predefined or a decision variable. The order of the disassembly operations in a specific disassembly process is called the *disassembly sequence.*

*Complete disassembly* is the separation of a product into all its components. *Incomplete disassembly* also involves the separation of a product but not all the components are separated from each other. The *disassembly depth* is the extent to which the disassembly process is carried out. *Selective disassembly* is a disassembly process that has to meet some specific criteria. It should be stressed that incomplete disassembly does not have a counterpart in assembly and hence is different from reverse of assembly, because assembly is aimed at obtaining a complete product and not just a partly complete product.

Disassembly sequences can be either sequential or parallel.

1. *Sequential sequences* (also called *one at a time* or *serial* sequences) are composed of disassembly operations in which at least one of the child subassemblies is a single component. Such an operation is called a *sequential operation.*
2. *Parallel sequences* include at least one disassembly operation in which both child subassemblies consist of multiple components. Such an operation is called a *parallel operation.*

### 1.4.3 The End-of-Life Process Chain

The chain of end-of-life processes includes:

1. *Disassembly*, which has already been discussed.
2. *Dismantling* (also called dismounting) is a process in which directed destructive operations are applied. These operations destroy one or more

components by breaking, sawing, cutting etc., thus devastating the component's value for reuse. Dismantling operations are frequently included in disassembly processes, especially if they enable further disassembly operations with the removal of components that obstruct accessibility and detachability of the desirable components. The removal of the roof of a car by cutting the bars is an example of such a destructive operation that facilitates the subsequent removal of the car's seats as a part of the selective disassembly process.

3. *Sorting* is the process in which various components and materials are divided into groups, which are called *clusters*. Each cluster meets some specific criteria on materials composition or component specifications. Examples of such groups include ferrous metals and printed circuit boards.
4. *Shredding* is an undirected destructive process, which breaks all the components into small pieces. The process is purely aimed at particle size reduction of products for increasing the materials homogeneity to enable the subsequent separation processes. Comparable processes such as milling, grinding, etc. are also included in shredding.
5. *Separation* is the postshredder process in which the shredder output is divided according to materials composition. The most frequently applied separation methods are *magnetic separation* and *eddy current separation.* A plethora of other physical and chemical separation methods are available, which originally come from the mining industry. The remainder after the chain of separation processes is called *shredder residue*, also called *shredder fluff* or *shredder light fraction.*
6. *Disposal* is the process of directing available residual waste to *discharge*, which includes incineration or landfill.
7. *Incineration* is aimed at waste volume reduction and energy recovery. Residual products of incineration are *filter, residue,* and *slag*.
8. *Controlled landfill* is the landfill under special conditions, particularly in the case of hazardous materials.

From the product's application point of view, the following processes are relevant:

Preconsumer phase:

1. *Rework* is aimed at transforming products, which are not produced according to the standards, into products that meet the standards of properly produced items.

Consumer phase:

2. *Maintenance* is aimed at enhancing the product's lifespan. It can include partial disassembly and reassembly, together with periodic replacement of components or modules.
3. *Repair* is aimed at restoring the product's functionality after its failure. It often includes partial disassembly, component replacement, and reassembly.

Postconsumer phase:

4. *Refurbishing* includes the processing of an end-of-life product such that its full functionality is restored. The resulting product is usually made available on the second hand market.
5. *Remanufacturing* is the composition of reconfigured products with components derived from end-of-life products. Usually, refurbishing of some of the components and modules of the original products is necessary.
6. *Reuse* is the employment of components and modules obtained from end-of-life products as spare parts or in other items.
7. *Recycling* is the recovery of materials out of scrap from end-of-life products.
8. *Recovery* refers to both reuse and recycling.
9. *Cascading* is the application of recycled materials for a lower-grade purpose than what it originally was used for.
10. *Downcycling* is the application of materials out of scrap for low-grade purposes, such as a filling agent in asphalt, an additive to cement kilns, or a basis for roads and buildings.

The sequence of processes in a chain is depicted in a directed graph, which is the graphical representation of a *product-process chain* with a *process flow diagram.* Its nodes represent processes and its arcs represent products.

## 1.5 PRODUCT MODELING

Two principal ways of product modeling are used in this book.

1. The *mechanical approach* considers a product starting from its geometric structure, such as the ones described in assembly drawings or virtual prototypes.
2. The *hierarchical tree approach* considers a product described in the form of an inverse product structure (see, for example, Gupta and Taleb, 1994).

The *mechanical approach* has been taken in assembly studies for a long time. It is comprehensive, as it accounts for the complete product geometry. This information is usually derived from assembly drawings or virtual prototypes. The mechanical approach is applicable for an arbitrary product structure.

The *hierarchical tree approach* accounts for a restricted number of components and modules only. Many other components, such as fasteners, casings, frames, etc., are only considered as obstruction for the removal of the targeted components, which are called *leaves.* A cluster of components held together is called a *root.* The relationships between roots and leaves are hierarchically organized, and can be depicted as a tree-shaped directed graph. The hierarchical tree approach also spontaneously appears if a product is disassembled according to some intuitive rules (Åkermark, 1997).

The hierarchical tree approach is simpler than the mechanical approach. There are relatively less decision variables considered in the hierarchical tree approach compared to the mechanical approach, and not all the theoretically possible disassembly sequences are included in the model. The advantage of the hierarchical tree approach lies in its ease of modeling and is specifically suited for the case where a family of products is considered.

## 1.6 SUMMARY OF THE CHAPTERS

The book is organized in three different parts, each of which is divided in several chapters.

### 1.6.1 Part 1 Disassembly Practice

This part of the book is devoted to the introductory aspects of disassembly process and its application to the end-of-life disassembly.

*Chapter 1* provides an overview of assembly and disassembly process and this book.

*Chapter 2* deals with the end-of-life processing of complex products. It starts with a general description of the philosophy of industrial ecology. The product life cycle and the processes that take place within the context of industrial ecology are discussed. Next the status of discarded complex products in the waste economy is expounded. The chapter concludes with the presentation of data derived from the literature on the composition of multiple waste streams and complex products.

*Chapter 3* discusses the disassembly process of complex products, particularly the technical aspects of product composition and fasteners. Typologies are presented. Environmental and financial costing methods are discussed here. The impact of product design on end-of-life processing is unfolded.

### 1.6.2 Part 2 Disassembly Sequencing

This part of the book deals with the theoretical analysis of disassembly sequencing. Both modeling and optimization procedures are considered.

*Chapter 4* discusses the theoretical foundations of disassembly and the basic mathematical and graphical tools used in disassembly theory. The most frequently applied methods and concepts are explained and these are applied to unconstrained products for determining the maximum search space in sequencing problems. These methods form the basis for the study of real world products.

*Chapter 5* is devoted to the restrictions on the product level, which are expressed via topological, geometrical, and technical constraints. The emphasis is on the geometrical constraints, which are converted to precedence relations that in turn can be transformed into selection rules. By this, the set of all possible disassembly sequences is derived and the different graphical representation methods: AND/OR graph, state diagram, and disassembly precedence graph, are discussed. Applications to different products are presented.

*Chapter 6* deals with movability and removability analysis. It introduces the theory in detail and also presents direction-oriented analysis. A relaxed version of the sequencing problem via obstruction diagrams and matrix methods is discussed. Real world topics on the physical level, including forces, are included in this chapter.

*Chapter 7* is devoted to optimization methods in disassembly sequencing using the exact methods of mathematical programming. A method for obtaining the optimum disassembly sequence, or a set of suboptimum sequences, based on different criteria, is discussed. Extensions are introduced for the incorporation of postdisassembly clustering processes, and for the optimization of processes with sequence-dependent disassembly costs. This chapter also presents the application of this method to disassembly-to-order problems, which include component commonality and multiplicity. In this approach, the hierarchical tree representation is applied.

### 1.6.3 Part 3 Disassembly Planning

This part of the book is devoted to disassembly planning. Because of the multi-objective nature of such problems, methodologies that specifically address multi-objective problems are presented.

*Chapter 8* extends the discussion on disassembly-to-order (DTO) problems. It highlights two multicriteria methods; goal programming and linear physical programming, to address the DTO problems.

*Chapter 9* discusses the disassembly line balancing problem and its principal differences with the assembly line balancing problem. A sampling of heuristic and metaheuristic approaches are presented to provide fast, good or near-optimal solutions to the multiobjective, nonlinear disassembly line balancing problems. Examples are considered to illustrate the approaches.

## REFERENCES

Åkermark, A.M., 1997, Design for Disassembly and Recycling. In: Krause, F.L. and Seliger, G. (eds.): Life Cycle Networks, *Proceedings of the 4th CIRP International Seminar on Life-Cycle Engineering*, 237–248.

Baldwin, D.F., Abell, T.E., Lui, M.M., De Fazio, T.L., and Whitney, D.E., 1991, An integrated computer aid for generating and evaluating assembly sequences for mechanical products. *IEEE Transactions on Robotics and Automation*, **7**, 78–94.

Batchelor, R., 1994, *Henry Ford*, Manchester UK; Manchester University Press.

Boothroyd, G., Poli, C., and Murch, L.E., 1982, Automatic assembly. In: G. Boothroyd, (ed.), *Manufacturing Engineering and Materials Processing*, Vol. **6,** New York: Marcel Dekker.

Bourjault, A., 1984, *Contribution à une approche méthodologique de l'assemblage automatisé: elaboration automatique des séquences opératoires*, (Contribution to a systematic approach of automatic assembly: automatic determination of operation sequences). Ph.D. Thesis, Besançon, France: Université de Franche-Comté, (in French).

Bourjault, A., Chappe, D., and Henrioud, J.M., 1987, *Elaboration automatique des gammes d'assemblage à l'aide de résaux de Petri* (automatic elaboration of assembly routines using Petri nets). RAIRO(APII), **21**, 323–342 (in French).

De Fazio, T.L. and Whitney, D.E., 1987, Simplified generation of all mechanical assembly sequences. *IEEE Journal of Robotics and Automation*, **RA-3**(6), 640–658.

Dewhurst, P., 1993, Product design for manufacture: design for disassembly. *Industrial Engineering*, **25**(9), 26–28.

Donnan, D.C. and Makan, K., 1983, Disassembly line balancing. *MTM Journal of Methods–Time Measurement*, **10**(2), 20–27.

Dornan, S.B., 1987, A status report: artificial intelligence. *Production*, **99**(4), 46–50.

EC, 2000, Directive 2000/53/EC of the European Parliament and of the Council of 18 September 2000 on end-of-life vehicles. *Official Journal of the European Communities*, L269/34, October 21st.

Encyclopedia Americana, 1965. New York: Americana Corporation.

Encyclopedia Britannica, 1994. London UK: Encyclopaedia Britannica.

Erdos, G. and Xirouchakis, P., 1999, Extended Petri-net modeling for re-manufacturing line simulation, *Proceedings of IEEE Conference on Ecodesign*, 83–88.

Feldmann, K. and Hopperdietzel, R., 1993, Environmental recycling by automated disassembly, *Zeitschrift für wirtschaftliche Fertigung und Automatisierung*, **88**(4), 148–150.

Fox, B.R. and Kempf, K.G., 1985, Opportunistic scheduling for robotic assembly. *Proceedings of 1985 IEEE International Conference on Robotics and Automation*, 880–889.

Gupta, S.M. and Taleb, K.N., 1994, Scheduling disassembly. *International Journal of Production Research*, **32**(8), 1857–1866.

Heemskerk, C.J.M., 1990, *A concept for computer aided planning for flexible assembly.* Ph.D. thesis. Delft: Technische Universiteit Delft. 276 p.

Homem de Mello, L.S. and Sanderson, A.C., 1990, And/or graph representation of assembly plans. *IEEE Transactions on Robotics and Automation*, **6**(2), 188–189.

Homem De Mello, L.S. and Sanderson, A.C., 1991, A correct and complete algorithm for the generation of mechanical assembly sequences. *IEEE Transactions on Robotics and Automation*, **7**(2), 228–240.

Isaacs, J.A. and Gupta, S.M., 1997, Economic consequences of increasing polymer content for the U.S. automobile recycling infrastructure. *Journal of Industrial Ecology*, **1**(4), 19–33.

Kanehara, T., Suzuki, T., Inaba, A., and Okuma, S., 1993, On algebraic and graph structural properties of assembly Petri net, *Proceedings of the 1993 IEEE/RSJ International Conference on Intelligent Robots and Systems*, Yokohama, July 26–30, 2286–2293.

Kilbridge, M.D. and Wester, L, 1961, A heuristic method of assembly line balancing. *The Journal of Industrial Engineering*, **12**(4).

Knapp, O. and Jansen, H., 1993, Environmental considerations in product development. *ETZ*, 114(22), 1386–1389 (in German).

Moore, K.E., Gungor, A., and Gupta, S.M., 1998, A Petri net approach to disassembly process planning, *Computers and Industrial Engineering*, **35**(1–2), 165–168.

Moore, K.E., Gungor, A., and Gupta, S.M., 2001, Petri net approach to disassembly process planning for products with complex AND/OR precedence relationships. *European Journal of Operations Research*, **135**(2), 428–449.

Moore, K.E. and Gupta, S.M., 1996, Petri net models of flexible and automated manufacturing systems: a survey. *International Journal of Production Research*, **34**(11), 3001–3035.

Prenting, T.O. and Battaglin, R.M., 1964, The precedence diagram: a tool for analysis in assembly line balancing. *Journal of Industrial Engineering*, **15**(4), 208–213.

Rekiek, B., Delchambre, A., Dolgui, A., and Bratcu, A., 2002, Assembly line design: a survey. *IFAC 15th World Congress*, Plenary Papers, Survey Papers, and Milestones, 131–142.

Seliger, G., Heinemeier, H.J., and Neu, S, 1993, Robot-guided disassembly and assembly of small transmissions for substitute production. *Zeitschrift für wirtschaftliche Fertigung und Automatisierung*, **88**(6), 245–248.

Tabucanon, M.T. and Somnasang, S., 1981, Line balancing in an automobile assembly plant: comparative applications of two techniques. *Industrial Management*, **23**(4), 18–23.

Tiwari, M.K., Sinha, N., Kumar, S., Rai, R., and Mukhopadhyay, S.K., 2002, A Petri net based approach to determine the disassembly strategy of a product. *International Journal of Production Research*, **40**(5), 1113–1129.

Wild, R., 1972, *Mass-production management*. London, UK: John Wiley and Sons.

Yuditskii, S.A., 1983, Discrete assembly-line processes and Petri nets. *Automation and Remote Control*, **44**(6, part 2), 806–810 (in English, translated from Russian).

Zussman, E., Zhou, M.C., and Caudill, R., 1998, Disassembly Petri net approach to modeling and planning disassembly processes of electronic products, *Proceedings of IEEE International Symposium on Electronics and the Environment*, 331–336.

# 2 Context of End-of-Life Disassembly

## 2.1 INTRODUCTION

In recent years, end-of-life disassembly has gained popularity because of both economic incentives and growing environmental concerns. Industrial ecology represents a firmly established systematic and integrated philosophy for dealing with the environmental aspects of the economy. The goal of industrial ecology is to promote sustainable development and affect a fundamental paradigm shift in thinking concerning the relationship between industry and ecology by efficient use of resources and by closing the loop of materials and energy flows in an effort to reduce their impact on the environment (Den Hond, 2000). Section 2.2 discusses the principal concepts of this method, with an emphasis on the materials life cycle, which is represented with the product-process chain. Section 2.3 focuses on the discussion of the different phases in the product-process chain of complex products. Section 2.4 places the waste from discarded complex products in the framework of the complete waste stream. Finally, section 2.5 deals with breakdown analyses of various complex products. Some concluding remarks are given in section 2.6.

## 2.2 INDUSTRIAL ECOLOGY

### 2.2.1 Introduction

As was previously mentioned, partial disassembly and reassembly for maintaining and repairing complex products have been practiced for a long time. Often, discarded products are used as a source for parts and materials for repairing and remanufacturing products. At times, the components of former products, such as buildings, cars, and machine tools, are reused in new ones, particularly in times of shortages. Items such as metals and textiles are also repeatedly recycled prior to their final disposal. The motivation for this, however, frequently stems from the value embedded within the components and materials rather than the environmental good. The upshot of this, at least in industrialized countries, has been to ignore any unprofitable processing of discarded materials resulting in the escalation of the amount of waste generated. In order to satisfy the appetite of corporate profits and the demand of consumers, more and more virgin resources are exploited on a larger and larger scale. In addition, as new types of products are developed, an increasing amount of chemically modified substances are interjected in the materials used that eventually enter the biosphere after they are discarded. Some of these are not only hazardous, but they are also long lasting and slow to biodegrade.

In an effort to (1) promote sustainability; (2) alleviate risk of future unavailability of natural resources; (3) offset continuous threat of depletion of natural resources; and (4) reduce the amount of waste (hazardous or otherwise), manufacturers and research communities have been compelled to approach the materials and energy flows in the industrial system in a more systematic and scientific manner. In the not so distant past, the lion's share of virgin resources were used only once before they were discarded. Evidently, no reasonable effort was made for energy and mass conservation. Although the environmental impact of emissions has been decisively reduced in the past century, thanks to end-of-pipe measures such as filters and wastewater treatment plants, the rate at which waste was added to landfills remained distressing. Apart from this, the massive reliance on fossil fuels of the industrialized world caused it to increase its dependence on politically unstable regimes. Of course, with a huge demand on natural resources, one has been left with decreasing quality of ores for materials reclamation. With the dawn of billions of people on the verge of entering the industrialized world, particularly in China and India, the problems facing materials use and discharge will be further exaggerated.

The desire to characterize materials flows and transformations together with development of tools to control them in an integrated fashion has led to the concept of *Industrial Ecology*, first proposed by Frosch and Gallopoulos (1989).

Industrial ecology, in part, addresses issues related to the technical-economic system, or *technosystem*. It also integrates the basic mechanisms of the technosystem to that of the natural environment or *ecosystem*. The ecosystem is considered sustainable because its energy is provided by sustainable energy sources (mainly the sun) and its material flows are organized in cycles, such as the carbon cycle and the nitrogen cycle. Assuming that the ecosystem is sustainable, industrial ecology seeks to organize the technosystem according to the basic principles of the ecosystem. Thus, industrial ecology seeks to exploit sustainable energy resources and to close the materials cycles. This takes place via waste reduction and the utilization of waste streams as a resource. For an extensive discussion of the basic concepts and many examples of industrial ecology see the books by Ayres and Ayres (1996), and Graedel and Allenby (1995).

Since this book focuses on materials flows in the technosystem, the following basic concepts of industrial ecology are noteworthy:

1. Product life cycle
2. Product recovery
3. Cascading

These are discussed in the subsequent subsections.

### 2.2.2 Product Life Cycle

#### 2.2.2.1 General

At least two interpretations of product life cycle appear in the literature, which include conceptual life cycle and physical life cycle.

1. The *conceptual life cycle* refers to the duration of time in which a product is considered viable in the market. The conceptual life cycle comprises the design phase, the production phase, the state-of-the-art (useful) phase, and finally the phase of product's decline. In this final phase, the product becomes outdated and needs replacing even though the product is technically sound. Depending on the product, the conceptual life cycle may encompass an appreciable period of time. On the other hand, electronic products such as computers and mobile phones tend to have substantially shorter conceptual life cycles. A shorter conceptual life cycle causes products to be discarded, even if they still are technically sound, thus preventing their components from being reused in new or remanufactured products.
2. The *physical life cycle* refers to the duration of time that spans from the production of a product up to the moment that it is discarded.

If the end of physical life cycle surpasses the end of conceptual life cycle, the product is still usable although it is technically outdated. This implies that end-of-life disassembly often involves a kind of industrial archeology. Even if the service that is offered by the product is still relevant, such as in a TV set, its components cannot be reused. Apart from this, some materials, particularly plastics, may have become obsolete. The fastener types may also be different from the current ones.

In some cases, a product could essentially disappear from the market. For example, mechanical and electric typewriters, duplicators, tape recorders, phonographs, centrifuges, manual sewing machines, matrix printers, and many other types of appliances are no longer produced or popular. The cathode ray tube is also following this trend. These products, however, have to be processed at the end of their lives. Note that special attention has to be paid to obsolete products that do not comply with current environmental regulations. For example, electric heaters containing asbestos and refrigerators containing hard chlorofluorocarbons (CFCs) fall into this category. These kinds of products require special handling (as dictated by the prevailing regulations because of their hazardous contents) as opposed to the modern version of the same product.

#### 2.2.2.2 Product-Process Chains

The product-process chains that describe the technosystem are similar to the process flow diagrams that are used for describing detailed production processes. However, product-process chains are presented at higher level of abstraction. The chain is represented by a directed graph. The directed arcs connecting the nodes represent products, which are considered as physical (mass and energy) flows. Every node represents a *process*, which is defined as a transformation of mass and energy flows. A stationary portrayal, with no changes over time, is commonly used. *Transformation* is defined as a change of a product physically or chemically (by processing), in time (by storing) and in place (by transporting). Transformations in time and place fall in the domain of logistics.

The following kinds of processes are recognized:

1. *Production processes* are transformations of physical flows aimed at increasing the economic and functional value of the product. The incoming flows include feedstock (raw materials or semifinished products) and ancillary materials such as containers and fuel. The outgoing flows include products and *process waste*. Process waste is the waste that originates from the process. Examples include emissions such as solvents and solid waste such as the metal that is left after a punch process.
2. *Consumption processes* produce services by consuming some product. Incoming flows include products and ancillaries. Outgoing flows include process waste and product waste. *Product waste* is the flow of discarded products, which is considered a feedstock for recovery processes.
3. *Recovery processes* can be considered as production processes. If discarded products are considered, these are also known as postconsumer processes.
4. *Disposal processes* prepare a product for discharge by decreasing its potential harm to the environment using some sort of technology (e.g., by volume reduction).
5. *Extraction processes* are production processes such as mining and agriculture that have materials from the environment as input.
6. *Discharge processes* finally bring the materials from the technosystem back to the environment. There are four types of discharges to the environment, viz., emissions to (i) the atmosphere, (ii) the surface water and (iii) the soil, and (iv) solid waste that is landfilled or stored.

Regardless of the process type, both mass and energy are conserved, which implies that the sum of the incoming mass flows equals that of the outgoing mass flows. The same holds true for the energy flows. Mass flows are expressed in multiples of kg/s and energy flows are expressed in multiples of Watt (W), which corresponds to Joules per second (J/s). Of course, the balancing of flows is only true in a stationary system, in which accumulation of matter and energy in the process units, such as changing stocks in a production plant, are neglected. Mass and energy flows are characterized not only by their quantity, but also by their qualitative properties such as composition as a measure for the quality of materials, and exergy content as a measure for the quality of energy.

A typical production process is depicted in Figure 2.1. It is a simplified version of the process as various flows are not explicitly depicted. Instead, the various flows are aggregated and represented by a single arrow. Typically, different types of products are produced in a single process, including *byproducts*, which are valuable, but not the intended products. This is an example of *divergence*. Many types of feedstock might also be required, which is particularly apparent in complex products. This is an example of *convergence*.

A typical aggregated product-process chain of the technosystem is depicted in Figure 2.2. It is assumed here that the system is linear, which means that the reuse and recycle loops are not included. Although the process waste also passes through

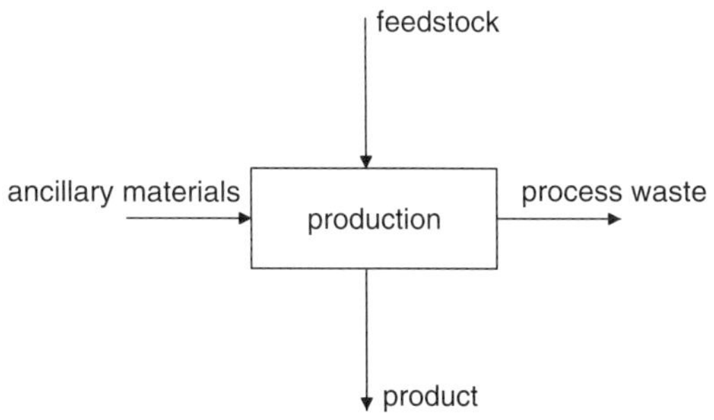


**FIGURE 2.1** Schematic representation of a typical production process.

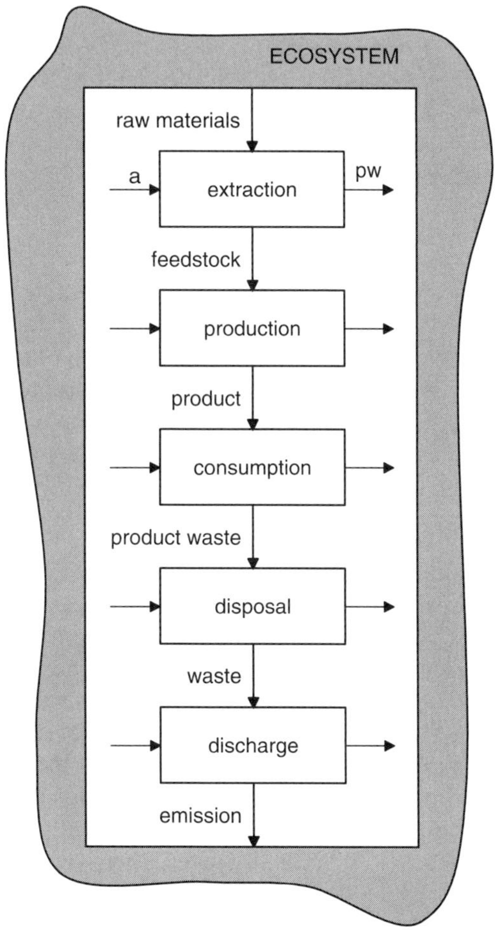


**FIGURE 2.2.** The linear product-process chain and its essential processes and mass flows. Note that ancillary mass flows (a) and process wastes (pw) are not elaborated in this figure.

the disposal and the discharge processes, these are not indicated in the diagram either as this is of lesser interest within the scope of this book.

The physical life cycle is a crucial concept in both the product design and the environmental evaluation of products and processes.

#### 2.2.2.3 Design for Life Cycle

Lately, there has been a tendency toward integrating all phases of a product's life cycle in its design. As was mentioned in chapter 1, there has been an evolution from design for functionality and cost, to design for assembly, to the current design methods that optimize products' performances over their complete life cycle. This evolution has taken place because a producer can actually be held responsible for any mishap that occurs during the complete life cycle of a product, including its end of life phase. In the literature this is referred to as *product responsibility* or *product stewardship*. Ishii et al.'s (1993) work was one of the first studies that addressed the concept of *design for life cycle*. As part of its objectives, design for life cycle aims to generate a high recovery rate, a high quality (and hence high value) of the recovered components and materials, and a minimum release of hazardous components to the environment all at minimum possible costs. Note that *Design for disassembly* is part of this concept.

#### 2.2.2.4 Life Cycle Assessment and Eco-Indicators

The environmental impact of a product for its complete life cycle is usually quantified via *life-cycle assessment* (LCA). This method results in the quantification of the different impact categories that are caused by resources consumption and by emissions and disposal. Combining these figures into a single quantity results in the eco-indicator, which is an aggregate figure. The ultimate value of this indicator strongly depends on the assignment of weight factors to the various impact categories.

### 2.2.3 Product Recovery

The linear version of the product-process chain can be expanded using vertical disaggregation of processes by adding postconsumer processing (or recovery). As was mentioned before, virtually no finite mineral or fossil resources are consumed in the ecosystem. In order to obtain a comparable state in the technosystem, materials flows have to be in a closed loop. This can be achieved by upgrading and recycling the waste streams (see Figure 2.3). Here, a *recovery process* is added between the consumption and disposal processes of the linear chain. A mass of *secondary feedstock* flows from the recovery process back to the production process. This partially replaces the *primary* (or *virgin*) *feedstock* that is extracted from the environment, thus reducing the dependence on virgin resources. There are many ways to achieve recovery; for example, by reusing components or recycling materials. Thus if secondary materials such as steel scrap were to be used as feedstock, it would reduce the burden on the environment by reducing the level of ore mining or iron production. Of course, one has to account for the environmental impact that is caused by the recovery processes, such as recollection, transportation, and melting and other

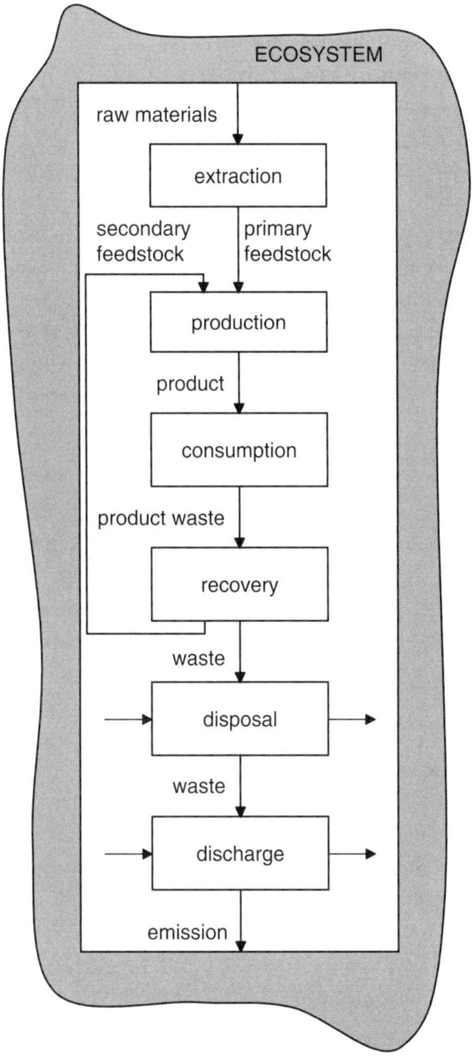


**FIGURE 2.3** Product-process chain with product recovery and a loop.

processing. However, the environmental impact of secondary materials recovery is normally much smaller than that of the production of the primary materials. In components recovery, this advantage is even more profound, because the cost of component production can be avoided here.

The principal feature of product recovery, therefore, is to avoid or at least mitigate the rate at which natural resources are depleted. This leads to substantial energy savings due to the reduction in primary materials production such as aluminum or plastics, which tend to be energy intensive. In addition, mineral resources are also saved. This is vital for materials with relatively restricted supply such as copper, and for materials that are only available as byproducts such as cadmium. Cadmium has virtually no natural deposits that are practical for mining it as a principal product. It is produced as a byproduct during the production of zinc. This puts restrictions on its production rate. Another example is iodine, which is produced from brines that are reclaimed as byproducts of natural gas extraction. Indium is produced as a byproduct of the processing of tin, zinc, and lead ores. It is used in solders, in the seals of cathode ray tubes, and as a phosphor in cathode ray tube (CRT) screens. Primary selenium is obtained as a byproduct of copper production only and is toxic. It is mainly used in diodes and by the photocopying industry. Recovery of such materials, therefore, needs to be practiced because of their limited supply and hazardous consequences. Data from 1989 reveal the following percentages of some secondary materials that are applied in the industrialized world (Potzschke, 1991): 27% of Al; 38% of Cu; 22% of Zn; 51% of Pb; and 18% of Sn.

### 2.2.4 Cascading

Even though Figure 2.2 was expanded into Figure 2.3, its scheme is still an oversimplification. For example, the materials and components of the discarded products are considered as one flow and the production processes are considered as one aggregated process, which obviously is an oversimplification, especially if complex products are considered. In addition, recollection of postconsumer products has to take place prior to recovery. The efficiencies of recollection and recovery processes can be defined as follows:

1. The *recollection efficiency*, $\eta_{recol}$, is the ratio between the weights of the products that are recollected for recovery and the total weight of the discarded products.
2. The *recovery efficiency*, $\eta_{recov}$, is the ratio between the weight of the materials that are actually recovered and the weight of the products that are offered for recovery to the recovery facility.

Consequently, the *cycle efficiency* $\eta_{cyc}$ is given by (usually quoted in percentage)

$$\eta_{cyc} = \eta_{recol} \cdot \eta_{recov} \tag{2.1}$$

Although frequently used, the calculations of these efficiencies assume that the processes in the product-process chains are static, i.e., changes over time are neglected. In practice, however, this assumption is not quite realistic, which undermines the workability of recollection efficiency. On the other hand, the recovery efficiency, which is more of a technical figure, is more firmly based. In addition, the aggregate efficiency does not account for the quality of the application of the recovered waste stream. The merit of product recovery thus cannot be completely expressed with one figure. High-grade applications have tight quality requirements. In most cases, upgrading metals to the original level is always possible. However, in the case of plastics, rubbers, glasses, and other compound materials, degradation often takes place because of the lack of information on exact composition of the materials used in the discarded product. For example, the broad spectrum of additives that are present in plastics include substances such as pigments, stabilizers, lubricants, fillers, antioxidants, and many more. These are no longer functional in discarded plastics, and are considered as contamination. Even if full information on the presence of additives were to be available, it would make the ultimate separation in completely homogeneous batches illusory in most of the cases because of economic and technical constraints. Destruction of essential structures, such as molecules and fibers, is another reason for degradation. Irreversible combination of different materials to composites is a third reason. Therefore, upgrading to a quality level that is same as the original feedstock is laborious and energy-intensive, if not impossible. This is particularly true for postconsumer waste, which is different from the homogeneous and well-defined batches of material. Therefore, waste materials are frequently considered as lower than original grade, which implies that their range of applications is restricted for processes that require lower than original quality standards. This downcycling can replicate itself several times if the material runs through the loop recurrently.

Evidently, materials exchange between different product-process chains may also occur. Steel from cars, for example, is recycled to steel that can have applications in various products, not just cars. But even in the same chain, different applications are possible. If plastics are considered, the following hierarchy is visible in cars. There are plastics in cars that have to comply with severe requirements on safety, strength, exterior, hardness, etc. Recycled versions of these plastics may find their way into other automotive components in which some of those requirements are somewhat relaxed. In the course of repeated application, degradation takes place from high-grade engineering plastic to low-grade filler in a component that does not have to meet any strict requirement. This kind of application thus is a sink into which all the plastic materials will finally wind up.

Another example of gradual downcycling is found in polyvinyl chloride (PVC), which can be used as a drinking water conduit in the highest-grade application, next as sewage conduit, next as drainpipe, and subsequently as a packaging material.

Paper can also be recycled, but a certain share (about 14%) of its fiber mass is lost to recycling because the fibers are too short. This share has to be rejected and is ultimately processed via incineration.

The process of gradual degradation is called *cascading* or *downcycling*. It ultimately brings about the material inevitably part of the chain, which leads to a process such as incineration or coincineration as in cement kilns. Energy may be recovered in such processes, but materials recovery is highly unlikely.

A sink that is frequently applied is the use of the material in bulk applications such as in road construction, as an addition to asphalt or in the basis of a road (sometimes called "linear landfill"). Characterizing this as materials recycling is questionable.

## 2.3 COMPLEX PRODUCTS

### 2.3.1 General

The theory in the preceding sections addresses all kinds of products and waste. However, complex products are of primary interest to us. *Complex products* are made up of multiple components and various materials. The life cycle of complex products can be understood by disaggregating the production and the recovery processes as presented in Figure 2.3.

### 2.3.2 Production of complex products

The production of complex products usually involves three main phases, each with its own characteristics.

1. *Materials production.* This refers to the processes where the physical and chemical intrinsic properties of the materials are tailored according to the manufacturers' requirements. Examples of the properties include strength, electrical conductivity, optical transparency, and chemical composition. This is the domain of *process industries.*
2. *Component production.* This refers to the manufacture of discrete components, which characterize their extrinsic properties, such as the geometry and the surface conditions of the components. This is the domain of the *manufacturing industries.*
3. *Product assembly.* This refers to the assembling of discrete components into modules, which in turn are assembled to produce a complete product. These processes define the product's functionality. This is the domain of the *assembly industries.*

### 2.3.3 Recovery of Complex Products

Although a precise picture of every possible product recovery process cannot be given as every product has its own attributes, we will highlight the foremost elements of postconsumer processing.

1. The first step is *recollection,* which consists of the logistic aspects of the end-of-life product acquisition.
2. Next, *repair* including *testing* may take place, after which the product can be reused as a whole. Occasionally, spare parts are needed for repairs.
3. If the whole product is not reused, a *selective disassembly process* may be carried out. The product is disassembled into modules and components up to some depth depending on which modules or components are targeted. The retrieved modules and components may need *testing* and possibly *repair.* They may be either reused as "new" parts in new products, or reassembled into *remanufactured* products, or used as spare parts.
4. Sometimes *selective dismantling* is done. Selective dismantling is a destructive process in which no effort is made to preserve the identity of a component or module. Therefore, the components or modules are only useful for their materials that are earmarked for recycling. Occasionally selective dismantling is done to facilitate selective disassembly of different components or modules, for example by removing obstructions. At times selective dismantling is preferred because it is cheaper and less labor-intensive than disassembly and workers with lower qualifications can perform it.

5. *Sorting* may take place to separate out different components or materials for recycling.
6. The remainder of the product after disassembly and dismantling is often *shredded*. Shredding can also be performed on disassembled and dismantled components or modules. The shredding process reduces the size of scrapped products into pieces that are small enough to be considered homogeneous or could be separated into homogeneous piles. This process can be accomplished in many ways, for example, hammering, cutting, milling, and grinding. It usually takes place in multiple steps for gradually obtaining smaller and smaller particles.
7. *Separation* process partitions the chunks obtained from the shredding process into appropriate materials categories.

   The methodology used for separation processes can be broadly classified into (i) physical methods and (ii) metallurgical and chemical methods. *Physical* methods essentially make use of the differences in densities and electromagnetic properties of materials to separate them. *Metallurgical and chemical* methods are used to decompose *compound materials* such as alloys and composites. These decomposition processes take place in specialized facilities. Since this book fundamentally deals with discrete products, further description of metallurgical and chemical methods is beyond the scope of the book.

   Most frequently used physical methods are magnetic and eddy current separation processes. *Magnetic separation* uses electromagnets for separating ferrous materials. *Eddy current separation* makes use of a rotating permanent magnet, thus inducing eddy currents in conductive materials. It is used subsequent to magnetic separation and is used to separate nonferromagnetic metals such as aluminum. *Electrostatic separation* makes use of electric fields and is used for the separating metallic content present in smaller particles, for example, recovery of metals and plastics from printed circuit boards, cables, and wires. *Air classification methods* and *jigs* are used to separate materials based on density differences. These are often used for separating the copper fraction from the light fraction that is left after the eddy current separation process. Sieves are used to separate materials according to the particle size.

   Although many other separation methods exist, they are not discussed here. Clearly, *manual sorting* is indeed an important method. Even in mechanized processes, manual sorting is frequently needed for removing large homogeneous chunks and large pieces that obstruct subsequent processing due to their composition or geometry. *Automated sorting* methods are also available, for example, *laser separation*, which makes use of the difference in surface textures of different kinds of plastic.
8. *Shredder residue* is what is left after the separation processes take place. The materials that leave the shredder are separated into ferromagnetic metal fraction and in one or more nonferrous metal fractions. Copper and aluminum rich fractions are frequently obtained. What remains is the shredder residue that consists of a mixture of plastics, rubber, glass, fabric, wood,

rust, paint, ceramics, sand, and possible contaminants such as oil. If additional separation steps are included, further fractions may be released reducing the final amount of shredder residue. Although technically possible, this often does not result in the recovery of valuable materials. Instead it generates mixtures that can only be used in low-grade applications such as filler.

A principal problem with shredder residue is its undefined composition. It may contain hazardous elements. Removal of hazardous components and materials before shredding could mitigate this problem. However, the composition of shredder residue cannot be guaranteed. Often, shredder residue is landfilled. However, tightening regulations that ban landfilling are propagating. Incineration under controlled conditions is considered a viable option. The *filter residue* and *slag*, which are residues of this process, can also contain hazardous substances. Therefore, their use in building or construction materials is prohibited. However, this is not true of the slag from municipal waste incineration. Consequently, incineration of shredder residue cannot be combined with that of ordinary household waste. Because hazardous waste can only be deposited in dedicated landfills, the disposal of shredder residue is expensive. Although this would ultimately encourage the introduction of more extended post-shredder separation processes, the disassembly of components that contain undesirable materials beforehand, may be more promising because of the quantity reduction and quality enhancement potential of the shredder residue. As an example, the removal of the seats from a car before shredding it can reduce shredder residue. Another measure that can be taken includes a ban on hazardous materials in the product design.

In light of the above discussion, the product-process chain of Figure 2.3 can be modified to the one shown in Figure 2.4. The selective disassembly process is visible here within its context. The products of this process include components intended for either reuse or recycling, materials aimed at recycling, and dismantled products that are intended to be shredded or separated. For a survey of articles published in the area of product recovery, see Gungor and Gupta (1999).

### 2.3.4 Selective Disassembly

Selective disassembly is generally used either for repair and maintenance of a product or for end-of-life disassembly. When used for *repair and maintenance*, some components and modules are removed to ensure accessibility to other components or modules for repairing, testing, and maintaining. Here selective disassembly is always followed by reassembly implying that damage to any disassembled component or module has to be prevented. That is, in most cases, only nondestructive disassembly is performed.

When selective disassembly is used for *end-of-life disassembly*, both nondestructive and semidestructive operations might be permitted. The selective disassembly process results in three types of outputs:

1. *Homogeneous components*, which cannot further be physically separated. Covers and casings, frames and parts removed from chassis of electronic products are typical examples of homogeneous components.

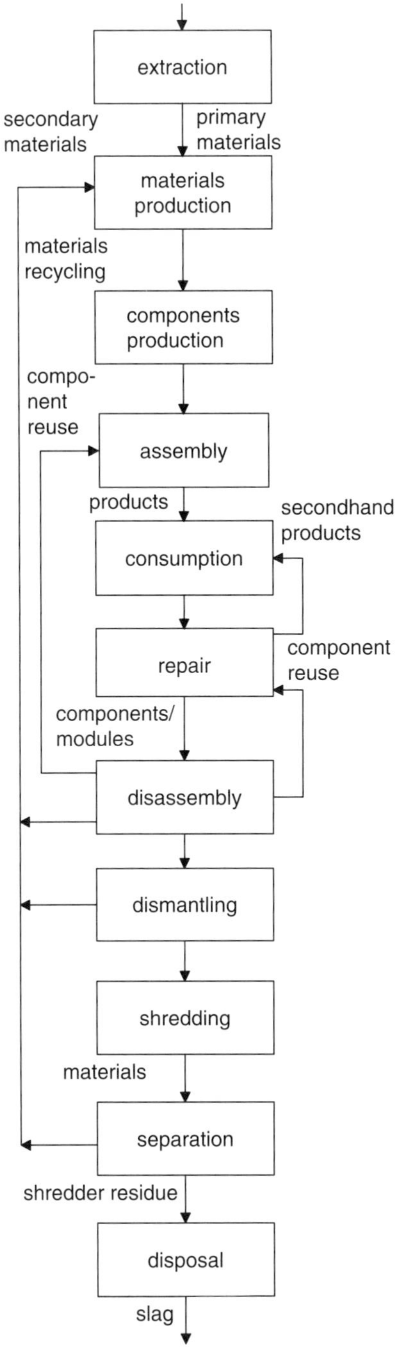


**FIGURE 2.4** Disaggregated product-process chain of a complex product, with reuse and recycling.

2. *Complex components* consist of several discrete homogeneous subcomponents but are normally not further disassembled because they are often connected via fasteners that require destructive disassembly for separation. Examples include cathode ray tubes, printed circuit boards, switches, rotors, stators, and transformers. Sometimes, individual electronic components, such as capacitors, also fall into this category.
3. *Modules* can normally be further disassembled, but sometimes are not as they possess their own functionality and thus may be reusable as such. Examples include electric motors, populated printed circuit boards, optical units, cables, engines, and batteries.

The presence of nonreversible connections in complex components is one of the reasons why end-of-life disassembly cannot be carried out to its full extent. A second reason is that the modules might be more valuable intact. When material recycling is strived for, the detachment of two homogeneous components that consist of the same or compatible material does not make sense. Finally, there is a tradeoff between costs and benefits, either environmental or financial, which results in an optimum disassembly depth that is different from complete disassembly.

End-of-life selective disassembly serves a variety of purposes, namely (Lambert, 1999):

1. Recovery of modules and components used for remanufacturing, spare parts, and secondary ("as new") modules and components for new products.
2. Removal of hazardous modules, components, and materials.
3. Regulatory requirement for removal of hazardous and nonhazardous components.
4. Removal of components that obstruct the removal of the components of interest.
5. Recovery of valuable materials.
6. Enhancement of the purity of materials.
7. Decrease of the quantity of shredder residue.
8. Increase of the quality of shredder residue.

It is obvious from Figure 2.4 that the process of selective disassembly influences the subsequent end-of-life processing steps. In particular, the quantity and quality of the shredder residue are crucial parameters that determine the extent of disposal costs. Tightening regulations have started to result in:

1. Avoiding the use of hazardous substances in the product.
2. Mandatory removal of hazardous substances (using selective disassembly) prior to shredding.
3. Ban on landfilling shredder residue.
4. Incineration of shredder residue separately from municipal and industrial waste.

5. Ban on the use of slag that results from incineration of shredder residue, in buildings and construction materials.
6. Obligatory disposal of incineration slag in landfills that are dedicated for hazardous waste.

It is noted that the disposal costs of shredder residue tend to increase over time, which encourages more and more use of selective end-of-life disassembly.

### 2.3.5 Bulk Recycling

Bulk recycling is a process in which an entire batch of products is *shredded* into a composite mixture of flakes and then *separated* into constituent components by one or more processes. Bulk recycling is frequently applied to electric and electronic scrap. Such processes have been modeled by several authors, see, for example, Weissmantel et al. (1997); Sodhi et al. (1999); Reimer et al. (2000); Sodhi and Reimer (2001); Stuart and Lu (2000ab); Knight and Sodhi (2000); Lu et al. (2000); Spengler et al. (2003).

#### 2.3.5.1 Modeling

Typical steps in modeling and analyzing bulk recycling processes consist of the following:

1. The *objective* of the model is stated.
2. A mass flow diagram consisting of acquisition, sorting, disassembly/dismantling processes, and bulk recycling process of the facility is developed (see Figure 2.5). The mass flow diagram duplicates the context in which the bulk recycling facility operates.
3. The bulk recycling process is examined. It typically is a chain of various shredding and eparation processes. Some loops are also there, which offer the possibility for *reprocessing*, i.e., a repeated application of shredding and separating processes on the materials in the same installation. An example configuration of a bulk recycling chain is presented in Figure 2.6.
4. The bulk recycling process is mathematically formulated. This includes formulation of constraints such as capacity limitations of the different shredding and separation processes.
5. Additional input data and assumptions used for the recovery efficiencies of the different separation methods are identified.
6. The model is used to obtain the optimal solution. Often, mathematical programming methods particularly linear programming and mixed integer linear programming are used for modeling. Comparative modeling, in which a discrete number of process configurations are compared with each other, is also applied.

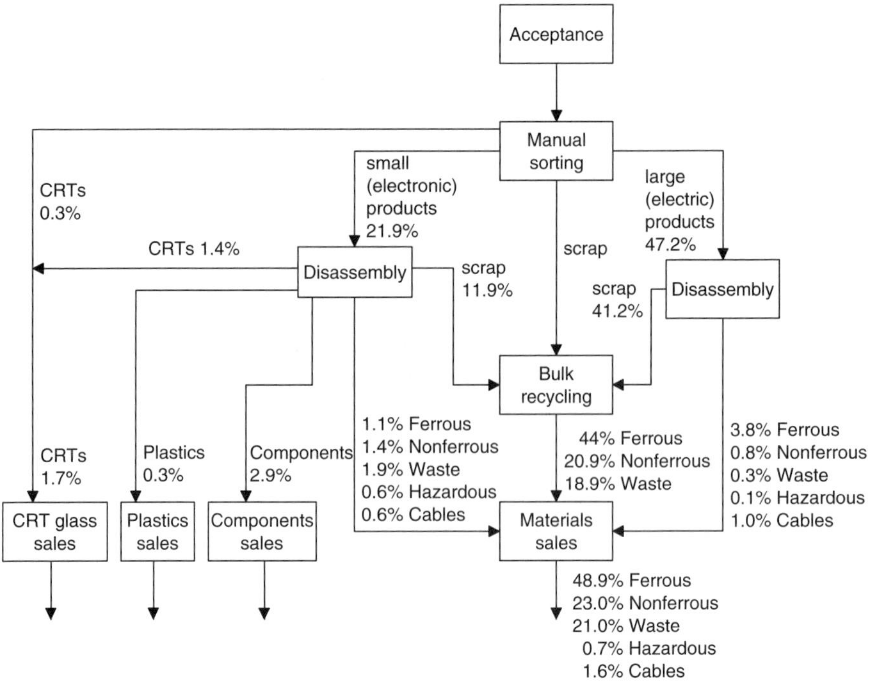


**FIGURE 2.5** Example of a scrap electronics recycling facility (Adapted from Spengler et al., 2003).

#### 2.3.5.2 A Typical Recycling Plant

A typical facility where a mix of electric household appliances, electronic products, and an assortment of electronic scrap are processed is depicted in Figure 2.5 (Spengler et al., 2003). The electronic scrap partly originates from professional appliances that typically contain a larger content of valuable materials than consumer goods. Process waste is also a major part of the input of many recycling companies, which often results in a higher recovery efficiency compared to the case in which only postconsumer products were to be processed. This may misrepresent the overall efficiency rates of such plants. In practice, a considerable quantity of never-used surplus products, intended for withdrawal from the market, is also processed. Because these represent a rather homogeneous and well-documented batch of products, this may also influence the overall figures.

Not every product will run through the complete line. The sequence of processes depends on factors such as the requirements of the suppliers and demand in the market. Some plants sell second hand components or remanufactured products. Normally, disassembly is carried out to a minor extent only. It is usually restricted to the removal of cables, hazardous components and fluids, CRT units, some plastic shells, and some valuable modules and complex components.

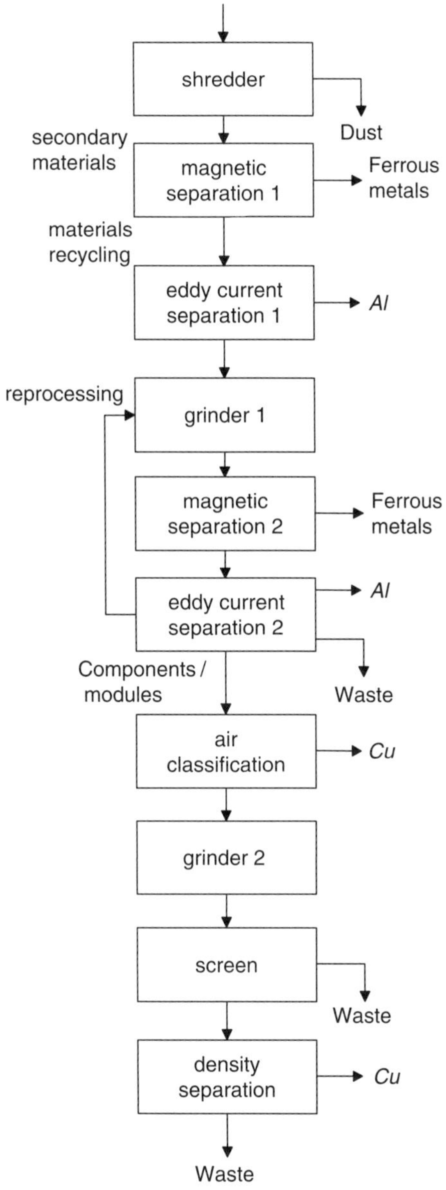


**FIGURE 2.6** Typical bulk recycling sequence in an electronics scrap recycling facility. (Adapted from Spengler et al., 2003).

The following data on capacity, taken from different facilities, indicate the order of magnitude of a large facility:

- Total processing capacity of the order of 100,000 t/yr
- Disassembly line capacity around 60,000 t/yr

- CRT processing line capacity about 40,000 t/yr
- Bulk recycling line capacity roughly 15,000–20,000 t/yr

#### 2.3.5.3 Typical Bulk Recycling Sequence

The supplied products are first separated in low-quality and high-quality items. Examples of high-quality products include computer systems, and telecom products. Examples of low-quality products include peripheral products, entertainment electronics, small white goods and what remains after monitors and TVs are dismantled. A simple bulk processing line consists of a shredder, a magnetic separator that removes the ferromagnetic fraction, and an eddy current separator that removes the light metals fraction (such as aluminum). What remains are a mixture of copper granulates and small fraction of plastics and ceramics. Typical proportions obtained from this process are:

- Ferrous fraction 42%
- Light metals fraction 8%
- Copper granulate 50%

The copper granulate contains about 18% Cu. Further increase in copper content can be established with air classification. The grade of precious materials is about 15–20 ppm Au; 300 ppm Ag; 5–10 ppm Pd for granulate from low-quality items and 70 ppm Au; 500 ppm Ag; 35 ppm Pd for granulate from high-quality items.

The bulk recycling sequence that is depicted in Figure 2.6 consists of several shredding processes, which produce a progressively reduced size of chunks, thus increasing the homogeneity of the individual particles. A reprocessing loop is intended for further reduction of the largest chunks for unlocking of the desired materials. A principal parameter for every separation process is its separation efficiency $\eta_{sep}$, which has to be defined for every separation step $j$ and every material $k$

$$(\eta_{sep})_{j,k} = \frac{\text{mass of separated material } k}{\text{mass of processed material } k} \tag{2.2}$$

Modelers are generally interested in the following problems:

1. Determining the sequence of the separation steps (when a mixture of materials has to be separated in the most efficient way) (Sodhi et al., 1999).
2. Comparing a discrete number of different sequences (Weissmantel et al., 1997).
3. Selection of reprocessing options under capacity constraints (Stuart and Lu, 2000ab).
4. Combined decision support on composition of supply, disassembly depth, and outsourcing, under capacity and demand constraints (Spengler et al., 2003).

These models are mentioned here because of their association with disassembly even though they do not consider the disassembly steps in detail. However, many of these models principally deal with scheduling issues. Therefore, they will not be discussed here in-depth.

### 2.3.6 Typical Costs and Revenues

In this section estimates of costs and revenues are compiled from numerous sources from the literature to provide a sense of their order of magnitude. However, it must be stressed that the data for different materials may fluctuate widely because of the volatile nature of the markets. In addition, the prices often depend on local circumstances. Other factors include purity, homogeneity, availability, supply and demand of the materials.

Different authors have assumed different prices for same scrap materials (see Table 2.1). Note that the prices are expressed in $/kg and are considered approximately equivalent to €/kg (other units were converted into $/kg). Negative figures in the tables mean that a fee had to be paid for the delivery of the respective material to the receiver.

Additional relevant data for components are compiled in Table 2.2. Typical costs for bulk processing adapted from the literature and compiled in Table 2.3.

Disposal costs are country dependent. Within large countries, such as the USA, considerable regional differences also exist. Estimates of disposal costs have been taken from the literature and from actual data. These are compiled in Table 2.4.

For electric and electronic postconsumer waste, the tariffs are different for different products. Some tariffs are compiled in Table 2.5.

### 2.3.7 Epilogue

As a general rule, disassembly costs are significantly higher than the revenues that can be obtained from nonprofessional end-of-life products. If only the minimum requirements for compliance with the law to remove hazardous components and fluids were to be met, a partly dismantled product would be left. Although further disassembly or dismantling is possible, the revenues anticipated may not offset the considerable cost associated with it. The existing practice of shipping significant amounts of nonreusable computer equipment to the Far East for further processing is estimated to be as high as about 80% of the U.S. electronic scrap production (Backthru, 2002). This is because the labor cost is still low in some of the Far Eastern countries. However, sound labor safety standards are not always assured in facilities that perform the disassembly/dismantling operations there. In addition, hazardous substances may find their way into those countries. Consequently, both labor forces and the local environment may be exposed to unacceptable risks, which is at odds with the intention of acting in an environmentally conscious and ethical way. Examples of possible hazardous substances include flame-retardants, mercury, heavy metals, lead in CRT units and polychlorinated biphenyls (BAN and SVTC, 2002). It is evident that strict monitoring, if not a complete ban, on the exportation and handling of product waste in developing countries is essential in dismissing such practices. Performing safe disassembly procedures should be promoted. In addition, design for disassembly must be encouraged to reduce the costs of disassembly operations.

**TABLE 2.1**
**Prices for Materials Scrap from Various Sources, in $/kg**

| Material | Åk | Smi | Gup | Das | Chen | Stu | Lu | Dini | Sodhi | RecW | SpM | Gao |
|---|---|---|---|---|---|---|---|---|---|---|---|---|
| Ferrous | 0.02 | 0.04 | 0.11 | 0.045 | 0.11 | 0.05 | 0.11 | 0.1 | 0.10 | 0.085 | | 0.02 |
| Stainless steel | | 0.33 | | 0.51/0.22[b] | 0.78 | | | | | 0.6 | | 0.23 |
| Nonferrous | | | 0.88 | | | 0.62 | | | | | | |
| Al | 0.60 | 1.89 | | 1.08/0.42[c] | | | | 0.61 | 1.06 | 1.57 | | 0.49 |
| Cu | 1.25 | 0.18 | | 2.20/0.45[d] | 0.79 | | | 0.31 | 2.16 | 2.25 | | |
| Brass | | | | | 0.96[e] | | | | | 1.15 | | |
| Ni | | | | | 2.78 | | | | 4.92 | | | |
| Pb | | | | | 2.26 | | | | 0.46 | 0.42 | | |
| Zn | | | | | 0.37 | | | | 1.06 | 2.20 | | |
| Sb | | | | | 1.34 | | | | 5.18 | | | |
| Precious | | | | | 174.00 | | | | | | | |
| Au | | | | | | | | | 8,566.00 | | 10,200 | |
| Ag | | | | | | | | | 75.80 | | 143 | |
| Pd | | | | | | | | | 11,065.00 | | 10,200 | |
| Pt | | | | | | | | | | | 19,000 | |
| Plastics | 0.05 | | | 0.045 | | | | 0.02 | 0.22 | 0.02[f] | | 0.11 |
| ABS | | 0.33 | | | | | | | | | | |
| PC | | | | | | | | | | 0.57 | | |
| PE | | | | | 0.30 | | | | | | | |
| PS | | 0.04 | | | 0.36 | | | | | | | |
| POM | | 0.04 | | | | | | | | | | |
| PVC | | | | | 0.24 | | | | | | | |
| Mix | | | | –0.24 | | –0.05 | | | | | | |
| Foam | | | | 0.00 | 0.08 | | | | | | | 0.06 |
| Glass | | | | | | | | 0.15 | | | | |
| CRT glass | | –0.25 | | | | –0.19 | –0.19 | | | –1.00[g] | | |

**TABLE 2.1**
**(Continued)**

| Material | Åk | Smi | Gup | Das | Chen | Stu | Lu | Dini | Sodhi | RecW | SpM | Gao |
|---|---|---|---|---|---|---|---|---|---|---|---|---|
| PWBs | | 0.08 | | 2.07 | | 1.67 | | 0.00 | | 2.20 | | 0.50 |
| Cables | | | | 0.40 | | 0.11 | | | | | | 0.18 |
| Batteries | | | | –0.50 | | | | | | –1.65 | | |
| Waste | | | | –0.11 | | | | | | | | |
| PC[a] | | | | | | | | | | –0.02 | | |

The sources of the data on prices have been indicated in the columns with abbreviations as follows:
Åk: Åkermark (1997) (with 10 SEK = 1 €)
Smi: Smith et al. (1995)
Gup: Gupta and Isaacs (1997)
Das: Das and Matthew (1999)
Chen: Chen (2001) (with 1 HK$ = 0.13 US$ = 0.13 €)
Stu: Stuart and Lu (2000a)
Lu: Lu et al. (2000)
Dini: Dini et al. (2001) (with 1936 ITL = 1 €)
Sodhi: Sodhi and Reimer (2001)
RecW: Recycler's World (2002)
SpM: Spot Market
Gao: Gao et al. (2002)

[a] Whole PC scrap
[b] Mixed
[c] Mixed
[d] Mixed
[e] Alloys included
[f] Computer shells
[g] CRT scrap. 1 CRT unit = 9 kg (exclusive deflection unit)

**TABLE 2.2**
**Prices that are Paid by Buyers of Car Subassemblies for Further Processing**

| Car Components and Modules[a] | Price ($/kg) |
|---|---|
| Car body | 0.065 |
| Motor block | 0.065 |
| Transmissions | 0.18 |
| Radiator | 0.82 |
| Battery[b] | 0.11 |
| Catalytic converter[c] | – 0.25 |
| **Electronic Components[d]** | **Price ($/kg)** |
| Clarkboard | 4 |
| Connector | 0.94 |
| CRT | 0.2 |
| Flatscreen | 0.5 |
| Keyboard | 0.23 |
| Magnet | 0.02 |
| UPS | 0.02 |

[a] Recycler's World, 2002.
[b] The mass of one battery has been assumed to be 13 kg.
[c] The mass of one catalytic converter has been assumed to be about $4^1/_2$ lbs or 2kg. It consists of stainless steel and Pt as a catalyst.
[d] Gao et al., 2002.

**TABLE 2.3**
**Typical Bulk Processing Costs, Compiled From the Literature**

| Activity | Cost ($/kg) | Source |
|---|---|---|
| Shredding small devices | 0.1–0.15 | Furuhjelm et al., 2000 |
| Shredding | 0.05 | Stuart and Lu, 2000a |
| Shredding hulk | 0.019 | Isaacs and Gupta, 1997 |
| Grinding | 0.02 | Stuart and Lu, 2000a |
| Transportation[a] | 0.045 | Stuart and Lu, 2000a |
| Storage[b] | 0.001–0.005 (Fe) | Stuart and Lu, 2000b |
| | 0.008–0.04 (Plastic) | Stuart and Lu, 2000b |

[a] Depends on distance and shipping method. Average distance has been assumed.
[b] Depends on storage time. Cost is inversely proportional to specific gravity, because of volume effects.

**TABLE 2.4**
**Typical Disposal Costs, Compiled From the Literature and Local Tariffs**

| Activity | Cost ($/kg) | Source | Country |
|---|---|---|---|
| Landfill of nonhazardous waste | 0.026 | Dini et al., 2001 | Europe |
| Landfill of nonhazardous waste | 0.033 | Isaacs and Gupta, 1997 | USA |
| Landfill of nonhazardous waste[a] | 0.16–0.2 | | Germany |
| Landfill of nonhazardous waste[b] | 0.03–0.083 | Stuart and Lu, 2000b | USA |
| Landfill of hazardous waste | 0.2 | Isaacs and Gupta, 1997 | USA |
| Landfill of hazardous waste | 0.3–0.5 | Stuart and Lu, 2000b | USA |
| Landfill of shredder residue | 0.12 | LAP, 2002 | |
| Incineration of shredder residue | 0.1 | LAP, 2002 | |
| Incineration of shredder residue | 0.15 | actual tariff (2002) | Netherlands |

[a] Tariff of an arbitrary German landfill.
[b] The authors took tariffs of different U.S. landfills.

**TABLE 2.5**
**Disposal Tariffs and Fees for Some Electric and Electronic Products**

| | Amount ($/item) | | | | | |
|---|---|---|---|---|---|---|
| Product | Landfill(Be) | Incin. (Be) | Fee(Be) | Fee(Ge) | Fee(Ne)[a] | Fee(US) |
| Monitor | 3 | 2.6 | 5 | 38 | n.a.[c] | 5.5–15 |
| TV | 6 | 5 | 6 | 38 | 8 | |
| PC without CRT | 4 | 3.5 | 6 | 13 | n.a. | |
| Printer | 4 | 3.5 | 3 | 13 | n.a. | |
| Washing machine | 14 | 12 | 32 | 38 | 5 | |
| Dishwasher | 8 | 7 | 36 | 38 | 5 | |
| Drying machine | 7 | 6 | 36 | 38 | 5 | |
| Cookery | 10 | 9 | 15 | 38 | 5 | |
| Stove | 8 | 7 | 18 | 38 | 5 | |
| Refrigerator[b] | | | | | 17 | |
| Small white | | | | 1.5 | 1 | |

These Figures are adapted from the following sources:
Belgium: L'Église et al., 2000
Germany: Tariff of a municipal recycling center
The Netherlands: NVMP, 2002
USA: Reimer et al., 2000.

[a] In The Netherlands, one has to pay take back fee when a product is bought. Take back takes place via the retailer and municipal recyling centers.
[b] Refrigerators contain CFCs that have to be captured.
[c] Take back and processing of ICT products (computers, monitors, peripherals, and telephones) are paid by producers and retailers in The Netherlands, in contrast with household appliances and electronic entertainment products.

## 2.4 COMPLEX PRODUCTS WASTE

### 2.4.1 General Waste Streams

Waste is a major problem in modern society. The common practice of disposing waste in landfills is increasingly discouraged in industrialized countries because of the risk of hazardous substances causing harm to the environment, for example, leaching of toxins from the waste could contaminate ground water. Besides this, appropriate areas for new landfill sites are virtually exhausted, particularly in densely populated regions.

It is difficult to give a precise definition of waste. However, virtually all materials that have been modified or diverted by humans (including moved soil) and subsequently discharged to the environment, can be considered waste, even if a negligible risk potential is associated with this. However, all wastes have some risk. For example, the vast quantities of processed soil and rock, which are associated with ore mining and refining (tailing), can sometimes be extremely hazardous. Then again, only those materials are considered wastes for which statistical records are kept. For example, in The Netherlands, statistics is kept for such waste streams as surplus manure, and contaminated sludge. A large contribution of these waste streams consists of water, sand, and other inert materials. If restricted to the solid waste, some typical waste categories are documented in the statistical data from industrialized countries, which are significant from a materials recycling point of view (Lambert, 2001). A major part of these streams consists of process waste, packaging materials, glass containers, and other wastes that originate from simple products.

In order to get a sense of solid waste production in the industrialized countries, some data is presented in Table 2.6. The quantities listed in the table have been derived from U.S. German, Dutch, and U.K. statistics. However, it must be noted that although statistics are not always reliable because, in some cases they might be up to 10 years old, contradictory, incomplete, not clearly defined, and not fully comparable, they, nevertheless provide information, at least in the order of magnitude. For attuning the data, some calculations were necessary. For example, if data were available on sales figures, the waste had to be estimated by multiplying those figures with the average mass of the products.

**TABLE 2.6**
**Principal Solid Waste Production in the Industrialized Countries**

| Waste Stream | Quantity (kg/capita-yr) |
|---|---|
| Industrial process waste | 1000 |
| Construction and demolition waste | 500 |
| Ordinary municipal waste[a] | 500 |
| Coarse municipal waste[b] | 50 |
| Discarded cars | 50 |
| Electric/electronic/small mechanical products | 25 |

[a] Included are the separately recollected wastes: glass, paper, organic waste, and fabrics.
[b] Furniture, carpets, mattresses, etc.

From this table it can be observed that, from a purely quantitative point of view, the waste generation from the end-of-life complex products is relatively modest. A large amount of this quantity originates from mechanical products, particularly from cars. This type of product has typical recovery efficiency up to 80% by weight, mainly due to its relatively high metals content and the extent of its nonmetallic components. The presence of valuable components makes the disassembly of such products relatively profitable as compared to that of the smaller products.

German data on waste sent to landfills is given by Weber (1991), in percentage by weight. These include 30% industrial waste; 60% municipal waste; 9% coarse waste; and 2% shredder residue. The shredder residue consists of 44% dust; 30% plastics; 9% metals; 9% fibers; and 8% rubber. The metals can be recovered, if desired, by additional processes, which include a further decrease of the particle size, and additional separation stages in bulk recycling.

## 2.4.2 Data on Complex Products Waste

In order to get a sense of the extent of complex products waste, data collected from various sources is presented in this section.

### 2.4.2.1 Introduction

There are four important categories of complex products that add to solid waste:

- Buildings and related infrastructures
- Ordinary complex products, such as furniture
- Vehicles
- Electric, electronic, and small mechanical products

We will briefly discuss these categories, with an emphasis on the last two categories.

### 2.4.2.2 Buildings and Infrastructure

In general, there are two types of waste streams associated with buildings and related infrastructures:

1. *Construction waste*, which represents the process waste that results from the creation of buildings and related infrastructure.
2. *Demolition waste*, which results due to modernization and demolition of buildings.

According to the available data, U.S. generates about 11 Mt/yr of construction waste and 128 Mt/yr of demolition waste. Data from the year 2000 suggest that Flanders and the Netherlands generate about 7 Mt/yr and 15 Mt/yr, respectively of construction and demolition waste. If these figures are analyzed based on the number of inhabitants, the resulting figures suggest that about 500 to 1000 kg/capita-yr of waste is generated in these countries, which is of the order of the data presented in Table 2.6.

The composition of construction and demolition waste in Flanders (Fl) and The Netherlands (Ne) is shown in Table 2.7

**TABLE 2.7**
**Composition of Construction and Demolition Waste in The Netherlands and Flanders**

| Component | Ne (%) | Fl (%) |
|---|---|---|
| Concrete | 42 | |
| Brick | 26 | |
| Various debris | 6 | |
| Stony materials | | 80 |
| Asphalt | 20 | 15 |
| Noninert[a] | 6 | 5 |

[a] Wood, fabric, paper, plastics.

The origin of construction and demolition waste in The Netherlands is listed in Table 2.8.

The recycling efficiency of these materials is up to 90% in The Netherlands and is aimed at 80% in Japan (Gao et al., 2001).

The composition of demolition waste from domestic buildings in the Upper Rhine Valley, a region that is partly French and partly German, is given in Table 2.9 (Spengler et al., 1997).

Within houses, the technical installations, such as the heating system, can be considered as mechanical or electric products. The house itself can also be considered as a complex product, but with a significantly different structure and materials composition than what is present in electric, electronic, and mechanical products.

### 2.4.2.3 Ordinary Complex Products

These products consist mainly of wood, fabrics, and steel. A detailed analysis of these products is not a subject of this book.

**TABLE 2.8**
**Origin of Dutch Construction and Demolition Waste**

| Origin | Contribution (%) |
|---|---|
| Residential | 20 |
| Industrial | 49 |
| Infrastructural | 31 |

**TABLE 2.9**
**Composition of Demolition Waste of Residential Buildings**

| Component | Share (%) |
|---|---|
| Brick | 50.2 |
| Concrete | 25.5 |
| Gypsum | 9.2 |
| Wood | 13.4 |
| Steel | 0.9 |
| Nonferrous materials | 0.2 |
| Paper and plastics | 0.6 |

*Source:* Spengler et al., 1997

### 2.4.2.4 Vehicles

Of the many types of vehicles that exist, only automotive vehicles are considered here such as trucks, vans, and cars. Trucks usually have a long life and refurbishing is frequently practiced. In recent years, recycling and take back of trucks is being advocated, particularly in the European Union (EU), along the lines of cars, which are already established. Recovery of cars for their valuable components, steel, and other materials has been practiced for a long time. The infrastructure for recycling cars in the U.S. (consisting essentially of two stages; dismantling and shredding) has been operating economically for about three decades (Isaacs and Gupta, 1977). Similar infrastructure exists in the EU where there are more than 200 shredder plants in operation.

Data from the EU countries[1] on new car registrations (European Shredder Group, 2002) show a gradual increase in the number of new car registrations from 11,671,000 units in 1995 to 14,460,000 units in 2001. In the EU, there were 160,784,000 cars in use in 1995, and 175,749,000 in 1999. U.S. figures show about 130,000,000 cars in use and a disposal rate of about 10,000,000 cars per year (Bandivadekar et al., 2002). The total number of cars in the world is estimated to be 475,000,000. Assuming an average life of 11 years results in a worldwide disposal rate of over 40,000,000 cars per year.

An average mass of 900 kg/car and a recovery efficiency of 80% would result in about 3.5 Mt/yr of automotive shredder residue (ASR) in the EU alone. The estimated figure for the U.S. is about 5.5 Mt/yr, mainly from cars and white goods. The global figure of ASR production is estimated to be at least 10 Mt/yr (Van Roosmalen, 2000). In the past ASR was landfilled. However, since ASR was heavily contaminated with oil, gasoline, sulfuric acid, and other working fluids, it threatened the groundwater and the environment.

[1] Austria, Belgium, Denmark, Finland, France, Germany, Greece, Ireland, Italy, Luxembourg, The Netherlands, Portugal, Spain, Sweden, and the U.K.

Therefore, draining of working fluids and battery removal prior to shredding is now obligatory in most industrialized countries. Detachment of large nonmetallic components prior to shredding is also stimulated aimed at reducing the production of ASR. The latter is vital since there has been a strong tendency toward substituting metal parts with other parts made with lighter materials (Isaacs and Gupta, 1997; Boon et al., 2000, 2003).

#### 2.4.2.5 Electric and Electronic, and Small Mechanical Products

It is difficult to organize the large variety of products that are included in this category into precise subgroups. Apart from the variety, the products may be professional or household appliances. Professional machines are often refurbished and reused. In some regions, small electromechanical tools, such as electric drills, are subjected to take back obligations as they contain electric motors, batteries, transformers, power electronics, and related components. Even completely mechanical tools may be contaminated with oil. Heaters, boilers, and other heat generating appliances are other types of products that have to be processed before disposal. Unfortunately, except for sales statistics, data on the recollection of these types of products are not readily available.

Electric and electronic products are broadly subdivided into *white goods*, *brown goods*, and *grey goods*. White goods, made up mostly of electric products, can be further divided into *large white goods* (e.g., refrigerators, washing machines, dryers, dishwashers) and *small white goods* (e.g., electric irons, vacuum cleaners). Brown goods customarily are made up of electronic entertainment products such as TVs and audio/video equipment. Grey goods typically include information and telecommunication products such as CRTs, personal computer (PC) systems, notebooks, and telephones.

Grouping the vast variety of complex products into categories can have a profound practical use, because disassembly facilities are usually interested in a restricted range of products only. These categories of related products are called *disassembly families* (Das et al., 2000).

The contribution of electric, electronic, and small mechanical products to the waste stream is continuously increasing. Table 2.10 presents the contribution of some principal household appliances to this stream. As before, the data in this table are coarse approximations only.

**TABLE 2.10**
**Principal Product Categories of the Electric, Electronic, and Small Mechanical Products**

| Waste Stream | Quantity (kg/capita-yr) |
|---|---|
| White goods (refrigerators, washing machines, etc.) | 10 |
| Electronic appliances (excluding monitors and TV sets) | 2 |
| Monitors and TV sets | 2 |
| Computer systems, keyboards, peripheral devices | 1 |
| Printed circuit boards (PWBs) | 0.5 |

From the table it is clear that a major part of this category consists of white goods, a large percentage of which is ferrous metals. Electronic appliances, monitors, TV sets, computers, and related products are quite different from that. In the first place, these are smaller. A mass that ranges from one to a few kg is typical. A large portion of small but complex electronic appliances, including mobile phones, digital cameras, global positioning systems (GPS), clocks, and palm computers typically have a mass well below one kg. Apart from the casings and frames, such products may include PWBs, CRT units, cables and wiring, connectors, and electromechanical units such as electric motors, transformers, and switches.

There are a lot of complex components present in almost all electronic products. The components contain a broad spectrum of materials including precious metals such as silver, valuable materials such as copper, and hazardous materials such as lead and batteries. Further breakdown via dismantling methods might be possible, but in most cases the only economically viable recycling method is bulk recycling.

Most of the materials used in electronic products are of the lasting type, which means that degradation takes place at a very slow rate. However, hazardous substances and fluids are susceptible to leaching into the ground.

The demand for electronic products is constantly increasing, which means that the number of end-of-life products is also increasing. The impressive replacement market created by newer models of products forces most of the existing products to become obsolete within a few years, even if they are in perfect working order. In addition, new types of application and technology are continuously being introduced in the market.

The growth in volume and variety of the products is, however, counteracted by miniaturization and dematerialization. Miniaturization is beneficial from technological, economic, and environmental points of view. Because of the resulting cost reduction, the frequency and variety of applications increase. Electronic devices are rapidly becoming part of virtually every product, from washing machines to cars to houses.

Dematerialization also results in positive outcomes. For example, instead of using a physical chip, chips printed on virtually every product, including packaging materials is now theoretically possible that could provide valuable product information that should be beneficial to the recovery industry. In some instances, virtualization of some interfaces is also practiced, for example, a virtual keyboard on a desktop that is projected and detected via laser light could be used instead of a physical one. These kinds of developments will inevitably have an effect on the recycling industry.

Of course, rapid technological changes could quickly make the existing recycling techniques obsolete, for example, the market for secondary materials might suddenly collapse. This phenomenon could be seen in CRT glass recycling where the demand for recycled glass has dropped sharply because CRTs are rapidly being replaced by liquid crystal display (LCD) and plasma screens. Other applications of the materials obtained from recycled CRTs are economically less viable.

Therefore, it is clear that the recovery of end-of-life complex products is extremely important yet challenging. Its solution will be beneficial to the society and the environment, and will trigger further technological development, combining both well-designed products and sophisticated new technologies.

### 2.4.2.6 Statistics on Electric and Electronic Products

Detailed statistics on electric and electronic end-of-life products are available from Germany (Ge), The Netherlands (Ne) and the Belgian City of Brussels (Br). For comparison purposes, note that the approximate number of inhabitants in Germany, The Netherlands, and the City of Brussels are 80, 16, and, 1 million, respctively. German data is published by the German Ministry for the Environment (Umweltbundesamt) (also see Knapp and Jansen (1993) and Angerer et al. (1993)). The Dutch figures are due to Van Bezooijen (2000). The Belgian figures were adapted from L'Èglise et al. (2000). (See Table 2.11). Note that further explication of the German data led to a yearly scrap supply of 18 kg/capita while the Dutch figures show a yearly supply of 8.3 kg/capita. The difference is partly due to the way the contributions of professional goods are reported. On the other hand, it also reflects the vagueness of available statistical data.

The German Ministry for the Environment also publishes statistics on scrapped investment goods. The supply of these goods is approximately 40% of the total scrap supply (see Table 2.12).

Additional data has also been reported. The U.S. had an annual supply of 200 kt of discarded PCs in 1995 (with an average mass of the complete system of 33 kg), which corresponds to about 6 million units. The global supply of discarded PCs amounted to about $25 \times 10^6$ units in 1995. Estimates for the year 2005 include $63 \times 10^6$ discarded computers in the U.S. and $150 \times 10^6$ PCs worldwide (Arner, 1997). For other estimates, see Boon et al. (2002). Belgian estimates show figures that increase from 100,000 scrapped computers in 1997, to 200,000 in 2000 to 400,000 in 2003 to 600,000 in 2006 and to 800,000 in 2010 (De Fazio et al., 1997). The Belgian population is about 10 million people. For the City of Brussels, with

**TABLE 2.11**
**Estimated Amount of Domestic Electric and Electronic Product Waste**

| | kt/yr | | $10^3$ items/yr | | kg/item | kt/yr |
|---|---|---|---|---|---|---|
| **Product** | **Ge(1993)** | **Ge(1999)** | **Ne(1998)** | **Br(2000)** | **Mass (est.)** | **Ne(1998)** |
| **White** | 585 | 630 | | | | |
| Refrigerator | | | 600 | 45 | 34 | 20 |
| Other large | | | 1,100 | 147 | 60 | 70 |
| Small white | | | 4,700 | | 4 | 18 |
| **Brown** | 270 | 400[a] | | | | |
| TV | | | 475 | 280[b] | 30 | 15 |
| Audio | | | 500 | | 4 | 2 |
| **Grey** | | | 1,300 | 320 | 8 | 10 |
| Telecom | 75 | | | | | |
| Computer | 75 | 110 | | | | |

[a] Telecom included.
[b] Monitors included.

**TABLE 2.12**
**German Statistics on Professional Electric and Electronic Scrap**

| Product | kt/yr Ge(1993) | kt/yr Ge(1999) |
|---|---|---|
| PWBs | 25 | |
| Lamp screw-caps | 80 | |
| Professional | 600 | |
| Office equipment | | 140 |
| Industrial electronics | | 360 |
| Medical equipment | | 50 |
| Domestic + Professional | 1,500 | 1,800 |

its one million inhabitants, a yearly supply of 17,000 t of electric and electronic scrap translates into a yearly supply of 17 kg/capita (L'Église et al., 2000).

For making valuable estimates on the future supply of end-of-life products, data on production rate, amount of units in use, annual growth rate, and average lifetime distribution are needed. Some indicative figures are as follows:

- The *global production* of monitors was expected to be 60,000,000 units/yr in 1998 (Smith et al., 1995).
- The *number of electric and electronic appliances* in German households was about $900 \times 10^6$ in 1999, of which $40 \times 10^6$ were TV sets.
- The German Ministry for the Environment estimates the growth rates for the annual supply of discarded items for the period from 1998 to 2005 as shown in Table 2.13.
- Data on *average lifetime* distribution are due to De Fazio et al., (1997). These refer to computers and indicate that 10% are discarded after 6 years. Subsequently, 20% are discarded after 7 yr, 40% after 8 yr, 20% after 9 yr, and 10% after 10 yr.

**TABLE 2.13**
**Estimated Growth Rate of the Supply of Discarded Products, (1998–2005)**

| Product | Growth Rate |
|---|---|
| Refrigerators | + 20% |
| Other large white | + 26% |
| TV sets | + 40% |
| Audio + small white | + 75% |
| ICT | + 270% |

*Note:* The statistics compiled here simply gives an order of magnitude of the quantities that are available or may become available for recovery. Recent developments, such as notebooks, mobile phones, and LCD screens are virtually not included in any of the statistics. Japanese data estimated an annual turnover of about $11 billion for large LCD screens (Thin film transister-LCD, i.e., TFT-LCD) and of about $22 billion for CRT units for the year 2000 (JTEC, 1992). The turnover for TFT-LCD screens is rapidly increasing and that of CRT units is stabilizing. Other figures indicate that the sales of DVD players are increasing and that of the video players are decreasing. These trends make the allotment of products into categories rather blurry. Nonetheless, for policy and management decisions, we have to rely on estimates and extrapolations.

## 2.5 BREAKDOWN ANALYSES OF SOME PROMINENT COMPLEX PRODUCTS

### 2.5.1 INTRODUCTION

There are three types of breakdown analyses:

1. *Component analysis* breaks down a product into its principal component categories such as casings, chassis, fasteners, PWBs, and CRTs. Some of these are homogeneous but many are complex consisting of multiple materials.
2. *Materials analysis* breaks down the product into its principal materials categories such as glass and nonferrous metals, without going into too much details.
3. *Elemental analysis* breaks down the product into its elements, as provided in the periodical table. Such an analysis is particularly important if trace elements have to be detected. This type of analysis is usually carried out via spectral analysis.

In practice, however, an analysis frequently has a *hybrid* character: By carrying out a set of disassembly operations, a product breaks apart in some homogeneous components that can be assigned to some materials category or some easily classifiable components category. If materials analysis is required, destructive operations such as crushing and subsequent separation are needed prior to the materials analysis. If this is not feasible, one can make use of estimates or standard values, such as an average for PWBs. The elemental analysis is more complex and requires technologies that are available in laboratories.

Available statistics on the supply of discarded products are not sufficient to estimate the materials that can be expected from the recovery of end-of-life products. It is, however, informative to gain insight in the composition of the different product categories. In the subsequent subsections, we present and discuss data on product compositions that have been compiled from the literature.

**TABLE 2.14**
**Breakdown Analysis of a House into its Construction Elements**

| Element | Share (%) |
|---|---|
| Walls | 37 |
| Ceilings | 29 |
| Foundations | 16 |
| Floor covering | 4 |
| Roofing | 3 |
| Wall paneling | 3 |
| Ceiling covering | 2 |
| Stairs | 2 |
| Roof frame | 2 |
| Doors, windows | 1 |

*Source:* Spengler et al., 1997

## 2.5.2 Houses

Houses vary in weight and composition. Typical weights that are mentioned in the literature vary from 50 t for a Japanese wooden house (Gao et al., 2001) to 220 t for a freestanding house that is mainly erected from stone (Spengler et al., 1997). The breakdown analysis of a house is given in Table 2.14.

The mass of the mechanical and electrical components in a house are given in Table 2.15. It is interesting to observe that the technical components of the house add up to 1100 kg or about 0.5% of the weight of the house.

## 2.5.3 Cars

The recycling of end-of-life cars has been practiced economically for several decades. Cars are of many brands and types with observable differences between

**TABLE 2.15**
**Mechanical and Electric Components of a House**

| Component | Mass (kg) |
|---|---|
| Plumbing | 400 |
| Fastening elements | 400 |
| Heating and ventilation | 210 |
| Sanitary | 82 |
| Electric installation | 8 |

*Source:* Spengler et al., 1997

the typical U.S., Japanese, and European cars. The composition of cars continues to change in favor of lighter materials such as plastics and nonferrous materials to enhance fuel economy and other addition of ancillary systems such as airco, airbags, and board electronics. The content of ferrous metals declined from 70 to 61% during the period from 1990 to 2000 in European cars (Samel, 2001). During the same period, the share of nonferrous metals increased from 5 to 9.5% and that of plastics from 9 to 11.5%. However, the average mass of a European car remained at about 900 kg. Weight savings via the use of plastics and light metals have been compensated by ancillary systems. Similar trend influenced the American cars with the average mass in the low 1300 kg (Isaacs and Gupta, 1997; Boon et al., 2000). The anticipated waste stream generated due to the introduction of hybrid or *clean vehicles* to the market may also have an effect on the current recycling infrastructure (Boon et al., 2003).

Some materials breakdown analyses of the U.S. and European medium-sized cars are given in Table 2.16.

Typically, aluminum, lead, copper, and zinc constitute the major portion of the nonferrous metals in cars. German data from 1986 suggest that a typical car has 4.5% Al; 1% Pb; 1% Cu; 0.5% Zn; and 0.03% Ni (Potzschke, 1991). With the nonmetallic fraction of about 30% in a car, the same data reveals its distribution as 4.7% glass; 9.3% plastics (including 2.1% PUR foam); 9.6% rubber; 3% coating; 2% textile; 1.5% wood. It is a common practice nowadays in the European countries to selectively disassemble cars according to the directive 2000/53/EC of the European Parliament and Council. In The Netherlands, the materials and components shown in Table 2.17 are always removed from an end-of-life car.

The masses in Table 2.17 refer to a standard car with a mass of about 900 kg. Selective disassembly along the Dutch covenant results in a reduction of about 90 kg of mass in addition to the working fluids. With a metals content of about 71% and a potential share of shredder light fraction of 29%, the reduction in the mass of the

**TABLE 2.16**
**Materials Breakdown Analyses of U.S. and European Cars**

| Material | US(1990)[a] | US(1997)[b] | European(1990) | European(2000)[c] |
|---|---|---|---|---|
| Ferrous (%) | 73.4 | 68.8 | 70.0 | 61.0 |
| Nonferrous (%) | 8.5 | 7.6 | 5.0 | 9.5 |
| Plastics (%) | 7.7 | 7.7 | 9.0 | 11.5 |
| Glass (%) | 2.9 | | 5.0 | 3.0 |
| Rubber (%) | 4.6 | | 5.0 | 5.0 |
| Other (%) | 2.9 | 16.9 | 6.0 | 10.0 |
| Mass (kg) | 1340 | 1316 | 906 | 901 |

[a] Samel, 2001.
[b] Isaacs and Gupta, 1997.
[c] ARN, 2002.

**TABLE 2.17**
**Materials and Components Removed from Cars in The Netherlands, Prior to Shredding**

| Name | Standard Mass (kg) |
|---|---|
| **Working Fluids** | |
| Gasoline | 5.0 |
| Cooling fluid | 3.6 |
| Lubricant | 4.9 |
| Window spray fluid | 0.9 |
| Brake fluid | 0.3 |
| **Hazardous Component** | |
| Battery | 13.6 |
| **Other Components and Materials** | |
| Tires | 27.3 |
| Windows | 25.4 |
| Rubber strips | 7.7 |
| Polyurethane (PUR) foam (seats) | 6.3 |
| Bumpers | 5.2 |
| Lights | 1.4 |
| Coir fiber | 0.9 |
| Grill | 0.8 |
| Hubcaps | 0.7 |
| Safety belts | 0.4 |
| Tubes | 0.2 |

shredder residue because of selective disassembly is about 35%. The contamination is significantly reduced because of the draining of the working fluids beforehand. Further reduction in the amount of shredder residue is possible by increasing the disassembly depth. This way, recycling percentages of over 92% can be achieved as has been observed in pilot plants (Hendricks, 1992). In practice, the share of shredder residue has decreased from about 20% in 1990 to about 13.5% in 1998, due to selective disassembly. Often, additional selective disassembly is carried out to harvest valuable parts to be used as spares. A list of the major nonmetallic components in a medium-sized car is given in Table 2.18.

The mass of the components in Table 2.18 add up to 170.92 kg. Although not listed, it should be mentioned that the components with a mass smaller than 1 kg add a further 17.216 kg to this amount, which results in an aggregate of about 188 kg of nonmetallic components. There are an additional 13.8 kg of electric cables. The power train amounts to about 14% of the car's mass consisting of less than 2% of plastics and rubber, 52% of cast iron, 27% of nonferrous metal and 19% of steel.[2] Ebersperger (1995) presented similar figures for a medium sized car with a mass of

[2] Ferrão et al., 2002.

**TABLE 2.18**
**Major Nonmetallic Parts of a Medium-Sized Car**[a]

| Component(s) | Mass (kg) |
|---|---|
| Wheels, hubcaps, tires, tubes | 26.025 |
| Dashboard + heater | 22.348 |
| Console + chairs + mats | 12.005 |
| Front window | 11.680 |
| Front doors + rubber | 8.114 |
| Fuel tank | 7.902 |
| Rear doors | 7.273 |
| Front chairs | 7.230 |
| Trunk lid | 6.560 |
| Rear window | 6.215 |
| Front bumper | 5.205 |
| Rear bumper | 4.882 |
| Rubber strips | 4.435 |
| Back rear bench | 4.413 |
| Trunk lining | 4.100 |
| Seat rear bench | 4.030 |
| Headlights | 3.247 |
| Steering unit | 2.573 |
| Small rear windows | 2.535 |
| Steering wheel | 2.285 |
| Skirts | 2.208 |
| Drop-in mats | 2.120 |
| Side parts trunk | 1.868 |
| Fog lights | 1.740 |
| Casing rear lights | 1.635 |
| Door case | 1.255 |
| Midconsole | 1.245 |
| Engine protection shield | 1.200 |
| Side fenders | 1.164 |
| Wheel houses | 1.154 |
| Triangular windows, rear | 1.142 |
| Air filter | 1.133 |

[a] Adapted from "Project Goes," which was carried out in The Netherlands in 1992.

1070 kg. The mass of the engine was 101 kg, that of the gearbox was 40 kg and the generator weighed 14 kg. The naked hulk of this specific car type had a specified mass of 257 kg. The measured mass of the hulk after selective disassembly was about 315 kg. The difference between the standard and the observed mass is due to sealed mats (7.3 kg), lacquer and coating (21 kg), sand, and some components that are fastened to the hulk. The presence of coatings and contamination makes a

**TABLE 2.19**
**Materials Breakdown of Supply of Electric and Electronic Products from Germany**

| Material | Content (%) |
|---|---|
| Ferrous metals | 47 |
| Nonferrous metals | 9.5 |
| Plastics | 20 |
| Glass | 8.5 |
| Miscellaneous | 14 |

recycling efficiency of 100% unrealistic. Metal parts other than the hulk (chassis parts, doors, etc.), and the working fluids, make up the residual 353 kg of this car.

Further data on materials breakdown analyses, including models for studying the effects of the introduction of plastics and light metals, can be found in the following papers: Samel, 2001; Bandivadekar et al., 2002; Isaacs and Gupta, 1997; Boon et al., 2000, and Ferrão et al., 2002.

## 2.5.4 Electric and Electronic Products

A review paper on end-of-life processing of electronic products is due to Moyer and Gupta (1997).

### 2.5.4.1 Aggregate Figures

Two aggregate sets of data on electric and electronic products are presented here. Even though they represent rough estimates, they do shed some light on their material content. German figures are presented in Table 2.19 (Schubert, 1996).

A breakdown analysis of the electric/electronic scrap for the Belgian City of Brussels is presented in Table 2.20 (L'Èglise et al., 2000).

Next, a compilation of the composition of the various electric and electronic products from various sources is presented.

### 2.5.4.2 White Goods

Some figures on the composition of white goods are compiled from the literature. These include washing machines, refrigerators and freezers, and some small household appliances such as vacuum cleaners, coffee machines, and toasters. Although many of these figures are imprecise, they nevertheless provide a valuable insight of the product composition (Table 2.21).

The following products are included in the table:

S Stove (Weber, 1991)
C Cooker hood (Weber, 1991)

**TABLE 2.20**
**Breakdown Analysis of Electric/Electronic Scrap**

| Material or Complex Component | Content (%) |
|---|---|
| **Materials:** | 76 |
| Ferrous metals | 36 |
| Stainless steel | 2 |
| Nonferrous materials | 0.2 |
| Plastics | 15 |
| CRT-glass | 17 |
| Non-CRT glass | 2 |
| Concrete | 3 |
| Wood | 1 |
| **Complex Components:** | 24 |
| Electron gun (CRT unit) | 0.1 |
| Deflector (CRT unit) | 1 |
| Wires | 3 |
| PWBs | 11 |
| Microprocessors | 0.03 |
| Power units | 5 |
| Transformers | 2 |
| Electric motors | 1 |
| Pump, compressors | 0.3 |
| Capacitors | 0.02 |

*Source:* L'Église et al., 2000

D Dishwasher AEG, 1989 (Potzschke, 1991)
W1 Washing machine (Ebersperger, 1993)
W2 Washing machine (Miele, 1989 and Potzschke, 1993)
R1 Refrigerator (Weber, 1991)
R2 Refrigerator, average (Coolrec, 1997)
Fre Freezer (Coolrec, 1997)
V1 Vacuum cleaner (Kaiser and Hirsch, 1995)
V2 Vacuum cleaner (Weber, 1991)
Cm Coffee machine (Kaiser and Hirsch, 1995)
Toa Toaster (Åkermark, 1997)

### 2.5.4.3 Brown Goods

Quantitative data on brown goods composition are not widely available. Some coarse data are compiled in Table 2.22. Data on recent products, such as CD players and midi sets, are not available.
The following products are included in the table:

Gr Gramophone (Anonymous, 1997)
Vi Video recorder (Anonymous, 1997)

**TABLE 2.21**
**Breakdown of Some Large and Small White Goods**

| Material | S | C | D | W1 | W2 | R1 | R2 | Fre | V1 | V2 | Cm | Toa |
|---|---|---|---|---|---|---|---|---|---|---|---|---|
| | Content (%) | | | | | | | | | | | |
| Ferrous metals | 73 | 48 | 56 | 53.9 | 61 | 63 | 36.1 | 35.3 | 50.7 | 25 | 14 | 56 |
| Stainless steel | | 16 | | | | | | | | | | |
| Nonferrous metals | | 3 | 3 | 2.7 | 5 | | | | 7.4 | 8 | 18 | |
| Cu | | | | | | 3 | 1 | 1 | | | | 5 |
| Al | | | | | | 5 | 7.5 | 11.7 | | | | 2 |
| Light fraction | | | | | | 26 | | | | | | |
| Plastics | 2 | 22 | 12 | 5.5 | 8 | | 12.6 | 8.6 | 35.2 | 60 | 68 | 36 |
| Foam | | | | | | | 9.9 | 11.9 | | | | |
| Resin | | | | | | | 3.7 | 3.4 | | | | |
| Seal | | | | | | | 0.2 | 0.1 | | | | |
| Rubber | | | | 3.4 | | | | | 4.9 | | | |
| Glass | 7 | | | 1.5 | | | | | | | | |
| Wood | | 8 | | | | | | | | | | |
| Dust | 5 | | | | | | | | 1.5 | | | |
| Concrete | | | | 21.6 | | | | | | | | |
| Working fluids | | | | | | 3 | 1.1 | 1.8 | | | | |
| CFCs | | | | | | | 1.1 | 1.0 | | | | |
| Oil | | | | | | | 1.4 | 1.2 | | | | |
| Electric materials | | | | 11.4 | | | 25.3 | 24.1 | 0.3 | | | |
| Compressor | | | | | | | | | | | | |
| Other [a] | 12 | 3 | 29 | | 25 | | | | | 7 | | |
| Product mass (kg) | 70 | 43 | | 80 | | 35 | 33 | 36.7 | 6 | 9 | 1.73 | 1.1 |

[a] "Other" is composed of the materials that are not explicitly quantified.

Cd Cassette deck (Anonymous, 1997)
TV1 TV set (Anonymous, 1997)
TV2 TV set (De Ron and Penev, 1995)
TV3 TV set, aggregate Dutch supply (Van der Hoek, 1994)
TV4 TV set (Dini et al. 2001)
TV5 TV set 14" from the year 1985 (Thomas et al., 1998)

#### 2.5.4.4 Grey Goods

The compositions of some grey goods are listed in Table 2.23. The following products are included in the table:

M1 Monitor (Anonymous, 1997)
M2 Monitor 17" (Smith et al., 1995; Menad, 1999)
Pr Matrix printer (Anonymous, 1997)
C1 PC configuration with system, monitor, and keyboard (Roberts, 1999)
C2 PC configuration with system, monitor, and keyboard (Zhang and Kuo, 1996)

**TABLE 2.22**
**Materials Composition of Some Brown Goods**

| | Gr | Vi | Cd | TV1 | TV2 | TV3 | TV4 | TV5 |
|---|---|---|---|---|---|---|---|---|
| **Material** | | | | **Content (%)** | | | | |
| Ferrous metals | 11 | 54 | 43 | 5 | 19 | 2.7 | 7.8 | 16 |
| Nonferrous metals | | | | | | | | |
| Cu | 7 | 7 | 5 | 3 | 6 | | 4.9 | 1.3 |
| Al | 1 | 1 | 1 | | 1 | 0.7 | 2.0 | 0.8 |
| Light fraction | | | | | | | | |
| Plastics | 67 | 20 | 38 | 5 | 20 | 8.9 | 34.2 | 24 |
| Glass | | | | 68 | 31 | | 46.5 | 55 |
| Wood | | | | 11 | 16 | 17.0 | | |
| PWB | 11 | 16 | 11 | 6 | | 7.1 | 4.5 | 2.9 |
| CRT unit | | | | | | 56.6[a] | | |
| Other | 3 | 2 | 2 | 1 | 7 | 5.9[b] | | |
| Product mass (kg) | 2.1 | 3.8 | 2.25 | 22 | | | 24.5 | 10 |

[a] Conus glass contributes to 8.8% of the complete mass; screen and funnel glass to 38.5%; ferrous and nonferrous metals to 8%.
[b] 3% deflection unit; 1.8% wire; 1.0% speakers; 0.13% elcos (electrolytic capacitors)

S PC system (Anonymous, 1997)
K PC keyboard (Kaiser and Hirsch, 1995)
Te Telephone set (Kaiser and Hirsch, 1995)
Mp Mobile phone

### 2.5.4.5 Elemental Analysis of a PC Configuration

An elemental analysis of a PC configuration (BAN and SVTC, 2002) has been carried out by Handy and Harman Electronic Materials Corp. The analysis refers to a configuration with a mass of 31.8 kg. It consists of a system, a monitor, a keyboard, cables, and a mouse. Table 2.24 is compiled on the basis of this research.

### 2.5.4.6 Complex Components in Electronic Products

A considerable part of an electronic product's mass consists of complex components, which cannot be disassembled other than via crushing or similar extremely destructive methods. The following categories can be found:

1. *PWBs*. These components have rapidly evolved over time. Those of the transistor era were rather massive. With the introduction of surface mount technology (SMT) and integrated circuits, the size and strength of the board has decreased, and the complexity of the PWB has increased. In small electronic devices and notebooks, the board has been

**TABLE 2.23**
**Materials Composition of Some Grey Goods**

| | M1 | M2 | Pr | C1 | C2 | S | K | Te | Mp |
|---|---|---|---|---|---|---|---|---|---|
| **Material** | | | | | **Content (%)** | | | | |
| Ferrous metals | 25 | 15.4 | 39 | 42 | 23[a] | 19 | 42[b] | 31[c] | 3 |
| Nonferrous metals | | | | | | | | | |
| Cu | 4 | 8.5 | 13 | 1 | | 1 | | | 15 |
| Al | | 5.1 | | 11 | | 4 | | | 7[d] |
| Light fraction | | | | | | | | | |
| Plastics | 46 | 17.6 | 36 | 17 | 23 | 41 | 31 | 40 | 49 |
| Glass | 21 | 42.5 | | 13 | | | | | |
| PWB | (23) | 10.6 | | 16 | 10 | 25 | | | |
| CRT unit | | | | | 29 | | | | |
| Cables | | | | | 5 | | | | |
| Other | 4 | | 10 | | | 10 | 27 | 29 | 25[e] |
| Product mass (kg) | 8 | 21.4 | 7 | | | 3.67 | 1.643 | 0.615 | 0.154 |

[a] With Al and Cu included.
[b] All metals included.
[c] All metals included.
[d] Ni, Zn, and Ag included.
[e] 9% epoxy and 16% ceramics.

replaced with a flexible foil. A diversification in functionality has also taken place. Most massive PWBs are the uninterruptible power supply (UPS) boards, which are populated with transformers, power transistors that are provided with heat sinks, coils, and rectifiers. There are also a lot of electrolytic capacitors (elcos) on the UPS board. There are boards, which have electric motors and others are populated with microchips and connectors. Therefore, strong individual differences between PWBs exist.

2. *CRT units*. These are present in TV sets and monitors. Although LCD and plasma screens are gradually replacing them, CRTs are still frequently used because they are cheap and technologically mature. From an environmental point of view, the CRT is unfriendly, because of its energy use and harmful contents. The CRT unit consists of a tube, with deflection coils sealed on it, an anode loop, and some additional electronics. The tube consists of glass, an implosion protection, an electron gun and a mask molded in it, and fluorescent powder. The glass is different in composition. 72% of it is screen glass. *Screen glass* typically contains 14% BaO which corresponds to 12.5% Barium. Other figures reveal a content up to 7% BaO and 10% SrO, which corresponds to 6.3% Ba and 8.5% Sr (Menad, 1999). 28% of the CRT glass is neck glass. *Neck glass* typically contains 18% PbO which corresponds to 16.7% Pb.

**TABLE 2.24**
**Elemental Analysis of a PC Configuration**

| Material | Content (%) | $\eta_{rec}$(%) | Use | Location |
|---|---|---|---|---|
| Silica | 24.88 | — | Glass; ceramics | CRT; PWB |
| Plastics | 22.99 | 20 | All organic materials | Casing; mechanical |
| Iron | 20.47 | 80 | Structure; magnetism | Casing; CRT; PWB |
| Aluminum | 14.17 | 80 | Structure; conductivity | Chassis; wire; PWB |
| Copper | 6.93 | 90 | Conductivity | Wire; windings |
| Lead | 6.30 | 5 | Radiation protection; joining | CRT; PWB |
| Zinc | 2.20 | 60 | Battery; phosphor | PWB; CRT |
| Tin | 1.001 | 70 | Joining | PWB; CRT |
| Nickel | 0.850 | 80 | Structure; magnetism | Casing; CRT; PWB |
| Barium | 0.0315 | — | Vacuum | CRT |
| Manganese | 0.0315 | — | Structure; magnetism | Casing; CRT; PWB |
| Silver | 0.0189 | 98 | Conductivity | PWB; Connectors |
| Tantalum | 0.0157 | — | Capacitor | PWB; UPS |
| Beryllium | 0.0157 | — | Thermal conductivity | PWB; Connectors |
| Titanium | 0.0157 | — | Pigment | Casing |
| Cobalt | 0.0157 | 85 | Structure; magnetism | Casing; CRT; PWB |
| Antimony | 0.0094 | — | Diodes | PWB |
| Cadmium | 0.0094 | — | Battery; phosphor | Casing; PWB; CRT |
| Bismuth | 0.0063 | — | Wetting agent | PWB |
| Chromium | 0.0063 | — | Surface; hardener | Casing |
| Mercury | 0.0022 | — | Battery; relay | PWB |
| Germanium | 0.0016 | — | Semiconductor | PWB |
| Indium | 0.0016 | 60 | Transistor; rectifier | PWB |
| Gold | 0.0016 | 99 | Conductivity | PWB; Connectors |
| Ruthenium | 0.0016 | 80 | Resistance | PWB |
| Selenium | 0.0016 | 70 | Rectifier | PWB |
| Gallium | 0.0013 | — | Semiconductor | PWB |
| Arsenic | 0.0013 | — | Semiconductor | PWB |
| Palladium | 0.0003 | 95 | Conductivity | PWB; Connectors |
| Vanadium | 0.0002 | — | Phosphor | CRT |
| Europium | 0.0002 | — | Phosphor | CRT |
| Niobium | 0.0002 | — | Welding | Casing |
| Yttrium | 0.0002 | — | Phosphor | CRT |
| Terbium | < 0.0001 | — | Phosphor | CRT |
| Rhodium | < 0.0001 | 50 | Thick film conductor | PWB |
| Platinum | < 0.0001 | 95 | Thick film conductor | PWB |
| Mass (kg) | | | 31.8 | |

3. *Transformers, coils*, and *relays* appear in different sizes. Large units contain mainly copper for the windings, iron for the core, and some insulating materials. Small units may also contain ceramics as a core.
4. *Electric motors* are of many types and sizes. There are *AC* and *DC* motors, with or without commutator, with coils, a copper cage, or a permanent magnet as a rotor.

**TABLE 2.25**
**CRT Composition**

| Material | Content (%) |
|---|---|
| Glass | |
| • Screen glass | 62.8 |
| • Neck glass | 24.4 |
| Ferrous metal | 12.0 |
| Electron gun | 0.4 |
| Seal | 0.4 |
| Fluorescent powder | 0.04 |

5. *Cables, wires*, and *connectors*. These consist of plastics, which are frequently PVC, and nonferrous materials, mainly Al and Cu. Precious materials may be present in the connectors.
6. *Switches* are often molds of plastics and nonferrous metals.
7. *Batteries* are presently often of the nickel-metal hydride (Ni-MH) type.
8. *Electronic components.*
9. *Mechanical units* such as gearboxes.
10. *Optical units.*

CRTs are well documented, as extensive research has been carried out on the reuse of the glass (Menad, 1999; Smith et al., 1995). The composition of CRTs (without deflection coil and anode loop) is given in Table 2.25.

Some coarse standard data on components in electronic products are given in Table 2.26 (Anonymous, 1997).

PWBs are frequently subjected to elemental analyses (Table 2.27). These data, however, do not provide much more than ad hoc information because the variety of PWBs is substantial and rapidly evolving. The data in different columns are due to

**TABLE 2.26**
**Provisional Data on Material Content of Complex Components**

| Material | PWB | Electric Motor | Trans-Former | Wire Cable | Wire Cable[a] |
|---|---|---|---|---|---|
| | Content (%) | | | | |
| Ferrous metal | — | 75 | 65 | — | — |
| Copper | 12 | 15 | 25 | 40 | 36 |
| Aluminum | — | 10 | 5 | — | 18 |
| Plastics | 70 | — | — | 60 | 45 |
| Other | 18 | — | 5 | — | — |

[a] Krikke et al., 1998.

**TABLE 2.27**
**Elemental Analyses of PWBs**

| | Sodhi | Brodersen | Angerer | Angerer |
|---|---|---|---|---|
| **Material** | | | **Content (%)** | |
| Silica | 30.2 | 49 | — | — |
| Plastics | 30.2 | 19[a] | — | — |
| Bromine | — | 4 | 2.7 | — |
| Iron | 8.1 | 6 | 10.8 | 5–10 |
| Copper | 20.1 | 7 | 3.7 | 10–20 |
| Aluminum | 2 | — | 4.8 | 1 |
| Tin | 4 | 1 | 3.1 | 2 |
| Nickel | 2 | 3 | 0.32 | 1–3 |
| Lead | 2 | — | — | 1–5 |
| Zinc | 1 | 2 | 1.45 | 0.3 |
| Silver | 0.2 | — | 0.08 | 0.05–0.3 |
| Gold | 0.1 | — | 0.01 | 0.0003–0.001 |
| Manganese | — | — | 2.15 | — |
| Antimony | — | — | 0.45 | — |
| Barium | — | — | 0.36 | — |
| Chlorine | — | — | 0.19 | — |
| Sodium | — | — | 0.18 | — |
| Chromium | — | — | 0.16 | — |
| Cadmium | — | — | 0.04 | — |
| Tantalum | — | — | 0.02 | — |
| Palladium | 0.005 | — | — | 0.004–0.003 |
| Other metals | — | 9 | — | — |

[a] Additives excluded.

Sodhi and Reimer (2001) adapted from Sum (1991); Brodersen et al., (1994); and Angerer et al. (1993) with data from Ansems and Esmeyer (1989) and with data from Schlag (1991). These data refer to the 386 and 486 processor era. Angerer et al. (1993) list Ti, Si, Ge, K, Ga, Pd, Rh, and Zr, as additional trace elements. Rhodium (Rh) is used in contacts. Bromine is used in flame-retardant, such as polybrominated diphenylethers (PBDE). These are present as additive in plastics such as in casing, wiring, and PWBs.

### 2.5.5 Hazardous Materials in Electronic Products

Data on the composition of electronic products gives an idea of the many substances that are present in them. Some of these substances are directly harmful to humans while others are directly harmful to the environment. Inadequate processing, such as uncontrolled incineration, results in additional harm to both people and the environment. For this reason, many substances have been banned. However, they remain resident in older products and thus will be around for many years. Some examples of such substances are:

- Asbestos in heating elements
- CFCs in cooling equipment
- Mercury in relays
- Cadmium in pigments
- Lead in solder
- $Cr^{6+}$ in corrosion protection layers
- PBDEs, which are widely used as a flame retardant in plastics

Materials such as Cd, Ni, Zn, Pb, and Hg are present in batteries. Diazo compounds are present in liquid crystal displays (LCDs). These might be toxic. Fluorescents are present in CRT screens. In addition, many components are contaminated with oil, toner, and with dust and dirt. Table 2.28, which is adapted from Angerer et al. (1993), presents a nonexhaustive survey of hazardous materials.

Additional information on hazardous substances in electronic products can be found in reports such as HDP (2000) on environmentally compliant electronics; EC (2000) on EC directives on hazardous materials in electronic products; and DEPA (1999) on brominated flame retardants.

**TABLE 2.28**
**Hazardous Substances in Electronic Products**

| Substance | Application |
|---|---|
| Heavy metals | |
| Cd; Ni; Zn; Pb; Hg | Batteries; fluorescent tubes |
| Sn; Pb; Cd | Solder |
| Ba; Sr; Pb | CRT glass |
| Cd; Y; Eu; Se; Zn | Fluorescent powder |
| Hg | Relays |
| Semiconductors | |
| B; Ga; In; As | Integrated circuits |
| GaAs | LEDs; photovoltaic cells |
| Se; Ge | Diodes |
| Se | Photocopying drums |
| Organic compounds | |
| PCBs | Capacitors |
| PBDEs | Flame retardants |
| Mineral oil | Lubricant |
| Additives in plastics | |
| Cl | PVC |
| Cd; Pb; Ni; Ti; Sb; Diazo compounds | Pigment |
| Pb; Ba; Cd; Sn | Stabilizer |

## 2.6 CONCLUSION

Extensive data on complex products were compiled and presented in this chapter. The data provides an order of magnitude and complexity of what is present in the end-of-life complex products as well as estimates of parameters that should prove useful in conducting quantitatively oriented studies dealing with end-of-life processing of complex products. The recovery process considered in this chapter emphasizes process technology. The next chapter will deal with the discrete character that is inherent to complex products.

## REFERENCES

Åkermark, A.M., 1997, Design for disassembly and recycling. In: Krause, F.L. and Seliger, G. (eds.): Life Cycle Networks, *Proceedings of 4th CIRP International Seminar on Life-Cycle Engineering*, 237–248.

Angerer, G., Bätcher, K., and Bars, P., 1993, *Upgrading of electronic scrap* (Verwertung von Elektronikschrott). Waste management in research and practice (Abfallwirtschaft in Forschung und Praxis), Vol. 59. Berlin: Erich Schmidt. (in German)

Anonymous, 1997, Results of some breakdown experiments, performed at a Dutch recycling company.

Ansems, A. and Esmeyer, F., 1989, *Scheiding en recycling van non-ferrometaal/kunststofcombinaties* (Separation and recycling of non-ferrous/plastics combinations). Apeldoorn (NL): TNO. Report (in Dutch).

ARN, 2002 Auto Recycling Nederland, Statistics.

Arner, R., 1997, Computer recycling in the U.S. http://www.computerecycleforeduc.com/recycling.html

Ayres, R.U. and Ayres, L.W., 1996, *Industrial Ecology: Closing the Materials Cycle*. Cheltenham: Edward Elgar.

Backthru, 2002, New York: Back thru the Future Micro Computers, Inc. http://www.thegreenpc.com/recyclin.htm

BAN and SVTC, 2002, *Exporting Harm: The high-tech trashing of Asia*. Seattle, WA: Basel Action Network; San Jose, CA: Silicon Valley Toxics Coalition. Report, p. 52.

Bandivadekar, A.P., Gunter, K.L., and Sutherland, J.W., 2002, A model for material flows and economic exchanges within the U.S. automotive life cycle chain and its sensitivity to systematic changes. *Proceedings of the 9th CIRP Life Cycle Engineering Seminar*, Erlangen, Germany.

Boon, J.E., Isaacs, J.A., and Gupta, S.M., 2000, Economic impact of aluminum-intensive vehicles on the U.S. automotive recycling infrastructure. *Journal of Industrial Ecology*, **4**(2), 117–134.

Boon, J.E., Isaacs, J.A., and Gupta, S.M., 2003, End-of-life infrastructure economics for 'Clean Vehicles' in the U.S. *Journal of Industrial Ecology*, **7**(1), 25–45.

Boon, J.E., Isaacs, J.A., and Gupta, S.M., 2002, Economic sensitivity for end-of-life planning and processing of personal computers. *Journal of Electronics Manufacturing*, **11**(1), 81–93.

Brodersen, K., Tartler, D., and Danzer, B., 1994, Scrap of electronics: a challenge to recycling activities. *Proceedings of IEEE International Symposium on Electronics and the Environment*, 174–176.

Chen, K.Z., 2001, Development of integrated design for disassembly and recycling in concurrent engineering. *Integrated Manufacturing Systems*, **12**(1), 67–79.

Coolrec, 1997, Information Brochure. Eindhoven.
Das, S.K. and Matthew, S., 1999, Characterization of material outputs from an electronics demanufacturing facility. *Proceedings of IEEE International Symposium of Electronics and the Environment*, 251–256.
Das, S.K., Yedlarajiah, P., and Narendra, R., 2000, An approach for estimating the end-of-life product disassembly effort and cost. *International Journal of Production Research*, **38**(3), 657–673.
De Fazio, T.L., Delchambre A., and De Lit, P., 1997, Disassembly for recycling of office electronic equipment. *European Journal of Mechanical and Environmental Engineering,* M **42**(1), 25–31.
De Ron, A.J. and Penev, K., 1995, Disassembly and recycling of electronic consumer products: an overview. *Technovation*, **15**(6), 363–374.
Den Hond, F., 2000, Industrial ecology: a review. *Regional Environmental Change*, **1**(2), 60–69.
DEPA, 1999, *Brominated flame retardants: Substance flow analysis and assessment of alternatives*. Report of Danish Environmental Protection Agency, p. 225.
Dini, G., Failli, F., and Santochi, M., 2001, A disassembly planning software system for the optimization of recycling processes. *Production Planning & Control*, **12**(1), 2–12.
Ebersperger, R., 1993, *Energieoptimierte Nutzungsdauer von Waschmaschinen* (Energetically optimal life-cycle of washing machines). In: Kumulierter Energie- und Stoffbilanzen. Düsseldorf: VDI-Verlag. VDI-Berichte Vol. **1093**. pp. 215–233. (in German).
Ebersperger, R., 1995, *Beispiele für Zurechnungsverfahren des Energieaufwands bei Entsorgung und Recycling von Produkten* (Examples for the calculation of the Gross Energy Requirement in recollection and recycling of products). In: Kumulierter Energieaufwand. Düsseldorf: VDI-Verlag. VDI-Berichte Vol. **1218**. pp. 11–31. (in German).
EC, 2000, *Directive of the European Parliament and Council on discarded electric and electronic products*. Brussels, COM(2000) 347.
European Shredder Group, 2002. http://www.efr2.org/EFR-ESG3.htm.
Ferrão, P., Reis, I. and Amaral, J., 2002, The industrial ecology of the automobile: a Portuguese perspective. *International Journal of Ecology and Environmental Sciences*, **28**, 27–34.
Frosch, R.A. and Gallopoulos, N.E., 1989, Strategies for manufacturing. *Scientific American,* **261**(3), 144–152.
Furuhjelm, J., Yasuda, Y., and Trankell, R., 2000, Recycling of telecommunication products in Europe, Japan and USA. *Proceedings of IEEE International Symposium on Electronics and the Environment*, 143–148.
Gao, M., Zhou, M.C., and Caudill, R.J., 2002, Integration of disassembly leveling and bin assignment for demanufacturing automation. *IEEE Transactions on Robotics and Automation*, **18**(6), 867–874.
Gao, W., Ariyama, T., Ojima, T., and Meier A., 2001, Energy impacts of recycling disassembly material in residential buildings. *Energy and Buildings,* **33**, 553–562.
Graedel, T.E. and Allenby, B.R., 1995, *Industrial Ecology.* Englewood Cliffs: Prentice Hall.
Gungor, A. and Gupta, S.M., 1999, Issues in environmentally conscious manufacturing and product recovery: a survey. *Computers and Industrial Engineering*, **36**(4), 811–853.
Gupta, S.M. and Isaacs, J.A., 1997, Value analysis of disposal strategies for automobiles. *Computers and Industrial Engineering*, **33**(1–2), 325–328.
HDP, 2000, *Environmentally compliant electronics framework specification.* Report of High Density Packaging User Group International, Scottsdale AZ.

Hendricks, V.G.M., 1992, Selective demontage van auto's voor materiaalhergebruik (*Selective disassembly of cars aimed at materials recycling*). M.Sc.- thesis, Technische Universiteit Eindhoven, 1992 (in Dutch).

Isaacs, J.A. and Gupta, S.M., 1997, Economic consequences of increasing polymer content for the U.S. automobile recycling infrastructure. *Journal of Industrial Ecology*, **1**(4), 19–33.

Ishii, K., Eubanks, C.F., and Marks, M., 1993, Evaluation methodology for post-manufacturing issues in life-cycle design. *Concurrent Engineering*, **1**(1), 61–68.

JTEC, 1992, *Display technologies in Japan*. Report of Japanese Technology Evaluation Center, http://www.wtec.og/loyola/dsply_jp.

Kaiser, H. and Hirsch, S., 1995, *Recycling von elektrischen Hausgeräten gelöst* (Recycling of electric household appliances solved). *Abfallwirtschafts-Journal*, **7**(3), 1–4. (in German).

Knapp, O. and Jansen, H., 1993, Environmental considerations in product development. *ETZ*, **114**(22), 1386–1389.

Knight, W.A. and Sodhi, M.S. 2000, Design for bulk recycling: analysis of materials separation. *Annals of the CIRP*, **49**(1), 83–86.

Krikke, H.R., Van Harten, A., and Schuur, P.C., 1998, On a medium term product recovery and disposal strategy for durable assembly products. *International Journal of Production Research*, **36**(1), 111–139.

Lambert, A.J.D., 1999, Linear programming in disassembly/clustering sequence generation. *Computers and Industrial Engineering*, **36**, 723–738.

Lambert, A.J.D., 2001, Life-cycle chain analysis including recycling. In: Sarkis, J. (ed.), *Greener Manufacturing and Operations*. Sheffield, UK: Greenleaf Publishing. Chapter 2, pp. 36–55.

LAP, 2002, Landelijk Afvalbeheersplan: Milieueffectrapportage 2002–2012. Utrecht: Commissie voor de Milieueffectrapportage (in Dutch).

L'Èglise, T., De Lit, P., Delchambre, A., and Raucent, B., 2000, Recycling of electric and electronic end-of-life devices: economical assessment study in Brussels, Belgium. *Proceedings of IEEE International Symposium on Electronics and the Environment*, 149–154.

Lu, Q., Christina, V., Stuart, J.A., and Rich, T., 2000, A practical framework for the reverse supply chain. *Proceedings of IEEE International Symposium on Electronics and the Environment*, 266–271.

Menad, N., 1999, Cathode ray tube recycling. *Resources, Conservation and Recycling*, **26**, 143–154.

Moyer, L.K. and Gupta, S.M., 1997, Environmental concerns and recycling/disassembly efforts in the electronics industry. *Journal of Electronics Manufacturing*, **7**(1), 1–22.

NVMP, 2000, http://www.nvmp.nl/html/asp/productenlijst.asp: List of takeback fees for electric and electromechanical household products. Zoetermeer (NL): Nederlandse Vereniging Verwijdering Metalektro Producten. (in Dutch).

Potzschke, M., 1991, *Gebrauchsgüter als komplexe Rohstofquelle* (Consumption goods as a complex raw materials resource). In: Recycling: Eine Heurausforderung für den Konstrukteur (Recycling: a challenge for the designer). Düsseldorf: VDI. VDI-Berichte, vol. 906, p. 43–74 (in German).

Recycler's World, 2002, http://www.recycle.net/price: *Computer & Electronics Recycling Index; RecycleNet Scrap Metals Index; Andela Scrap Glass Index; RecycleNet Automotive Recycling Index; RecycleNet Plastic Recycling Index; RecycleNet Tire & Rubber Index*. August 28.

Reimer, B., Sodhi, M.S., and Knight, W.A., 2000, Optimizing electronics end-of-life disposal costs. *Proceedings of IEEE International Symposium on Electronics and the Environment*, 342–347.

Roberts, B., 1999, Mining cast-off PCs. *Electronic Business*, **25**(12), 47–48.

Samel, R., 2001, Materials in cars: options for change. *Materials Technology & Advanced Performance Materials*, **16**(1), 4–7.

Schlag, D., 1991, *Entsorgung von Elektronikaltgeräten* (Recovery of end-of-life electronic products). Stuttgart: Landesanstalt für Umweltschitz Baden-Württemberg. Sachstandsbericht (in German).

Schubert, H., 1996, *Electric/electronic waste in the Closed Cycle Economy* (*Elektro-/Elektronikschrott in der Kreislaufwirtschaft*), Fraunhofer Institute for Process Technology, Karlsruhe (in German).

Smith, D., Small, M., Dodds, R., Amagai, S., and Strong, T., 1995, Computer monitor recycling: A case study. *Proceedings of IEEE Conference on Clean Electronics Products and Technology* (CONCEPT), 124–128.

Sodhi, M.S., Young, J., and Knight, W.A., 1999, Modelling material separation processes in bulk recycling. *International Journal of Production Research*, 37(10), 2239–2252.

Sodhi, M.S. and Reimer, B., 2001, Models for recycling electronics end-of-life products. *OR-Spektrum*, **23**, 97–115.

Spengler, T., Püchert, H., Penkuhn, T., and Rentz, O., 1997, Environmental integrated production and recycling management. *European Journal of Operational Research*, **97**(2), 308–326.

Spengler, T., Ploog, M., and Schröter, M, 2003, Integrated Planning of Acquisition, Disassembly and Bulk Recycling: A Case Study on Electronic Scrap Recovery. *OR Spectrum*, **25**(3), 413–442.

Stuart, J.A. and Lu, Q., 2000a, A model for discrete processing decisions for bulk recycling of electronics equipment. *IEEE Transactions on Electronics Packaging Manufacturing*, **23**(4), 314–320.

Stuart, J.A. and Lu, Q., 2000b, A refine-or-sell decision model for a station with continuous reprocessing options in an electronics recycling center. *IEEE Transactions on Electronics Packaging Manufacturing*, **23**(4), 321–327.

Sum, E.Y.L., 1991, The recovery of metals from electronics scrap. *Journal of the Minerals, Metals and Materials Society* (JOM), **43**(4), 53–61.

Thomas, V., Caudill, R., and Badwe, D., 1998, Marginal emissions and variations across models: Life-cycle assessment of a television. *Proceedings of IEEE International Symposium of Electronics and the Environment*, 48–53.

Van Bezooijen, G. 2000, *EEE waste in the Netherlands*, data presented at Workshop on Refrigerator Recycling, held at Universitè Politechnique, Mons, Belgium, May 2000.

Van der Hoek, A., 1994, *De retourstroom van TV's*. M.Sc.-Thesis. Eindhoven: Technische Universiteit. (in Dutch).

Van Roosmalen, R. 2000, Hoogwaardige verwerking van shredderresidu (*High-quality processing of shredder residue*). Magazine Recycling Benelux, Issue 5, 2000, 12–15, (in Dutch).

Weber, R., 1991, *Die Entwicklung recyclinggerechten Automobile* (Design for recycling of cars). In: Recycling: Eine Herausforderung für den Konstrukteur (Recycling: a challenge for the designer). Düsseldorf: VDI-Verlag. VDI-Berichte Vol. **906** p. 75–88. (in German).

Weissmantel, H., Baire, C., Kaase, W., and Thomas, A.G., 1997, Benefit function for determining the optimum recycling option for products and parts. Life Cycle Networks. *Proceedings of 4th CIRP International Seminar on Life Cycle Engineering*. London: Chapman & Hall, 276–287.

Zhang, H.C. and Kuo, T.C., 1996. *Disassembly model for recycling-personal computer*. Dearborn, MI: Society of Manufacturing Engineers (SME). Technical paper MS96–132.

# 3 The Disassembly Process

## 3.1 INTRODUCTION

Chapter 2 focused on the materials aspects of a product and the significance of disassembly process within the process-product chain. This included a concise description of the bulk recycling process. This chapter takes a closer look at the disassembly process itself. It considers a complex product as a set of connected discrete components and the disassembly process as a sequence of (disassembly) operations. The key elements here are components and connections. The relationships between these elements are portrayed via a connection diagram, which forms the basis of theoretical disassembly research. Starting with this connection diagram, various approaches to disassembly operations are discussed and compared. The differences between component- and connection-oriented approaches are highlighted. The distribution of component masses in a product is analyzed. Typologies of materials, connections, fasteners, and disassembly tasks are also presented. A typology of disassembly tasks is one of the basic tools in cost metrics that is important for optimizing disassembly processes. All theory is illustrated with disassembly examples that have been taken from practice. These also demonstrate the hierarchical tree representation, which is frequently used in disassembly analysis and offers an alternative method in representing a product with a definite structure. Finally, an example of a car is considered to illustrate various aspects of disassembly.

## 3.2 THE PRODUCT

### 3.2.1 Introduction

Some basic terminology was presented in subsection 1.4. We use that as a starting point for a systematic discussion of disassembly processes. This section is dedicated to the product itself. A *complex product* is considered as a functional unit that consists of *components* that are related to each other via *connections*. Components and connections are more or less complementary to each other. Just as a product and a process-oriented approach can be distinguished, distinction can also be made between the component and the connection-oriented approach. Therefore, these issues will be introduced in subsections 3.2.2 and 3.2.3, respectively. Subsection 3.2.4 will deal with fasteners, which represent the materialization of connections. Subsection 3.2.5 presents some experimental data on components' mass distribution in complex products. New theory on this topic is discussed in subsection 3.2.6. Components and fasteners are the material objects that define the functionality of a product. Therefore, both are important in the design of the product. Design issues, with an emphasis on design for disassembly, are briefly outlined in subsection 3.2.7. Materials breakdown analyses are carried out in subsection 3.2.8. Although seemingly related only to bulk recycling, it is further discussed here aimed at assigning components to specific material categories, rather than at separation processes that are applied to flows of chunks or grains. Finally, subsection 3.2.9 introduces the connection diagram, which is a prerequisite for discussing disassembly processes from a theoretical point of view.

### 3.2.2 Components

A *component* is a material entity that can be separated from a product via disassembly operations, thus keeping the extrinsic properties of this entity intact, and cannot be further separated except via destructive operations. A *connection* forms a relationship between two components that restricts their relative movement. Material objects that are used to restrict the movements of two components are called *fasteners*.

Components fulfill one or more functions in a product, such as protection and transfer of forces. Functionality, however, is not the main objective of this book, which is devoted to disassembly.

Identifying a component introduces some arbitrariness, as disassembly may include some semidestructive actions. The following types of components are recognized:

1. Homogeneous components
2. Composite components
3. Complex components

*Homogeneous components* consist of homogeneous materials, which may be a mixture or an alloy. Casings and frames are typical examples of homogeneous components. Minor parts of different materials could also be present.

*Composite components* consist of multiple materials that are linked in an irreversible way, such as in sandwich structures. A tire can be considered a composite component.

*Complex components* are units that consist of a set of irreversibly connected homogeneous components. Examples of such components are printed wiring boards (PWBs), electronic components, electric motors, and cables. Some complex components can also be considered modules. A module is a functional set of connected components.

Components are linked to each other via *connections*. A connection is a relationship between two components that restricts their movement or detachability. A component is *moveable* with respect to a subassembly if it can be displaced with respect to this subassembly over, at least, an infinitesimal distance. This motion can be translational or rotational. If infinite displacement is possible, the component is considered *detachable*. Many products have some internal degrees of freedom, as movement of components with respect to each other might be essential for the product's functionality. Examples of this kind of product include switches, gears, and lids.

### 3.2.3 Connections

The separation of components or disassembly is closely related to the disestablishment of connections. However, prior to getting into disassembly task analysis, an exploration of connections and fasteners is needed.

There are a considerable variety of connection types that are used in complex products. These connection types offer varying degrees of difficulties in their ability to nondestructively release the fasteners, the amount of force required to undo the connections, the restriction in movement and the type of fastener used. The following are some typical types of connections that are present in different products. Special attention is given to the possibility of separating the connections without damaging the connected components:

1. *Mating connection.* In this type of connection, motion is restricted by mating surfaces. The extent of restriction depends on the shape of these surfaces, which is a subset of translational and rotational directions. The simplest mating connection is a planar contact. A variety of connections such as holes and pins (or rods), notches, and slots also fall into this category. Complete restriction of motion of a component by mating connection is possible. Forces, such as gravity and friction, keep the mated components in place. A special case of a mating connection is a *lock connection*. In this case, motion of a component is only enabled if a specific set of components is moved first.
2. *Bundling connection.* This type of connection is aimed at tying up some ducts, for instance, electric wires and cables. This is usually accomplished with the help of a discrete fastener such as a bundler or adhesive tape. This type of fastener usually has to be destroyed prior to detaching the bundled ducts.
3. *Spring connection.* In this type of connection, motion can be unrestricted or restricted, but detachment is prohibited because of the presence of a deformable fastener (i.e., a *spring*) that applies internal forces to the components that are in relative motion. Frequently, the deformable fastener can be reversibly detached, which has to occur prior to detachment of the component that is connected with the spring.
4. *Screw connection, bolt and nut connection*, etc. are reversible connection types that usually completely restrict the motion of the components involved, by applying dedicated fasteners, which are separate components. This type of connection is common in many products. The motion of fasteners can also be restricted or prohibited by additional fasteners such as lock washers or seals (for example as a protection against vibrations).
5. *Cotter pin, staple, and related connections* are established via fasteners that have to be deformed in a reversible or irreversible way prior to detaching the components.
6. *Snap fit connections* can be either reversible or irreversible. The fastener is not a separate component but rather a feature of the component. Snap fits are frequently used because they make it easy to establish the connection. However, this type of connection can sometimes be challenging to undo

and any attempt to accomplish it could harm the component. This is especially true if the product is designed for assembly without any regard to the disassembly process.

7. *Press fit connection* is established via pressure that is achieved by a minor deformation of the components involved. For disestablishing the connection, a definite friction force has to be surmounted. The amount of deformation, force, and the properties of the components influence the degree of reversibility and destructiveness of the disestablishment of the connection. *Shrink fit connections*, for example, often cannot be disestablished in a reversible way. Discrete fasteners are not involved in this type of connection. Electric cable connections of the *connector type*, such as pin and plug type connections, are examples of completely reversible press fit connections.
8. *Rivet connections* are established via fasteners that are deformed during the process. Disestablishment is only possible via the destruction of the fastener, for example, by drilling.
9. *Seam fold connections* are established via the deformation of a sheet material, exceeding its plasticity limit. The disestablishment of such a connection usually damages the sheet material, which impedes the reestablishment of such connection. Seam fold connections are frequently used in electronic components, batteries, electric motors, and mechanical devices.
10. *Glue and seal connections* are established via an agent that is applied to the components to be connected. Adhesion, chemical reactions and phase transition are the mechanisms that connect the components. The disestablishment of such a connection is usually irreversible. However, the principal components might remain intact. An important example is the deflection coil of a cathode ray tube (CRT) unit, which is sealed to the cathode ray tube. Glue is also frequently applied via discrete fasteners, such as adhesive tapes and foils, for example, securing the heat protective materials.
11. *Solder connections* are established with an agent that undergoes a phase transition. Although solder connections are considered irreversible, judicious heating can be applied to release the components and to recover the soldering agent.
12. *Weld connections* are also established with an agent that undergoes a phase transition. However, the phase transition also takes place in the materials of the components that are welded together. The disestablishment of weld connections, therefore, always damages the components to some extent. Since the welded components are often made of comparable materials, the disestablishment of those connections is often not required from a materials recovery point of view.
13. *Mould connections* are characterized by a phase transition of at least one of the components that are connected. Mould connections are established via melting, baking, etc. (e.g., metal parts in plastic, ceramic, or glass components). Examples of such connections are in electronic components and switches. Separation of the different materials is only practical via bulk processing, thus completely destroying the components due to shredding

and grinding. From a disassembly processing point of view, components that are mould connected have to be considered as a single component.

### 3.2.4 Fasteners

It is clear from the above discussion that there are many different types of fasteners that may be present in complex products. Fasteners can be discrete components such as screws, or nondiscrete material objects such as snap fits. If the fastener is a component, but not considered as such in modeling, it is considered a *quasi-component*. Even so, some ambiguity is present in the definition of a fastener, which may include every component that has a fastening function. Frequently, however, components have multiple functions, which makes the definition of a fastener a bit fuzzy. For example, connective strips, which connect two components, can also be considered as components as they also make up part of the construction. Das et al. (2000) presented a typology of fasteners. They presented a list of 13 different types of fasteners and included fasteners such as nails, zippers, and Velcro®, which are generally not present in mechanical, electric, and electronic products. Therefore, for such products, we propose the following typology of fasteners (the numbers in parentheses refer to the connection type listed in the previous subsection):

1. Discrete components, which are not deformed
   a. Bundler (2)
   b. Spring (3)
   c. Screw, bolt, nut, washer, lock washer, spacer (4)
2. Discrete components, which are reversibly deformed
   a. Cotter pin, staple (5)
3. Discrete components, which are irreversibly deformed
   a. Rivet (8)
   b. Adhesive tape (10)
4. Parts of components, reversible connection
   a. Surface (1, 7)
   b. Snap fit connection (6)
5. Parts of components, irreversible connection
   a. Surface (7, 13)
   b. Seam (9)
   c. Seal (10)
6. Virtual components
   a. Solder (11)
   b. Weld (12)

The principal advantage of the above typology is that any deviant type of fastener can also be accommodated within this scheme.

### 3.2.5 Distribution of Components' Masses in a Product

The recovery of large components (made of homogeneous materials) is crucial to the efficiency of selective disassembly. Even though the large components in a complex product are small in number, they nevertheless represent a substantial share of the product's mass. An efficient disassembly process intended for material recovery focuses on the retrieval of the most important homogeneous components. It is, however, not always possible to recover the desired components in the order of their importance because of the presence of precedence relationships between the components and the costs associated with the disassembly operations.

Often a complex product has a few large components and numerous small ones. Schuckert (1993) ascertained the mass distribution in a complex product with the help of a graph by ordering the components of a car according to their mass with the component number on the $x$-axis, and the component's mass on the $y$-axis. If the points are connected, a hyperbolic graph is revealed. The $x$- and the $y$-axes are asymptotes because there are no components with infinite mass and the number of components with zero mass can be assumed to be infinite! In general, there are many components with small masses and only a few components with relatively high masses. Schuckert noted that the points on the graph (representing a component) could be grouped in distinct domains, thus providing a tool for classification.

A complicating factor of Schuckert's method is, however, the existence of complex components. The result of the method strongly depends on the extent to which the complex components are decomposed. This depends on the type of disassembly operations used, thus introducing a certain degree of ambiguity. Some disassembly operations that destroy fasteners or that restrict damage to "components of interest" may be included as well. In practice, modules are often considered as composite components. Thus, for example, in disassembling PC systems, the drives are considered as components.

In this subsection, the disassembly of two different electronic products is discussed, regarding a notebook and a monitor. Only nondestructive and semidestructive operations are included. Therefore, populated PWBs are considered components, except for components that are screwed to it, or cables that are soldered to it. Also switches, cables, and cathode ray tubes are considered components here. Prying out glue connections is considered semidestructive and is permitted.

The distribution of components' masses of a Toshiba Satellite® Notebook PC and a Compaq monitor are depicted in Figure 3.1. The masses are weighed with a resolution of 1 g. Despite some differences with Schuckert's original approach, the overall picture here is similar to that of a car, albeit on a different mass scale and with a lesser number of components involved. If the figures are normalized by using a 0 through 100% scale on both axes, they are comparable at first glance. However, differences in the character of the product can give rise to a slightly different shape of the plot. For example, in the case of the notebook, the plot is influenced by the surplus of small components. This is due to the number of small fasteners and the hundreds of small components that are assigned to the 86 keys. The monitor shows a dominant contribution of the tube, and the appearance of relatively few components.

Although Figure 3.1 presents a general picture, it does not go into detail about the priority of components that have to be detached in selective disassembly. Intuitively, the component should be detached in order of its mass, but this neither accounts for

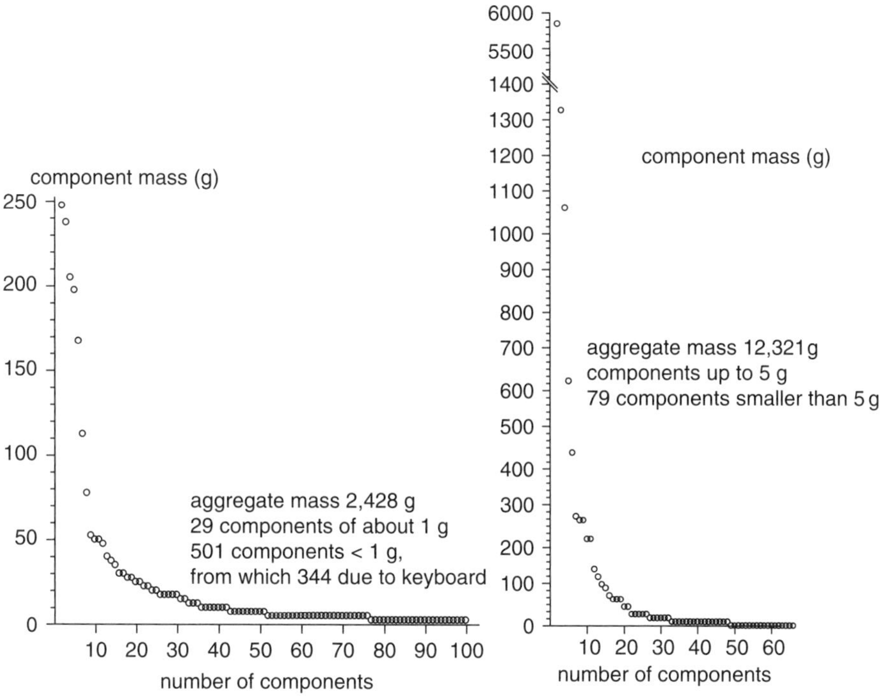


**FIGURE 3.1** Distribution of the component's masses in (a) a Toshiba Satellite Notebook PC and (b) a Compaq monitor.

the value of the component, nor does it address the safety and environmental issues. In addition, it completely ignores the precedence relationships, which frequently dictate the detachment of many lighter components prior to detaching the desired one. In the case of the notebook, the recovery of the heaviest components, which include the bottom protection and the motherboard, is not possible without detaching a multitude of other components.

Apart from mass distribution figures (such as Figure 3.1), pie charts are also used in studies on component mass distribution (Riller, 1991). In this alternative representation, the share of the aggregate mass of the components within a definite range is presented. Riller, for example, reported that in an electric dishwasher approximately 45% of the weight is due to components with a mass exceeding 200 g, 20% of the weight is due to components within the range 50–200 g, and 20% of the weight is due to components with a mass <50 g. Riller's method makes sense only if the components can be divided in some distinct classes. For the cases of notebook and monitor, such data is presented in Table 3.1.

As can be seen from Table 3.1, Riller's method is not appropriate for distinguishing between different classes of the product. The following subsection offers an aggregate approach that provides more insight into the regularities of component mass distribution of different products.

**TABLE 3.1**
**Mass Distribution of Components in a Notebook PC (*m* is the Mass of the Component)**

| Notebook | | Monitor | |
|---|---|---|---|
| Range (g) | Mass Contribution (%) | Range (g) | Mass Contribution (%) |
| $m > 100$ | 48 | $m > 1000$ | 68 |
| $100 > m \geq 25$ | 22 | $1000 > m \geq 200$ | 20 |
| $25 > m \geq 5$ | 20 | $200 > m \geq 50$ | 7 |
| $m < 5$ | 9 | $m < 50$ | 5 |

### 3.2.6 Aggregate Approach

As was mentioned before, the plots in Figure 3.1ab can be approximated with hyperbolic functions that roughly pass through the points depicted in the figures. In an ideal case, such curves can be represented with

$$y = \frac{a}{x} \tag{3.1}$$

where $a$ is a coefficient that represents mass in kg. The cumulative mass of the product is given, in the discretized case, by the expression

$$M = \sum_{n=1}^{N} \frac{a}{n} \tag{3.2}$$

where $M$ is the aggregate mass in kg, $n$ is the component number when the components are ordered according to decreasing mass, and has a value from 1 through the number of components $N$.
Expression 3.2 can be approximated with a continuous relationship as follows:

$$M - M_1 = \int_{x_1}^{x_N} \frac{a}{x} dx = a(\ln x_N - \ln x_1) \tag{3.3}$$

By putting $x_1 = 1$ and $x_N = N$, this can be further reduced to

$$M - M_1 = a \ln x_N \tag{3.4}$$

or,

$$\frac{M_n - M_1}{M - M_1} = \frac{\ln x_n}{\ln x_N} \tag{3.5}$$

where $M_n$ represents the cumulative mass of the $n$ most massive components.

In a logarithmic graph, if $x_n$ is replaced with $n$, and $x_N$ with $N$, it exhibits a straight line from the origin to the point (ln $N$, 1). Using this method, the characteristics of various products can be represented in one figure. The figure shows the extent to which the masses of the components of a complex product approximate a hyperbolic curve. This appears especially true for the components with intermediate mass. The graphs according to Expression 3.5 have been depicted in Figure 3.2 for the cases of the notebook and the monitor that were considered in the previous subsection.

The expression on the left-hand side of Expression 3.5 is plotted on the $y$-axis in Figure 3.2. A strong analogy between the plots of different products can be observed here. The different slopes of the plots are due to the different number of components (the number of components of the notebook exceeds that of the monitor by a factor of about six). In addition, the share of the most massive component to the complete product's mass, and the complete product mass $M$, have to be added to the product characteristics, as these are not represented in the plot. The heaviest component can be introduced, if desired, via plotting $\ln(n+1)$ instead of ln $n$ on the $x$-axis.

Figure 3.2 confirms the assertion that the component's mass distribution as depicted in Figure 3.1 indeed approximates a hyperbolic plot. This appears to be a universal property of a wide variety of complex electric and mechanical products. However, in the domain of the small components, there is a significant deviation of the hyperbolic structure. This is due to the overrepresentation of the small components in the product, which results in a decrease of the plot's slope in its upper part as is also visible in Figure 3.2 (for the case of the notebook). Note however, that the $y = 100\%$ line in these plots is not asymptotic, since the number of components is finite, and so is the mass of the smallest component. This feature appears in virtually every complex product. The hyperbolic structure and the deviation from it at large values of $n$ means that the number of smaller components is large compared to the number of larger components. For smaller values of $n$, which refer to the most massive components, there is also some deviation, as there are only a few of these. Actually, some of these properties are governed by accident as many of these components are composites and each composite is often treated as one component. Even so, it appears that the typical properties of component mass distribution, such as the ones depicted in Figure 3.2, are robust. This means that the properties are not significantly disturbed even if the data are slightly manipulated, for instance, by combining components into modules, or by considering specific components together, such as some plastic components. The only thing that happens is that it influences the slope of the plot. It should be noted that properties of these types of plots are completely different for mass series than other products' components, such as masses of bodies in the solar system. This accentuates the nontriviality of the properties in component mass distribution.

The method outlined here has also been validated using data from the literature (Gao et al., 2002). These data also refer to a notebook. However, the components in this case are more aggregated than in the notebook considered in Figure 3.1. Only 28 components are discerned here. The most massive component here is a battery cover, which is 35% of the notebook's mass. The screen is not disassembled and its mass is 427 g, which is the second massive component. Fasteners add up to an

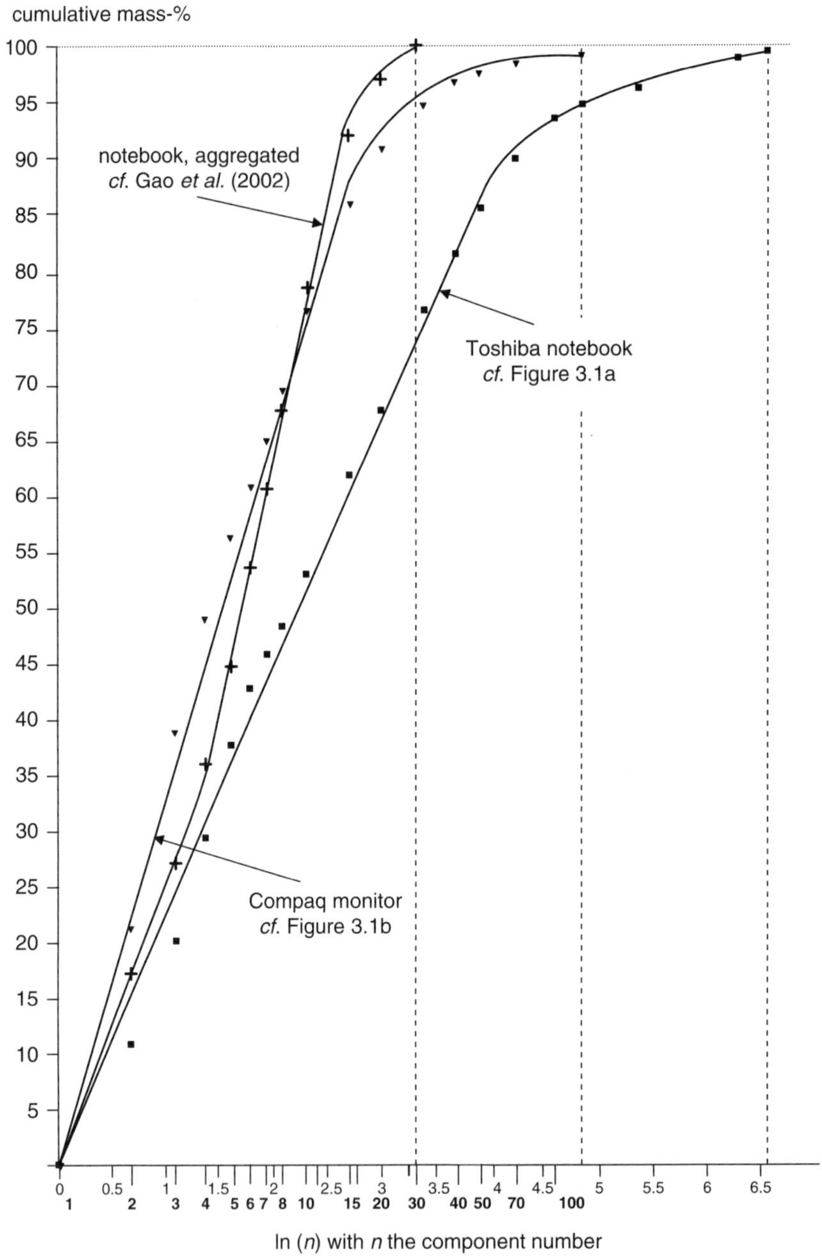


**FIGURE 3.2** Cumulative mass with the heaviest component excluded, as a function of the component number for two products, and an example from the literature.

aggregate of 41 g. Consequently, the graph behaves more compact, with a steeper slope, although showing an initial slope that is comparable to the other two examples of Figure 3.2, due to the contribution of covers and so forth which are not aggregated. The final part, with gradually decreasing slope, is less pronounced, which is also a consequence of the aggregation of the small components into groups or complex components.

The average ratio between the masses of two subsequent components may also provide relevant information. Expression 3.5 can be rewritten as

$$M_n - M_1 = \left(\frac{M - M_1}{\ln x_N}\right)\ln x_n = \alpha \cdot \ln x_n \qquad (3.6)$$

where $\alpha$ is a coefficient that is independent of $n$. From this, it follows

$$\frac{m_{n+1}}{m_n} = \frac{M_{n+1} - M_n}{M_n - M_{n-1}} = \frac{\alpha(\ln x_{n+1} - \ln x_n)}{\alpha(\ln x_n - \ln x_{n-1})} = \frac{\ln(n+1) - \ln n}{\ln n - \ln(n-1)} \quad (n \geq 3) \qquad (3.7)$$

where $m_n$ is the mass of the $n$th component in the list of components with decreasing mass.

Because the linear relationship between ln $n$ and $y$ is only approximate, and only valid in a finite range, it is obvious that the ratio between succeeding masses converges to 1, as long as these are in the linear section of the plot.

It follows from Expression 3.7, for example, that, $m_3/m_2 = 0.58$, $m_4/m_3 = 0.71$, $m_7/m_6 = 0.85$, $m_{20}/m_{19} = 0.95$. Because the components are ordered according to decreasing mass, the ratio between the masses of two subsequent components can never exceed the value of 1. However, real values of mass ratios can substantially deviate from their theoretical counterpart as can be observed in Figure 3.1 by the deviation of the points from the hyperbolic graph. Examples for the above-mentioned ratios for the notebook are 0.830, 0.572, 0.979, 0.825 and the ratios for the monitor are 0.866, 0.957, 0.693, 0.960.

Plots such as those in Figure 3.1 give an insight into the number of components that has to be detached in order to obtain a certain percentage of recovery. It has already been noticed, however, that this theoretical sequence is not practical because of precedence constraints.

In the subsequent chapters, particularly in chapters 5, 6, and 7, the theory of selecting the appropriate disassembly sequence will be treated in detail.

### 3.2.7 Design Issues

As products become more complex, disassembling them also becomes challenging. Even though it is always possible to perform both nondestructive and semidestructive disassembly operations in a safe and clean way that would normally result in a high recovery efficiency, the disassembly time and the associated costs can be substantial. It is for this reason that disassembly, even though desirable, is practiced only to a limited extent. Enhancing the ease of disassembly and thus reducing disassembly time (and cost) is one of the objectives of a rational product design process, frequently referred to as *design for disassembly.* Although a detailed description of the associated design principles is beyond the scope of this book, these principles are closely related to a thorough analysis of the disassembly process. Three topics have to be considered within this framework:

1. Appropriate materials composition:
   - Banned substances, such as Cd and asbestos, should be avoided.
   - Potentially hazardous substances should be avoided.
   - Substances that are difficult to recycle, such as composites, should be restricted.
2. Appropriate mechanical properties:
   - The product's structure should be transparent, which means that the fasteners to be loosened to gain access to the product's interior or for separating it in modules should be clearly indicated and not be hidden.
   - The product should have a hierarchical and modular structure, which means that it is easily separable into its main functional units.
   - The fasteners should be accessible and, if forces have to be applied, this should be facilitated.
   - Connections should be reversible as much as possible.
   - The components, as much as possible, should be made of homogeneous materials.
   - The number of applied materials should be restricted, particularly with respect to plastics.
   - The number of fastener types should be restricted.
   - Operations, as much as possible, should be carried out with one tool only.
   - The number of disassembly directions should be restricted.
3. Appropriate availability of information:
   - A code should be applied on the plastic components, to indicate their materials composition.
   - Product data sheets, including data on materials composition, mass, and geometry of components, should be made available.

Much of the above-mentioned criteria also comply with criteria for the ease of assembly. However, the reversibility criterion and the reduction in the variety of materials are different from pure *design for assembly* concept. From the many papers in the field of design for disassembly, those of Boothroyd and Alting (1992), Wang and Johnson (1995), Harjula et al. (1996), Jovane et al. (1997), Lee and Ishii (1997), Murayama et al. (1999), Shu and Flowers (1999), Veerakamolmal and Gupta (1999, 2000), Chen (2001) and Kuo et al. (2001) are mentioned here.

**TABLE 3.2**
**Breakdown Analyses of a Notebook and a Monitor**

| | Notebook | Monitor |
|---|---|---|
| | 2428 g | 12,321 g |
| Material | Percentage | Percentage |
| Fe | 15.4 | 12.9 |
| Fe-Ni | 2.2 | 0.9 |
| Al | 6.5 | — |
| Cu | — | 4.2 |
| Other nonferrous | 2.6 | 0.2 |
| ABS + PC | 33.5 | 22.4 |
| Various plastics | 8.6 | 1.6 |
| Nylon | 0.9 | — |
| Tapes; foils | 2.1 | 0.2 |
| Rubber | 0.2 | 0.1 |
| Glass | — | 42.3 |
| Ceramics (ferrite) | — | 5.2 |
| PWBs and components | 13.7 | 6.8 |
| Wire; cables; connectors | 2.1 | 2.3 |
| Other complex | 3.3 | 0.6 |
| Other hazardous | 9.0[a] | —[b] |

[a] Internal batteries (Ni-MH) 19 g; LCD screen with diazo compound 198 g; fluorescent tube (Hg-bearing) 3 g.
[b] Fluorescent powder.

### 3.2.8 Materials Breakdown Analyses

In subsection 2.5, three types of breakdown analyses were discussed, including component analysis, materials analysis, and elemental analysis. The first two of these are elaborated below in connection with the notebook and the monitor discussed in one of the previous subsections.

The breakdown analyses of a notebook and a monitor are given in Table 3.2.

Note that in Table 3.2, some complex component categories appear, such as PWBs and cables. These can be decomposed into materials, if desired, according to standard compositions. Although these can differ slightly from case to case, standards are useful as a first order of estimate (see Table 2.26 for an example of such estimate).

From Table 3.2, it follows that for a notebook a theoretical maximum of 72% by weight can be recovered as homogeneous materials via the detachment of homogeneous components, if only destructive and semidestructive operations are applied. From this, 27% is assigned to metals and 45% is assigned to plastics. No less than 33% by weight is assigned to homogeneous components of one kind of engineering plastic, which is typically ABS+PC. From the complex components, a contribution of 14% by weight is due to PWBs. These can be further processed via bulk recycling for reclaiming precious materials and some copper. Typical to notebooks is the liquid crystal display (LCD) screen, which accounts for 8% by weight.

The monitor is different from the notebook because it consists of larger components. The contribution of cathode ray tube (CRT) glass is typically over 40% by weight. Besides that, the monitor contains a considerable amount of coils that contribute to a large content of magneto ceramics (ferrite), and copper. The PWBs present in the monitor have lots of coils and transformers on it, which also contain copper. If required, these major components can be separated from the PWB by prying them out and considering them separately.

### 3.2.9 Connection Diagrams and Connectivity Matrices

As mentioned before, the components and their connections represent the basic elements of a product. The essential data for components include their weight, their dimensions, and their materials composition. The necessary information about the connections is the type of connection and the type of fastener.

The topological relationships between the components of a complete product are graphically represented using a *connection diagram* (also known as a *liaison diagram* or a *graph*). A connection diagram (Bourjault, 1984) is an undirected graph in which the nodes represent components and the arcs represent connections. Connection diagrams will be discussed here with the help of a simple assembly depicted in Figure 3.3. Three full components can be seen here, referred to as A, B, and C. A and B represent mating components that are connected with a strip C. The strip is connected to A via a weld D and to B via a screw E. Thus, components D and E are fasteners. However, D is a virtual component, and E is a discrete component (Navin-Chandra, 1994).

The connection diagram in its most extended form can be depicted as shown in Figure 3.4a. However, it can be argued that D is not a component at all. Consequently, the diagram can be reduced to the form shown in Figure 3.4b. Here, the virtual component D is incorporated within the connection between A and C. If recovery of the discrete component E (fastener) or its material is of minor importance (i.e., if E is considered a quasi-component), then the connections between the real components (B, C) and the quasi-component (E) are represented by dashed lines. If quasi-components are incorporated within the connection, the minimal connection diagram of Figure 3.4c

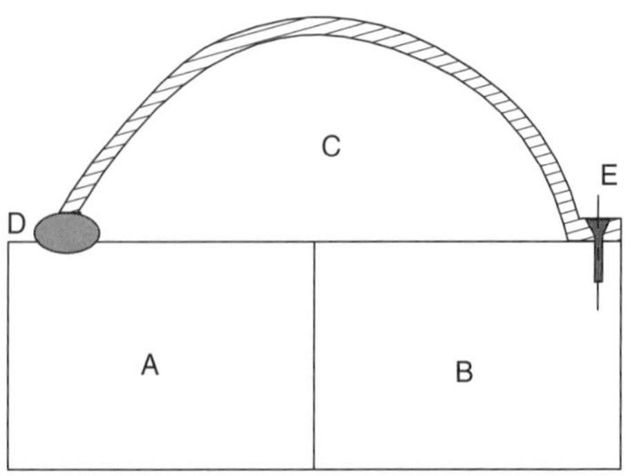


**FIGURE 3.3** Example product for demonstrating the connection diagram.

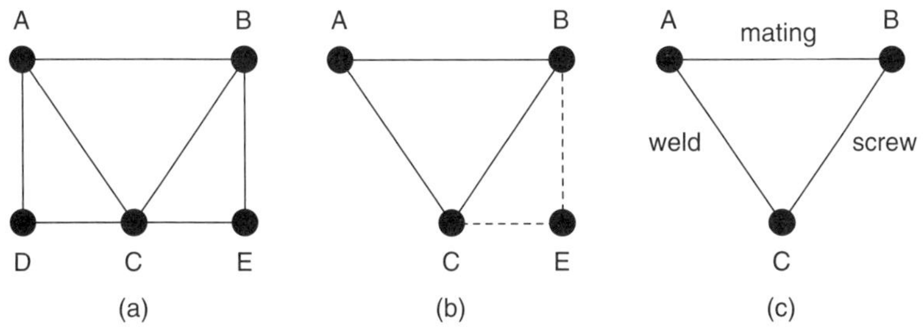


**FIGURE 3.4** Connection diagram of the product of Figure 3.3. (a) Extended; (b) reduced; (c) minimal.

results. In this figure, the type of connection has been added as an indicator. However, this is not obligatory in a connection diagram.

It is possible to further reduce the connection diagram. If the materials content of strip C is not considered relevant, and the strip is only attached to immobilize components A and B, it can also be considered as a quasi-component (Figure 3.5a). Destructive dismantling may also be permitted. This is usually described by separating the component, which has to be destroyed, into two subcomponents. In this case, cutting strip C can separate the product into its principal components A and B. Therefore, strip C is decomposed into two subcomponents C′ and C″ in the connection diagram (Figure 3.5b). Processes such as these are discussed by Armillotta and Semeraro (1997).

From the above discussion it is clear that there is some sort of flexibility as well as arbitrariness in generating the connection diagram. However, this also enables us to adapt the connection diagram for any desired purpose.

Since computers usually generate the disassembly sequences, it is necessary that the connection diagrams have a format that can be communicated with the computer. This can be done with the help of an $N \times N$ *connectivity matrix* C, where $N$ represents the number of components in the product. The value in a cell of the matrix equals 1 if the respective components are connected, and zero if they are not. One can restrict the matrix strictly to the lower triangle, because the square matrix is actually symmetric, which means that $C_{i,j} = C_{j,i}$. The diagonal elements $C_{i,i}$ have no topological

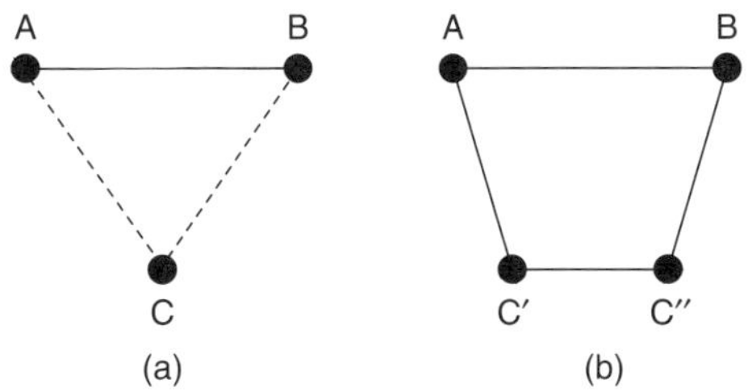


**FIGURE 3.5** Connection diagram with (a) strip C as a quasi-component, and (b) destructive operation of the strip.

| | A | B | C | D | E |
|---|---|---|---|---|---|
| A | | | | | |
| B | 1 | | | | |
| C | 1 | 1 | | | |
| D | 1 | 0 | 1 | | |
| E | 0 | 1 | 1 | 0 | |

**FIGURE 3.6** Connectivity matrix for the connection Figure 3.4a.

significance and can be left blank or put equal to zero. The connectivity matrix that corresponds to the connection diagram of Figure 3.4a is presented in Figure 3.6.

The maximum number of connections $K$ in a product with $N$ components is given by the following expression:

$$K = {}^{1}/_{2}\, N \cdot (N - 1) \tag{3.8}$$

If all the connections in a product are established, i.e., all the significant elements equal 1, the matrix represents a strongly connected product, i.e., a product in which all the components are mutually connected. A detailed discussion of this can be found in chapter 4.

## 3.3 THE PROCESS

### 3.3.1 BASIC APPROACHES

The connection diagrams can be used to define disassembly operations via the following two complementary approaches:

1. Connection-oriented approach
2. Component-oriented approach

In the connection-oriented approach, a typical operation involves the disestablishment of a connection, which is represented by a *cut* in the connection diagram (see Figure 3.7a). In the component-oriented approach, however, a typical operation is the division of a subassembly into two subassemblies, which divides the corresponding connection diagram into two separate connection diagrams. An additional

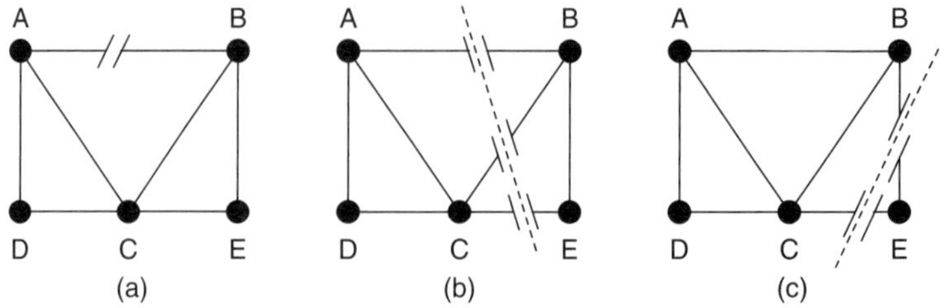


**FIGURE 3.7** Disassembly operations. (a) Connection-oriented; (b) cut-set; (c) cut-set for the detachment of a single component.

constraint here is that both connection diagrams must be connected themselves. This is represented by a *cut-set* in the connection diagram (see Figure 3.7b). In case one of the two resulting subassemblies is not connected (i.e., it consists of one component only), the resulting operation is termed *sequential* (see Figure 3.7c); otherwise the operation is termed *parallel*.

Note that the disassembly operations considered in the previous paragraph and depicted in Figure 3.7 were only presented to define some terms and may or may not actually be feasible. For example, in the case of the connection-oriented approach, the connection between A and B cannot be disestablished without previously disestablishing other connections (see Figure 3.3). In addition, the disestablishment of a single connection may not be possible without simultaneously disestablishing one or more other connections. For instance, the connections between B and E, and between C and E, have to be simultaneously disestablished. Similar constraints occur in the cut-set cases, i.e., not every cut-set is possible. For example, the cut-set in Figure 3.7b suggesting the detachment of subassembly BE, is actually infeasible, as can be seen in Figure 3.3. On the other hand, the cut-set in Figure 3.7c representing the detachment of component E, is feasible. Furthermore, it is also clear that E must be detached before the detachment of B. In other words: the detachment of E *must precede* the detachment of B. This type of constraint is called *geometric constraint*. Geometric constraints will be extensively discussed in chapter 5.

From the above discussion, the component-oriented approach appears to be preferable in dealing with geometric constraints. It will be demonstrated in chapter 5 that this results in a transparent formalism that can be applied in a broad range of product configurations. The option of parallel disassembly can easily be incorporated using this approach. Parallel disassembly, particularly the detachment of modules, is often beneficial because it enhances flexibility. The resulting subassemblies can be disassembled independently of each other.

### 3.3.2 Disassembly Task Typology

Disassembly operations can be subdivided into disassembly *tasks*. Formulating a typology of disassembly tasks is indispensable for disassembly costing. As noted above, the component-oriented approach is preferable in disassembly studies. However, the process of identifying different tasks may need some explanation. In the component-oriented approach, the separation of a parent subassembly into two child subassemblies has to be further subdivided into tasks. Tasks may not only include the disestablishment of connections, they may also include the movements necessary to transfer the subassemblies to a different location or disassembly station. In addition to the tasks that are intrinsic to disassembly, supplementary tasks such as cleaning, fixturing, tool exchanging, product reorienting (to guarantee access or stability), and testing must also be identified.

Other issues may also arise. For example, in both manual and robotic disassembly operations, there are times when a product to be disassembled has to be immobilized with a fixture. If the disassembly operation at hand results in two child subassemblies, one of them could still be left clamped in the fixture while the other one could be free to be moved. The one in the fixture is called the *static subassembly* while the one that can be moved is called the *dynamic subassembly.* In such cases, the time required to retrieve the static subassembly will be more than the time required to retrieve the dynamic subassembly! Another example involves testing. Testing results can influence the disassembly process as well. For instance, if a module is found to be sound, no further disassembly may be performed. On the other hand, if a module is defective, its disassembly into submodules may be required.

The following disassembly tasks are frequently present in disassembly operations:

1. Attachment to a fixture
2. Tool exchange
3. Product reorientation
4. Tool reorientation
5. Disestablishment of connections
6. Gripping the dynamic child subassembly
7. Moving the dynamic child subassembly
8. Releasing the static child subassembly from the fixture

Of course, tasks such as "attachment to a fixture," "tool exchange," and "product reorientation" may not be needed in every disassembly operation. On the other hand, some tasks may occur multiple times in performing just a single disassembly operation, e.g., the detachment of a subassembly may require the unscrewing of multiple screw types involving multiple tool exchanges and orientations.

Since disestablishment of connections is crucial in disassembly cost assignment, various authors have proposed typologies for these tasks. Dowie and Kelly (1994) discern 12 ways of breaking a connection; Kroll et al. (1996) define 10 types; Das et al. (2000) propose an elaborate listing of 13 types of unfastening processes, which include a listing of fasteners rather than a listing of processes. Apart from this, 16 types of disassembly processes are distinguished, which also include processes that belong to the bulk-recycling domain. Starting from this, and based on various experiments conducted, the following typology of disestablishing tasks is proposed, which combines the advantages of the above-mentioned typologies and lists the tasks according to their increasingly destructive character. Most of the disassembly tasks that occur in practice can be positioned within this typology.

| **Task type** | **Example** |
|---|---|
| *Reversible* | |
| Move: Unlock | Open lid, enabling accessibility or movability |
| Remove (only gravity force) | Disestablish mating connection(s) |
| Pull (friction force is present) | Disestablish press fit connection |
| Unscrew | Rotate screw, nut, against friction |
| Deform (reversible) | Release a clamp |
| Drain | Remove a fluid |
| *Irreversible* | |
| Deform (irreversible) | Unfold a seam fold |
| Peel | Remove a sticker or foil |

| | |
|---|---|
| Pry out | Detach a press fit, rivet, or glue connection |
| Shear cut | Cut a wire |
| Saw cut | Destroy a strip |
| Drill | Destroy a bolt |
| Break | Destroy by applying force |
| Melt | Release soldering connections |
| Crush | Destroy CR tube for accessing interior |

Some of the above-mentioned disassembly tasks can be assigned to specific connection types that are listed in subsection 3.2.3 (the numbers in parentheses refer to the connection type listed in subsection 3.2.3):

| Connection type | Disassembly task |
|---|---|
| Mate (1) | Remove |
| Lock (1) | Move |
| Bundler (2) | Shear cut |
| Spring (3) | Deform/pull |
| Screw (4) | Unscrew (drill) |
| Nut (4) | Unscrew |
| Lock washer (4) | Deform/pull |
| Cotter pin (5) | Pull |
| Snap fit (6) | Deform, pry out/pull |
| Connector (7) | Pull |
| Press fit (7) | Pull/pry out |
| Shrink fit (7) | Pull |
| Rivet (8) | Pry out/drill |
| Fold (9) | Deform |
| Glue (10) | Peel/pry out/break |
| Solder (11) | Shear cut/break/melt |
| Weld (12) | Saw cut/break |
| Mould (13) | Break |

### 3.3.3 Disassembly Tools

During the late 1990s, German and Japanese investigators advocated for robotic disassembly lines. Unfortunately, the flexibility and adaptiveness of currently available robots are inadequate to handle the requirements for fully automated disassembly lines, which, among other things require processing of different products with uncertainties in configurations, geometry, and conditions of fasteners. Even though state-of-the-art sensor technology can add some flexibility to robotic operations, most robots are primarily designed to execute predetermined tasks. This makes widespread practice of robotic disassembly lines less viable, although future developments, together with adequate product data management, may extend the possibilities of this type of operation. Note that there are disassembly lines that do use some robots but offer very little flexibility. Manual disassembly, on the other hand, is flexible albeit relatively expensive. Humans, particularly skilled ones, can adapt to the uncertainties that frequently occur in the supply of end-of-life products.

Both robotic and human-operated disassembly lines use a range of tools. These are intended for efficiently applying forces and guaranteeing stability to the product and its subassemblies. Robotic operations require several types of fixtures, jigs, and grippers etc., which have to mimic the flexibility of human hands and sensory capabilities. Manual operations, on the other hand, can be accomplished with relatively modest types of tools, although some additional amenities are needed to protect the health and the safety of the employees. These may include gloves, respiratory protection, hearing protection, etc. Protection is not only needed because of the debris generated (e.g., iron filings, glass shells) by the task at hand, but also because many end-of-life products are contaminated with working fluids, dust, butyric acid, soot, toner, and so on.

It can be demonstrated that manual disassembly (to a considerable disassembly depth) of many electric and electronic products requires only a restricted set of tools, such as a few types and sizes of screwdrivers (e.g., slotted head, Phillips, telephone, socket head), wrenches (e.g., socket head and open-ended), cutters and pliers, crowbars and hammers, and so forth. However, in some cases, tools that are powered with electricity or compressed air are preferable. It all depends on the magnitude of the forces that have to be applied, the volume of work that has to be accomplished, the accessibility of components, and the repeatability and frequency of tasks, whether or not more specialized tools are preferable. In some cases, machine tools for destructive operations, such as mechanical scissors, may also be needed.

### 3.3.4 Disassembly Process Analysis

Several reports on disassembly processes in the literature are devoted to the estimation of time and cost. In most of these cases, a predefined disassembly sequence, arranged as a *hierarchical tree structure*, which is similar to the bill of materials, is assumed. Åkermark (1997) presented a time analysis of the complete disassembly of a toaster. The registered time was subdivided into handling time and disassembly time. Further suggestions were made on the operator's need for additional information and on the need for specialized tools. As this study was only cost-oriented without any regard to the disassembly depth, no estimates of revenues were provided.

Typically, the hierarchical tree representation has a modular structure. This implies that parallel disassembly is permitted, and at times may even be essential, in such a model, which results in a divergent structure. As an example of this approach, some steps for disassembling a printer are depicted in Figure 3.8.

Additional information on the composition and mass of the components may be essential. The hierarchical tree may be complex, even for relatively simple products such as a monitor (see Figure 3.9). Minor components, such as fasteners, are not explicitly shown here. Semidestructive operations are avoided as much as possible. Disassembly tasks are assigned at the branching points.

Note that the disassembly task descriptions and the mass and composition descriptions are missing from Figure 3.9. An example of a disassembly task description for uncovering the monitor might include something like this: *unscrew* 4 *x sunken Phillips*

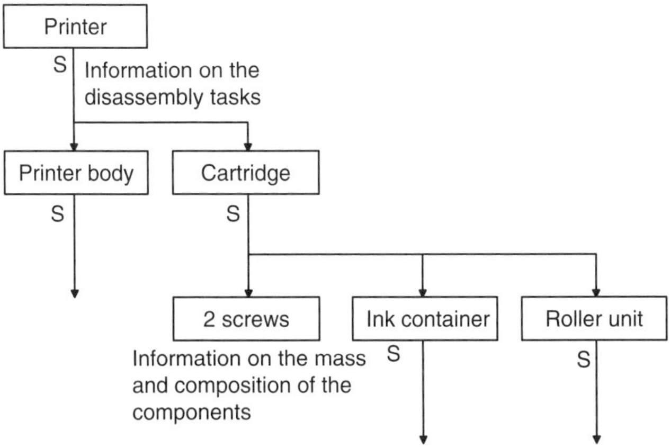


**FIGURE 3.8** Part of a hierarchical tree structure in disassembly study of a printer (the symbol S refers to a subassembly; it is absent in case of a single component).

*screws*. An example of a mass and composition description for the cover might look something like this: *mass equals* 1369 *g; materials composition is ABS* (*Acrylonitrile Butadiene Styrene*)*, with a minor amount of foam.*

The hierarchical tree structure is only appropriate for a product that has a distinct modular structure. If there is a demand for a particular component, the tree structure offers a unique way to get it. For example, to retrieve the "main PWB," the following sequence of disassembly operations is necessary:

- Detachment of the cover
- Detachment of the CRT unit
- Detachment of the screen module
- Detachment of the cable entry
- Detachment of the data cable
- Detachment of the main PWB

Evidently, a thorough analysis of disassembly processes by subdividing them into operations and tasks is the basis for sound cost metrics.

## 3.4 COST METRICS

### 3.4.1 Introduction

Two approaches for cost metrics are available in the literature, which include the technical approach and the work measurement approach. The technical approach (discussed in subsection 3.4.2) is based on robotic assembly whereas the work measurement approach (discussed in subsection 3.4.3) is based on human work analysis. The distinction between fixed and variable costs is discussed in subsection 3.4.4.

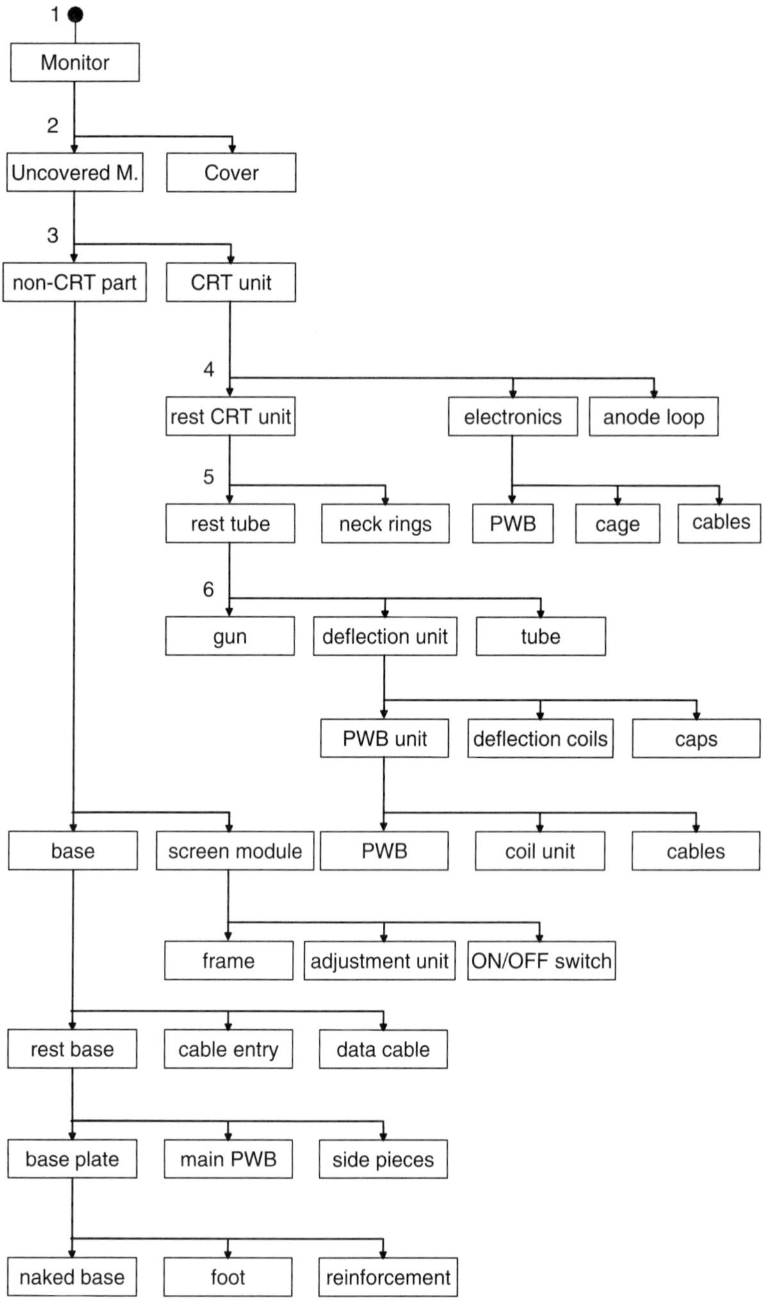


**FIGURE 3.9** Hierarchical tree structure for disassembling a monitor.

### 3.4.2 Technical Approach

In robotic assembly, the costs are mainly determined by technical factors that can be estimated with a high degree of precision. Examples of detailed deterministic approaches are due to Lee and Shin (1990), and Suarez and Lee (1997). These authors follow the connection-oriented approach starting with the connection diagram. Costs are assigned to every connection and, consequently, to every arc in the diagram. This is called a weighted connection diagram, which also addresses characteristics such as separability and required forces, manipulability, fastener type and so forth. Based on these, detailed formulae are derived, which reveal a metric for the complexity of the disestablishment of these connections. Despite the complexity of the formulae, the introduction of multiple weight coefficients cannot be avoided, which introduces new ambiguities. Therefore, this method, which was originally developed to automatically define modules in mechanical products based on heuristics, is not suitable for disassembility study because it suffers from over-exactness in a field in which a lot of uncertainty is inevitable.

Cost metrics that are based on the technical aspects of the disassembly process are due to Asiedu and Gu (1998), who also present a review on life cycle costing. Their method includes a formula for the disassembly time, which is based on the product structure. It is composed of the following components:

1. Time to remove components
2. Time to release fasteners
3. Time to release connections without discrete fasteners

In a slightly modified form, the formula is as follows:

$$t_D = \sum_{i=1}^{N} t_i + \sum_{k=1}^{n}\left(\sum_{f=1}^{F}(n_f \cdot t_f)_k + \sum_{p=1}^{P}(n_p \cdot t_p)_k\right) \tag{3.9}$$

where $t_D$ is the disassembly time of the product, $t_i$ is the time for removing component $i$, $t_f$ is the time for removing one item of discrete fastener type $f$, $t_p$ is the time for removing one item of nondiscrete connection type $p$, $N$ is the number of components, $n$ is the number of connections; $n_f$ is the number of items in one discrete fastener type $f$, $n_p$ is the number of items in one nondiscrete connection type $p$, $F$ is the number of discrete fastener types (screw, etc.) $P$ is the number of nondiscrete fastener types (spot weld, fold, etc.).

### 3.4.3 Work Measurement Approach

As mentioned before, robotic operations are not always suitable for disassembly. In such cases manual disassembly operations are necessary where costs are often based on direct labor charges, which, in turn, are proportional to operation times. Operation times (especially for assembly processes) are typically measured using standard work measurement tools such as MOST (Maynard Operations Sequence Technique), see Zandin (1980). However, these tools are not suitable for disassembly processes, as they fail to fulfill the assumption of repetitive and predictable tasks because of disassembly's inherently uncertain behavior.

Several modified work measurement methods have been developed especially for disassembly processes. Examples of such methods are due to McGlothlin and Kroll (1995); Boks et al. (1996); Kroll (1996); and Kroll and Carver (1999). These authors distinguish a standardized base time, which is based on the most efficient method for performing a task under average conditions. This is translated into a 1 to 10 dimensionless scale based on the difficulty of the task. The resulting value is increased with penalties based on factors such as accessibility, tool positioning, and force to be applied. Exchange of tools and additional hand movements are also incorporated. The set of values that are obtained by this are distilled into one figure. This is multiplied by a specific factor to reveal the disassembly time. In methods like this, the usefulness of the typology of disassembly operations is apparent, as it opens the way for collecting basic data on a restricted number of unit tasks.

Time metric is important in the optimization of both product and disassembly process design. In the approach of Kroll and his colleagues, who use a disassembly evaluation chart, the disassembly sequence is predefined. Their approach is primarily aimed at the evaluation and optimization of the product design. Alternative disassembly sequences can be evaluated on a one by one basis. However, they do not present a systematic optimization method.

Although methods that are based only on work measurement can generate estimates for disassembly time, this is not the only component that contributes to the disassembly cost. To this end Das et al. (2000) suggest an additional cost driver called the *disassembly effort index* (DEI). Additional points toward costs are given for:

1. The use of specialized tools
2. The use of specialized fixtures
3. The need for instructions or skilled workers
4. The need for safety measures, such as gloves or masks

The DEI represents indirect costs, which have to be added to the direct labor costs according to some distributive code.

In practice, the disassembly time can vary considerably because of *corruption* of connections and other *contaminations.* Although not included in the above-mentioned studies, these factors can be significant, particularly for those products that have been operated in aggressive environments, such as cars and washing machines. Corruption and contamination not only degrade the quality of components, they also contribute to more difficult and unsafe disassembly activities.

Scholz-Reiter and Scharke (1997), and Salomonski and Zussman (1999) address three causes of corruption: rust, deformation, and missing fasteners. These are incorporated in a predictive model that estimates disassembly time as a function of the amount of rust and deformation, and the presence or absence of fasteners, measured with the help of sensors. Methods like these, however, are only viable if a large number of similar products are processed repetitively using robots.

Considering direct labor time as the primary cost driver causes a problem in traditional cost metrics. With overhead cost assigned to direct labor cost, this method could unfairly distort the economic feasibility of manual disassembly. Therefore, instead of traditional costing methods based on direct labor, activity-based costing (ABC) appears to be more suitable for this situation in which direct labor contributes to a minor fraction (down to 5%) of the total cost, because of the increased capital and knowledge intensiveness of production. ABC allocates the overhead cost more appropriately to the activities that generate these costs than direct costing does (Bras and Emblemsvåg, 1995). Using this method, a redistribution of overhead cost takes place, which shifts the optimum for product and process design in a direction that is beneficial for efficient recovery of components and materials, thus reducing the environmental load of products during their end-of-life phase.

### 3.4.4 Fixed and Variable Costs

In the previous two subsections, it was argued that the knowledge of disassembly time and direct labor cost would facilitate the calculation of disassembly cost. Since each item processed will require a different amount of disassembly time, the disassembly cost obtained in this way is a variable cost. Although in traditional costing methods most of the overhead costs are incorporated into direct labor cost, and major cost components, which are independent of the disassembly time and are overlooked. Examples of such cost components include internal transportation costs, and storage costs, building costs, etc. These costs can be allocated equally to each item processed and would thus lead to the fixed cost portion of the total disassembly cost.

In the literature, multiple expressions are proposed for determining the variable and fixed costs portions of the total disassembly cost per item (see, for example, Armillotta and Semeraro, 1997 and Jovane et al., 1998).

The variable cost per item $C_v t$ is given by the following expression:

$$C_v t = \frac{1}{\eta}\left[c_D + \frac{c_I}{v} + \frac{1}{hd\tau}(c_E + c_B \alpha A)\right] t \tag{3.10}$$

where $C_v$ is the variable cost per hour, $t$ is the disassembly time per unit in hours, $d$ and $h$ are the number of working days in a year, and that of working hours in a day, respectively, $c_E$ and $c_B$ are the disassembly equipment cost and the building cost per unit area, respectively, $c_D$ and $c_I$ are the labor costs of the direct operator and supervisor, respectively, $v$ represents the number of operators per supervisor, $\tau$ is the depreciation time in years, $A$ is the area of the disassembly cell, and $\alpha$ is a correction factor attributed to materials handling.

The fixed cost per item $C_f$ is given by the following expression:

$$C_f = C_H + \frac{C_G}{V} \tag{3.11}$$

where $C_H$ is the handling cost per item which includes the handling of both the products and the resulting materials and is assumed to be independent of disassembly time, $C_G$ represents the general expenses per year that includes such things as building costs, administration costs and so forth, and $V$ is the number of items processed per year.

The total disassembly cost per unit $C$ can therefore be calculated as follows:

$$C = C_v t + C_f \tag{3.12}$$

## 3.5 REVENUE METRICS

### 3.5.1 Introduction

There are two fundamental drivers for end-of-life product disassembly, which include financial revenues and environmental benefits, both of which are discussed in the following subsections.

### 3.5.2 Financial Revenues

Financial revenues derived from end-of-life products depend on the mass and the composition of the different components and subassemblies. In order to estimate these revenues, the prices (which are a function of market conditions) of different modules, components, and materials should be known. In addition, their demands need to be appraised. Typically, the demands for modules and materials are restricted. These restrictions, however, are less pronounced for materials recycling, as the spectrum of their applications is much larger. Some estimates of materials prices were compiled in Table 2.1. Apart from market conditions, the prices are heavily affected by the tolerances on the impurities permitted by various buyers. At times, one has to pay a fee to dispose of some materials. This is treated as negative revenue.

Typical outputs generated by disassembly facilities consist of unprocessed products, modules, components, damaged components, and waste. After leaving the disassembly facility, these items are sent to one of the following (Figure 3.10): refurbishing, remanufacturing, reuse, recycling, and final processing (incineration and landfill).

In industrial settings, the detached components or subassemblies are usually *sorted* by placing them within a limited number of bins, each assigned to a particular type of material or module. The number of bins strongly depends on the type and quantity of products disassembled. More bins offer a better level of separation with respect to materials or component category, although there might be constraints (e.g., limited space) that would restrict the number of bins that can be accommodated. Some constraints might come from the buyers, for example they might want items sorted based on the degree of contamination or the minimum contents of a particular material, or the batch size or other criteria. Optimizing the sorting process via an

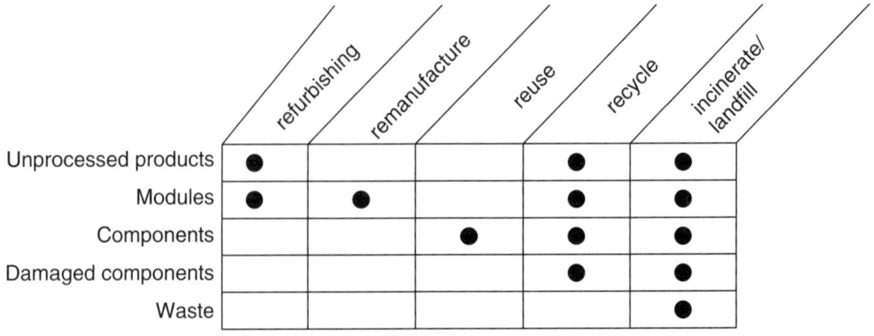


**FIGURE 3.10** Destination for the output flows of a disassembly facility.

appropriate definition of the streams, including further subdivision, or blending of these streams, belongs to the domain of *clustering* or *bin allocation problem*.

Some data on limits on impurities in various streams, and the corresponding materials prices, can be found in Das and Matthew (1999). The blending problem is more complicated as compatibility issues have to be accounted for. For example, some types of blends of plastics or metals may be more useful to materials recyclers. Reduction in the number of bins can be obtained if compatible materials are blended, which can proceed without a considerable loss in value. Jovane et al. (1993) present a compatibility matrix for engineering plastics. It showed, for example, that PE can be blended with PP; PVC with SAN and ABS; PC with PVC, SAN, and ABS; and PA with PS. Further details on blending of plastics are presented by Jorden (1991). Of late, the composition of plastics in electronic products is specified on major components, which facilitates appropriate sorting. Automatic methods for materials analysis, based on laser reflection techniques, are also available.

### 3.5.3 Environmental Benefits

#### 3.5.3.1 Introduction

Environmental benefits of disassembly processes are not easily quantifiable. The basic reason for this is that the environmental impact of a product or a process consists of various unrelated aspects that have to be combined in some way to get a sense of its benefit. Essential to validating the environmental merits of the product lifecycle and the various processes that are part of it, is complying with legislative standards. Legislation prescribes definite product design constraints (such as the absence of cadmium in the product), mandatory end-of-life processing steps (such as the removal of batteries), requirements on end-of-life processing (such as a minimum recycling efficiency), and restrictions on final processing steps (such as a ban on landfill for shredder residue).

#### 3.5.3.2 Life Cycle Assessment

Since ecological issues are one of the driving forces behind end-of-life processing, these should be incorporated in models that are aimed at optimizing disassembly operations. The basic tool for quantifying the environmental impact of products over their life cycle is life cycle assessment (LCA). LCA is a quantitative method that determines the complete environmental impact of a product during its life cycle, including the impact of ancillary processes, such as energy supply. This methodology is standardized via the ISO 14040 series. Detailed discussion of LCA is not within the scope of this book. However, the basic features of LCA are outlined here (Guinée et al., 1993ab). The method proceeds along five phases, viz.:

1. Goal definition
2. Inventory
3. Classification
4. Valuation
5. Improvement analysis

Phase 2 and Phase 3 are most important for this book's purpose. In the course of the inventory phase, all relevant processes and their associated mass and energy flows are quantified, including ancillary processes. Besides the extraction and production processes, the consumption process and the end-of-life process are also considered. The life-cycle chain has already been depicted in Figure 2.2. The *second-* and *higher-order processes* that are connected with the production of the ancillary mass and energy flows must also be incorporated.

The environmental impact of the product life cycle is related to (1) the depletion of natural resources and (2) the discharge of waste and emissions to the soil and groundwater, surface water, and the atmosphere. Apart from this, the solid waste, which is discharged to landfills, incinerated, or retained in a controlled depot dedicated for hazardous materials must be considered. Because typical inventories house a multitude of materials, these are assigned to a set of selected impact categories. Such analysis is called *impact assessment*. Although there may be other categories and subcategories, the most frequently used categories are:

1. Depletion of natural resources (minerals, ores)
2. Depletion of fossil energy carriers
3. Ozone depletion potential (ODP)
4. Global warming potential (GWP)
5. Eutrophication/acidification
6. Human toxicity
7. Ecotoxicity
8. Heavy metals
9. Photochemical oxidant formation

Each of these is affected by multiple components. GWP, for example, is not only boosted by carbon dioxide ($CO_2$)-emission, but also by the emission of other gases, such as methane ($CH_4$). Prior to combining the various contributions to a particular impact category, they have to be normalized to the same unit, using a weight factor. An appropriate unit for the GWP, for example, is the ton $CO_2$-equivalent.

Apart from the above-mentioned effects, which are directly connected with materials flows, other effects, known as disturbances, are also extremely important. These effects are related to the deterioration of ecosystems, the occupation of area, and so forth. Unfortunately, this kind of effect is not easy to quantify. It is, therefore, often underestimated in LCA analyses and related studies.

The LCA method is supported by software tools that include databases containing the impact of many frequently occurring processes and materials. A serious drawback of standard LCA methodology is that it insufficiently supports the end-of-life phase. Current software tools usually support a user-defined mix of definite end-of-life options such as landfill and incineration. Secondary materials are treated by assigning a negative environmental impact to the flow, which is the reverse of the impact caused by producing the equivalent amount of virgin material. Of course, the environmental impact that is caused by the recycling processes has to be added as a positive value. This scheme fails in the presence of complex recycling loops.

A discussion of the techniques that are available for dealing with this problem is due to Tillman et al. (1994). These are nonstandard and extend LCA more or less to network analysis.

#### 3.5.3.3 Eco-Indicator

As mentioned previously, aggregating the impact categories via LCA requires a certain degree of agreement on the needed weight factors. Other easy to use tools such as the *eco-indicator* are frequently used. Consider *eco-indicator 99*, for example (Goedkoop and Spriensma, 2000). Using this tool, the aggregation takes place in multiple steps, which reveals three aspect categories:

1. Damage to the mineral and fossil resources.
2. Damage to the ecosystem.
3. Damage to human health.

These three categories are combined via additional weight factors into one single estimator called the eco-indicator. The weight factors are established by a committee of advisors, and have to be periodically reestablished, because of changing policies and scientific insights.

#### 3.5.3.4 Environmental Constraints

The principal environmental benefits that are tied to reuse and recycling of end-of-life products are:

1. Reduction in waste and emission that is related to the discharge of the end-of-life products.
2. Reduction in the use of mineral and fossil resources via substituting primary products by secondary products.

From an environmental point of view, these beneficial effects are counteracted via the energy use and emissions that are related to the end-of-life processing chain itself, including additional transport, processing, and residual waste. Fortunately, in many cases, the net environmental benefit of partly closing the cycle is positive.

In disassembly modeling, two complementary methods are available for incorporating the environmental effect. The first is a *legislation-oriented approach.* Its premise is that compliance with the regulation should be guaranteed. For example, a hazardous material such as mercury should be recovered because of legislation, which is considered a "hard" constraint. No further discussion on the effects of this substance is then required. The second approach is an *optimization-oriented approach.* The goal here is to reach maximum environmental benefit at minimum expense. In contrast with the legislation-oriented approach, the constraints here are "soft." For example, the recycling percentage of steel, which is not subjected to any legislation, will be according to the outcome of an optimization procedure.

## 3.6 ECONOMIC OPTIMIZATION

### 3.6.1 INTRODUCTION

In the last two sections, costs and revenues were treated as separate entities. In this section, they are treated together (economic modeling). In the evaluation of recovery processes, economic models are used to determine a tradeoff between costs and revenues, in an attempt to maximize the profit. When used to evaluate the disassembly process, economic models are instrumental in determining the optimal disassembly depth of a product. Sometimes additional processes such as bulk recycling are also included in the model proving trade-off between even more variables further influencing the disassembly depth (for an example of the analysis of disassembly and subsequent bulk recycling of a refrigerator, see Cagno et al., 2000). The profit function is discussed in subsection 3.6.2. The return on investment and profit rate are the subjects of subsection 3.6.3.

### 3.6.2 THE PROFIT FUNCTION

The economic modeling of disassembly processes is discussed in many papers, see, e.g., Chen et al. (1993); Dowie and Kelly (1994); Smith et al. (1995); Harjula et al. (1996); Knight (1996); Armillotta and Semeraro (1997); Asiedu and Gu (1998); Jovane et al. (1998); Sodhi and Knight (1998); Veerakamolmal and Gupta (1998, 1999); Knight and Sodhi (2000); Reimer et al. (2000); Dini et al. (2001); Kongar and Gupta (2002).

The most elementary relationship that defines the profit $P$ is as follows:

$$P = R - C \tag{3.13}$$

Where $P$, $R$, and $C$ are the profit, revenue, and cost per item, respectively.

The profit of a disassembly process depends on two principal parameters, namely the disassembly depth, which is expressed by the *state* of the processed product, and the *disassembly sequence*, which represents the sequence of disassembly operations, starting from the initial product, that is required to reach this state. The state $s$ of a product is given by the set of subassemblies that is obtained after the selective disassembly process has been carried out. The disassembly revenue depends on the state of the product once the intended disassembly depth has been obtained, and the disassembly cost depends on the disassembly sequence that has been followed. This will be illustrated with an example.

#### EXAMPLE 1

If a product consists of six components (A, B, C, D, E, and F), it is represented by ABCDEF. A *state* of this product at a particular disassembly depth provides a list of various subassemblies available. For example, ABD+CE+F means that subassemblies ABD, CE, and F are present at this particular disassembly depth. Obviously, all the components within each of the subassemblies have to be connected, and any particular state should be obtained via disassembly operations only. Any particular state can be reached via one of many possible *disassembly sequences*. In the example considered here, a possible sequence is as follows

$$\text{ABCDEF} \rightarrow \text{ABCDE} + \text{F} \rightarrow \text{ABD} + \text{CE} + \text{F}$$

Another possible sequence is as follows

$$\text{ABCDEF} \rightarrow \text{ABDF} + \text{CE} \rightarrow \text{ABD} + \text{CE} + \text{F}$$

If $N$ represents the number of components in a product, then the number of possible states and disassembly sequences increase rapidly with both $N$ and the disassembly depth. The theory behind this issue will be discussed in the next three chapters.

Let $S$ be a state, which is a set of subassemblies $\sigma$ that corresponds to a partition of the set of components of the original product. Let the set of disassembly sequences that result in state $S$ be $Q_S$. Let $q_S$ be an element of $Q_S$. Consequently, $q_S$ is an ordered set of disassembly operations $j_{q_S}$. The revenue $R_S$ is the sum of the partial revenues $R_\sigma$ of every subassembly that is an element of $S$. The cost $C_{q_S}$ is the sum of the costs of every individual disassembly operation that is an element of $q_S$. The profit, which depends on both the state $S$ and the particular disassembly sequence that results in this state, is given by the expression

$$P_{S,q_S} = R_S - C_{q_S} = \sum_{\sigma \in S} R_\sigma - \sum_{j \in q_S} C_j \tag{3.14}$$

Optimization procedures identify both the state and the appropriate disassembly sequence that result in the maximum profit. This problem can be solved with the help mathematical programming.

The profit that is obtained in the initial state, which is represented by $S = 0$, is given by the following relationship, with both terms on the right being negative:

$$P_0 = R_0 - C_0 \tag{3.15}$$

Here the index 0 refers to the initial operation, which simply is the availability of the product. From the viewpoint of the disassembly company, the cost $C_0$ is negative if the supplier pays a fee to the disassembler. At this state the revenue is also negative, because the disassembler has to pay a fee to get rid of the product.

The subassemblies that are generated during the course of the disassembly process can produce either positive or negative revenues. Negative revenues occur, for example, if a subassembly has no demand or if its materials cannot be recycled. We use the convention of assigning the revenues to the subassemblies, and the costs to the operations.

The revenue from a subassembly depends on how it is further processed. Therefore, for any subassembly $i$, the following relationship for the revenue holds

$$R_i = \text{Max}(R_{lf}, R_{inc}, R_{rec}, R_{reu})_i \tag{3.16}$$

This relationship states that the revenue obtained from any subassembly equals the maximum value that can be obtained from one of four choices, including landfill, incineration, materials recycling, and subassembly reuse, denoted by the indices *lf*, *inc*, *rec*, and *reu*, respectively.

Although the materials composition of the initial product is given, its redistribution over multiple subassemblies strongly influences the total revenue. By applying one or more disassembly operations to the product, it separates into two or more subassemblies. The materials composition of each of these subassemblies is different from that of the original product, which makes the sum of the individual revenues of these subassemblies different from the initial product's revenue. This usually has a positive effect as homogeneous materials or fractions with enhanced contents of valuable materials may be obtained. In addition, hazardous materials can often be isolated. This is illustrated with the help of the following two examples:

### Example 2

A product with mass $M$ consists of a hazardous component $X$ that makes up 20% of the product's mass, and a useful component $Y$ that makes up 80% of the product's mass. The revenue from the useful component is \$0.20/kg and that from the hazardous component is – \$0.40/kg. The separation cost $C_1$ = \$0.30 $M$/unit product. The complete product $XY$ is considered hazardous waste. The product is made available at zero cost, thus $C_0 = 0$.

Thus, the profit $P_0$ is equal to the revenue before separation $R_0$ = – \$0.40 $M$/unit product. If, on the other hand, disassembly were performed on the product, the revenue after separation, $R_1 = R_X + R_Y = (0.40 \times 0.2 + 0.20 \times 0.8)\ M$ = \$0.08 $M$. Consequently, the profit has increased to $P_1 = R_1 - (C_0 + C_1)$ = – \$0.22 $M$/unit product. For making the process profitable, the resulting negative profit will have to be compensated, for example, by imposing a fee that the supplier must pay to the recycler (i.e., make $C_0$ negative enough to affect this, instead of keeping $C_0 = 0$). In other words, impose a fee greater than \$0.22 $M$/unit product.

### Example 3

A product consists of components $A$ and $B$. Component $A$ is considered useful and component $B$ is hazardous. If $B$ is not removed, the complete product is considered hazardous waste. The supplier pays the recycler \$2 to acquire product $AB$. If this product is not disassembled, it can only be incinerated, which costs \$3. If disassembled, component $A$ can be sold to a refurbishing plant for \$4. The hazardous component $B$ can be disposed of at the cost of \$1. The maximum cost of the disassembly operation: $AB \rightarrow A + B$ for guaranteeing a minimum profit $P_{min}$ = \$1 can be calculated as follows:

Note that state $S = 0$ corresponds to product $AB$ before disassembly and state $S = 1$ corresponds to the status after the disassembly operation (i.e., A + B). Therefore, the costs are given by $C_0$ = – \$2 and $C_1 = x$, where $x$ is what needs to be calculated. Also, $R_{AB}$ = – \$3, $R_A$ = \$4; $R_B$ = – \$1.

Therefore, profit, if no disassembly is done, $P_0 = R_{AB} - C_0$ = –\$5 and profit, if disassembly is performed, $P_1 = (R_A + R_B) - (C_0 + C_1) = (4 - 1) - (-2 + x) \geq P_{min}$.

Consequently, $x \leq 4$. Therefore, the cost for disassembling the product should not exceed \$4 in order to keep the minimum profit at the desired level.

For a specific disassembly sequence, the cost and revenue can be plotted against the disassembly time, thus revealing a $(C,t)$ plot and an $(R,t)$ plot. Cost is considered proportional to the disassembly time, and the revenue exhibits a stepwise function. This is because a new state is entered only when a disassembly operation is completely finished, which is a discrete process. Profit is obtained by subtracting cost from revenue. The construction of the $(P,t)$ plot from the $(C,t)$ plot and the $(R,t)$ plot is illustrated in Figure 3.11. It exhibits a saw tooth shape because, as time elapses, the additional cost keeps increasing proportional to it, which causes the $(P,t)$ plot to

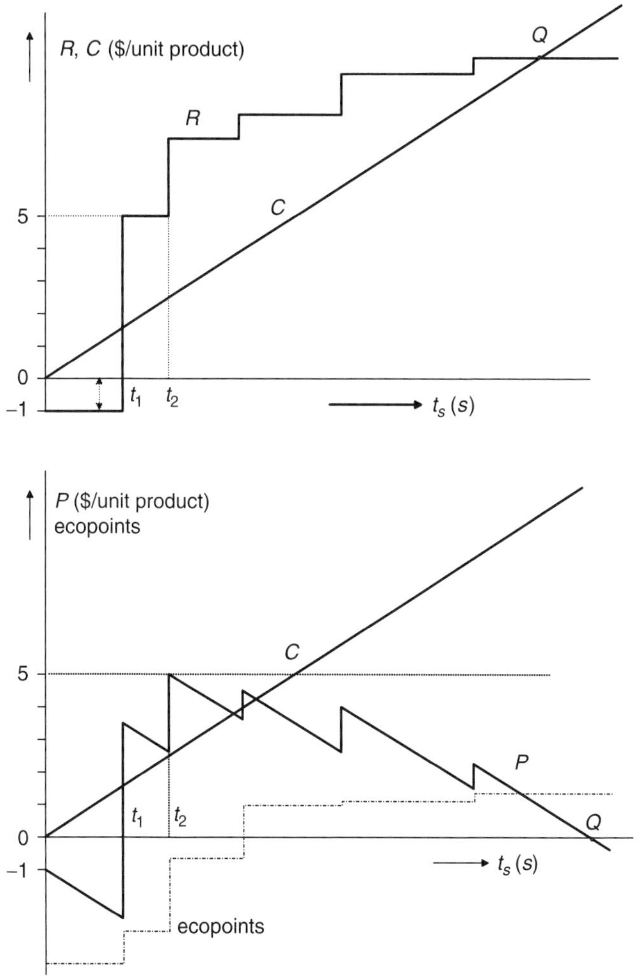


**FIGURE 3.11** The construction of the $(P,t)$ plot from the cost and revenue plots. ($P$ profit: $C$ cost: $R$ revenue: $Q$ profit break even point).

decrease with time after starting from a maximum value. Ultimately, the profit becomes negative, which makes further disassembly uneconomical.

Note that not every individual disassembly operation needs to be profitable, because there are times when some unprofitable disassembly operations may be necessary to enable subsequent profitable ones. For example, when geometric constraints occur, obstructing components may have to be disassembled prior to the disassembly of the desired components.

### 3.6.3 Return on Investment and Profit Rate

In addition to the absolute profit, one is often interested in the return on investment and the profit rate. The *return on investment* (*ROI*) of the disassembly process ($S \neq 0$) can be expressed as follows:

$$ROI_{S,q_S} = \frac{P_{S,q_S}}{C_{q,S} - C_0} = \frac{\sum_{\sigma \in S} R_\sigma - \sum_{j \in q_S} C_j}{\sum_{j \in q_S} C_j - C_0} \tag{3.17}$$

Note that the denominator represents the disassembly cost of the product.

**Example 4**

For the data given in example 3, the *ROI* for state 1 can be calculated as follows:

$$ROI_1 = \frac{(R_A + R_B) - (C_0 + C_1)}{(C_0 + C_1) - C_0} = \frac{(4-1)-(-2+x)}{(-2+x)+2} = \frac{5-x}{x}$$

The *disassembly profit rate* $U_s$ is as follows:

$$U_{S,q_S} = \frac{P_{S,q_S}}{\tau_{S,q_S}} = \frac{\sum_{\sigma \in S} R_\sigma - \sum_{j \in q_S} C_j}{\sum_{j \in q_S} t_j} \tag{3.18}$$

where $\tau_{S,q_S}$ is the disassembly time, and $t_j$ is the time required to perform disassembly operation $j$.

($U,t$)-plot can be derived from the ($P,t$)-plot by dividing it by $t$ (Figure 3.12).

In Figure 3.12, a nonlinear saw tooth shape is observed, which follows straight from the saw tooth in Figure 3.11. The nonlinearity is due to the presence of the variable $t$ in the denominator of $U$ (see Expression 3.18). This decrease is represented with a hyperbolic curve, which has the following general shape:

$$\frac{a_j - b_j t}{t} \tag{3.19}$$

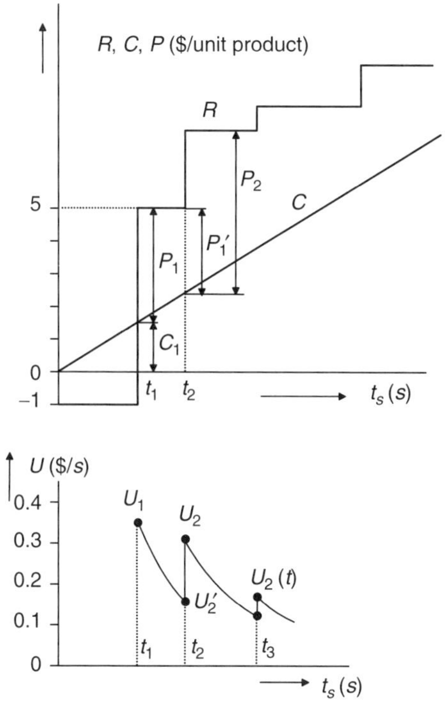


**FIGURE 3.12** The construction of the $U(t)$-curve. $P$ profit: $C$ cost: $R$ revenue: $U$ profit rate.

where $a_j$ and $b_j$ are constants within each interval $j$ that represents the interval between discontinuities $j-1$ and $j$, which can be interpreted as the disassembly operation $j$. Note that the disassembly profit rate for $t < t_1$ is not plotted because it has no practical significance.

If cost is assumed to be proportional to $t$, i.e., $C(t) = ct$, where $c$ is the cost per unit of disassembly time, the following relationship holds:

$$U_j(t) = \frac{R_{j-1} - ct}{t}, \text{ for } t_{j-1} \leq t \leq t_j \tag{3.20}$$

The maximum disassembly profit rate for $t = t_j$ is given by

$$U_j = \frac{P_j}{t_j} \tag{3.21}$$

The minimum disassembly profit rate for $t = t_j$ is given by

$$U_j' = \frac{R_{j-1} - ct_j}{t_j} \tag{3.22}$$

Consequently

$$U_j - U_j' = \frac{R_j - R_{j-1}}{t_j} \tag{3.23}$$

In practice, the disassembly profit rate tends to decrease quickly with time.

**Example 5**

Let $c = 0.15$, $t_1 = 10$, $t_2 = 16$, $t_3 = 25$, $R_1 = 5$, $R_2 = 7.4$, and $R_3 = 8.05$.

The profits (which are $R - C$) can now be calculated and are as follows:

$$P_1 = 3.5,\ P_2 = 5,\ P_3 = 4.3.$$

Similarly, the disassembly profit rates can be calculated and are as follows:

$$U_1 = 0.35,\ U_2 = 0.313,\ U_3 = 0.172.$$

$$U_2' = 0.165,\ U_3' = 0.146$$

## 3.7 ECONOMIC-ECOLOGICAL (ECO-ECO) MODELS

Economical-ecological (eco-eco) models are used to study the trade-offs between economic costs and environmental benefits. As can be observed from Figure 3.11 and Figure 3.12, if the cost per unit of disassembly time $c$ is high (which is often true), the added value due to the disassembly process is modest. However, disassembly processes are not carried out just for economic reasons; environmental issues (such as isolating hazardous substances and reducing the amount of shredder residue) also play a crucial role.

The ecopoint method, which is based on the philosophy that is similar to the methods discussed in subsection 3.5.3, provides a tool for relating cost with ecological merit in the course of a disassembly process. Ecopoints are depicted in Figure 3.11 with a dotted line. Points on this line start with negative values, because disposal of the complete product has a negative impact, resulting from both the loss of resources, and the waste that is generated. If hazardous materials were present in the product, this negative value would be even more pronounced. The subsequent disassembly operations separate recyclable materials and components from nonrecyclable or hazardous items, thus reducing the amount discharged to the environment and decreasing the environmental burden. However, care should be taken to account for the additional consumption of resources and the generation of emissions due to recycling-related processes. This can be illustrated by considering the example of chlorofluorocarbons (CFCs) in refrigerators. If inadequately disposed, CFCs will escape from the discarded products to the atmosphere and ultimately affect the stratospheric ozone layer. If CFCs are captured during the end-of-life processing and subsequently destroyed (Stoop and Lambert, 1998; Lambert and Stoop, 2001), the environmental impact of the discarded product will become far less negative, although the containment and treatment will consume additional energy.

From an environmental point of view, those actions that contribute to a rapid gain in ecopoints should be performed at an early stage of the disassembly process (Harjula et al., 1996). Within this framework, maximizing the quotient of environmental improvement and additional cost can be taken as an objective.

Note that the net ecological impact of a product during its complete life cycle will almost always be negative. Exceptions are products that are specially designed for the benefit of the environment, such as water treatment plants. If a product is designed for life-cycle then the impact of a conventional product must be taken as a reference. It includes the environmental consequences of production, consumption, and end-of-life processing (Harjula et al., 1996). If only the end-of-life phase is considered, discarded products can actually have a positive environmental effect because they can be mined for secondary materials, which reduces the consumption of virgin resources.

## 3.8 EXAMPLE: DISASSEMBLY OF DISCARDED CARS

Among the many products, cars represent one of the earliest examples of large-scale end-of-life processing. The processing essentially involves selective disassembly of reusable components and modules (to be used as spare parts) and hazardous parts (to comply with environmental laws), and bulk recycling for retrieving ferrous and nonferrous materials. In the past, materials recovery via bulk recycling was relatively simple and profitable, due to the high metal contents in end-of-life cars. A recovery efficiency of over 80% was common, obtained via shredding and subsequently extracting ferrous metals by magnetic separation, followed by separating nonferrous metals via eddy current techniques. The remaining shredder residue, which consisted of a contaminated blend of light materials, was landfilled (Isaacs and Gupta, 1997).

Since the 1970s, two opposite developments have taken place. First, the governmental regulations gradually tightened, including in some cases, a ban on certain substances (e.g., Pb, Hg, Cd, and $Cr^{6+}$) used in car designs, minimum requirements for recovery efficiencies, enforced draining of working fluids and removal of hazardous components, professionalization of the car recycling industry, and requirements for final processing of the shredder residue. In this framework, the directive 2000/53/EC of the European Communities is relevant (EC, 2000). Apart from the above-mentioned requirements, it required the removal of tires, major plastic components, and glass, via selective disassembly, and encouraged design-for-life-cycle practices, including the development of recycling techniques for major nonmetal components.

Second, a gradual decrease in the amount of ferrous metals started to take place in favor of nonferrous metals, plastics, and complex ancillary equipment. The replacement of steel with engineering plastics and aluminum was used for both cost and mass reduction. In practice, however, some of this mass reduction was counterbalanced by the addition of an increasing amount of ancillary equipment, thus further complicating conventional end-of-life processing. Ancillary equipments are used for safety (airbag, safety belt), comfort (airco, servo-systems), information (navigation system, sensors), and entertainment. This equipment introduces new types of hazardous materials, such as cooling agents and explosives, to the cars. In the course of a 15-year period, the share of ferrous metals in cars decreased from 73.5 to 61%. The share of nonferrous metals increased from 7 to 9.5%, plastics from 6.5 to 10% and that of various other materials from 6.5 to 10%. This trend will continue, which implies that the content of easily recyclable materials will further decrease in the future. If complete nonmetal fraction were to enter the shredder residue, it could impede economically feasible shredding and threaten the existence of the car-recycling infrastructure (Isaacs and Gupta, 1997; Boon et al., 2000, 2003).

In 1992, a project involving cars was undertaken in the Netherlands aimed to get an accurate assessment on selective disassembly. This was called "project Goes," named after the town where it was carried out. This project included the comparison of different selective disassembly strategies for establishing the optimum disassembly depth and recycling percentage. A set of identical cars that were previously used as test vehicles, were dismantled to different depths. Although, a lot of real life uncertainties and variability were avoided by the experimental setup of this project, it nevertheless provided valuable information.

The used cars involved in the project were of the Volvo brand and had an average mass of 1010 kg. After disassembling these cars, only 996 kg of the components and substances were left. The difference could be assigned to some spillage to the ground that could not be weighed, and the various working fluids that were captured separately.

After full dismounting, a carcass with a mass of 315 kg remained. The catalog mass of this was supposed to be 257 kg. The discrepancy was attributed to contaminants such as lacquers and coatings (21 kg), some hard to remove rubber components that were sealed to the body, and other residual waste. Part of the contaminants end up in the shredder residue while part of it enters the steel mill as a contaminant in the recycled steel, where it is incinerated or appears in the slag. As the difference between the mass that is actually obtained and the mass that it should be (according to the catalog) is significant, it can be concluded that recycling grades above 95% are difficult to obtain.

The metals portion of the cars was estimated to be 70.5% (or about 684 kg) of the total mass, which was considered as the initial recycling efficiency, assuming, of course, that perfect magnetic and eddy current separation processes could be carried out. For enhancing the recycling efficiency, selective disassembly of the nonmetal fraction was carried out, and the masses of these fractions were determined. These fractions were considered recyclable and thus were assumed to improve the recycling efficiency, regardless of how they were further processed.

Disassembly started with the preparatory work (compulsory) of draining the working fluids. The average amount of these fluids was 10 kg. The battery (14 kg) and the fuel filter (0.5 kg) also had to be removed (see Table 2.17). The time for carrying out these processes was 27 min, which included preparatory tasks (7 min) such as pulling the car on a bridge, and draining plus related disassembly operations (20 min). Next, other major nonmetallic components were disassembled, such as the bumpers and the dashboard. The principal nonmetallic components of a car are listed in Table 2.18. The detachments of modules that mainly consist of metals, such as the engine, were not included in improving the recycling efficiency. Instead, the engine was assumed to be completely reusable, as a remanufactured engine.

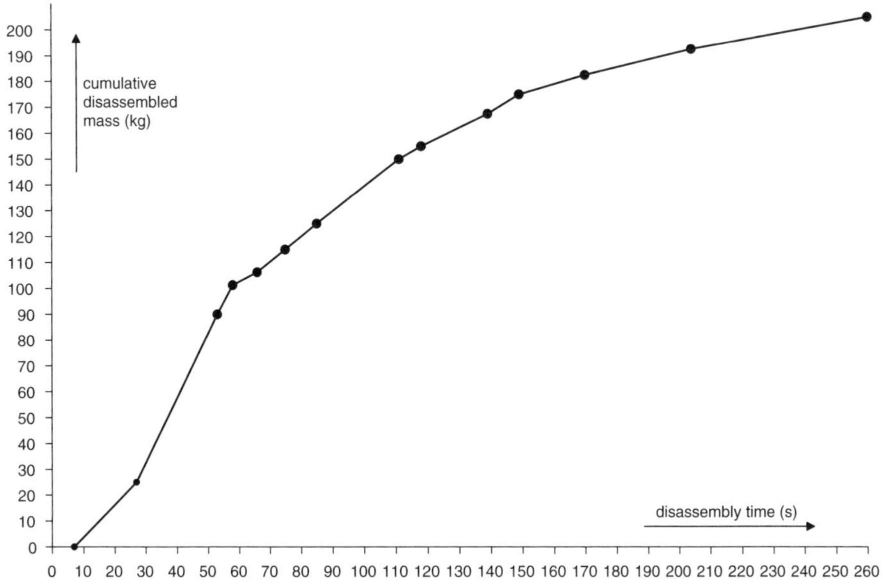


**FIGURE 3.13** The amount of disassembled nonmetal components in kg, as a function of the cumulative disassembly time in sec.

A graph depicting the cumulated mass of the disassembled nonmetal components that resulted from the experiments vs. the disassembly time is presented in Figure 3.13. After a disassembly time of 260 min, in a process layout with two employees involved, the maximum technically feasible recycling efficiency of 92.3% by weight was obtained, corresponding to the detachment of about 204 kg of nonmetallic components.

This curve is representative for typical disassembly processes. The slopes gradually decrease in the course of the disassembly process. In this particular case, the steepest slope corresponds to the second operation, and not the first, because of the technical reasons of accessibility. The curve flattens over time, which indicates a decreasing efficiency of the subsequent disassembly operations. This prompts the disassembler to stop the process, resulting in incomplete disassembly. Economic constraints further reduce the disassembly depth and, consequently, the recycling efficiency.

## 3.9 CONCLUSION

This chapter, which concludes the first part of this book, focused on the disassembly process, which acts on discrete components and consists of discrete operations. This is in contrast with chapter 2 that emphasized bulk recycling and the process technological aspects of end-of-life recycling. In this chapter, the disassembly costs and the component revenues were analyzed. These are important parameters in disassembly sequencing problems. Aspects of disassembling real life products were discussed and the representation via connection diagrams was introduced. The significance of discrete components was expounded.

Although this part of the book focused on end-of-life disassembly, the basic principles employed here are also applicable to the maintenance and the repair process. In addition, they are useful in assembly studies. With the framework that is outlined in the preceding chapters as a starting point, the theoretical aspect of disassembly sequencing will be treated in the next part, culminating in solving optimization problems.

## REFERENCES

Åkermark, A.M., 1997, Design for disassembly and recycling. In: Krause, F.L. and Seliger, G. (eds.): Life Cycle Networks, *Proceeding of the 4th CIRP International Seminar on Life-Cycle Engineering*, 237–248.

Armillotta, A. and Semeraro, Q., 1997, Assessment of destructive operations in disassembly process planning. *International Journal of Flexible Automation and Integrated Manufacturing*. **5**(1–2), 57–78.

Asiedu, Y. and Gu, P., 1998, Product life cycle cost analysis: state of the art review. *International Journal of Production Research*, **36**(4), 883–908.

Boks, C.B., Kroll, E., Brouwers, W.C.J., and Stevels, A.L.N., 1996, Disassembly modelling: two applications to a Philips 21" television set. *Proceedings of IEEE International Symposium on Electronics and the Environment*, 224–229.

Boon, J.E., Isaacs, J.A., and Gupta, S.M., 2000, Economic impact of aluminum-intensive vehicles on the U.S. automotive recycling infrastructure. *Journal of Industrial Ecology*, **4**(2), 117–134.

Boon, J.E., Isaacs, J.A., and Gupta, S.M., 2003, End-of-life infrastructure economics for 'Clear Vehicles' in the United States. *Journal of Industrial Ecology*, **7**(1), 25–45.

Boothroyd, G. and Alting, L., 1992, Design for assembly and disassembly. *Annals of the CIRP*, **41**(2), 625–636.

Bourjault, A., 1984, *Contribution à une approche méthodologique de l'assemblage automatisé: elaboration automatique des séquences opératoires* (Contribution to a systematic approach of automatic assembly: The automatic generation of operation sequences), (Besançon, France: Faculty of Science and Technology, Université de Franche-Comté), Ph.D. Thesis, 12 November (in French).

Bras, B. and Emblemsvåg, J., 1995, The use of activity-based costing, uncertainty, and disassembly action chards in demanufacture cost assessments. *Proceedings of 1995 ASME Design Engineering Technical Conferences*, **1**, 285–292.

Cagno, E., Di Giulio, A., and Trucco, P., 2000, Planning the end-of-life management within the product design process. *Proceedings of SPIE International Conference on Environmentally Conscious Manufacturing*. Belligham, WA: SPIE. Vol. **4193**, 299–308.

Chen, R.W., Navin-Chandra, D. and Prinz, F.B., 1993, Product design for recyclability: a cost benefit analysis model and its application. *Proceedings of IEEE International Symposium on Electronics and the Environment*, 178–183.

Chen, K.Z., 2001, Development of integrated design for disassembly and recycling in concurrent engineering. *Integrated Manufacturing Systems*, **12**(1), 67–79.

Das, S.K. and Matthew, S., 1999, Characterization of material outputs from an electronics demanufacturing facility. *Proceedings of IEEE International Symposium on Electronics and the Environment*, 251–256.

Das, S.K., Yedlarajiah, P., and Narendra, R., 2000, An approach for estimating the end-of-life product disassembly effort and cost. *International Journal of Production Research*, **38**(3), 657–673.

Dini, G., Failli, F., and Santochi, M., 2001, A disassembly planning software system for the optimization of recycling processes. *Production Planning & Control*, **12**(1), 2–12.

Dowie, T. and Kelly, P., 1994, *Estimation of disassembly times*. Technical report. Manchester UK: Manchester Metropolitan University, Dept. of Mechanical Engineering, Design and Manufacture.

EC, 2000, Directive 2000/53/EC of the European Parliament and of the Council of 18 September 2000 on end-of-life vehicles, *Official Journal of the European Communities,* October 21st, p. L269/34.

Gao, M., Zhou, M.C., and Caudill R.J., 2002, Integration of disassembly leveling and bin assignment for demanufacturing automation. *IEEE Transactions on Robotics and Automation*, **18**(6), 867–874.

Goedkoop, M. and Spriensma, R., 2000, *The eco-indicator 99. A damage oriented method for Life Cycle Impact Assessment: Methodology report.* Den Haag: Ministry of Housing, Spatial Planning and the Environment (2nd edition). Available on the Internet on http://www.pre.nl

Guinée, J.B., Udo De Haes, H.A., and Huppes G., 1993a, Quantitative life cycle assessment of products. Part 1: Goal definition and inventory. *Journal of Cleaner Production*, **1**(1), 2–13.

Guinée, J.B., Heijungs, R., Udo De Haes, H.A., and Huppes G., 1993b, Quantitative life cycle assessment of products. Part 2: Classification, valuation and improvement analysis. *Journal of Cleaner Production*, **1**(2), 81–91.

Harjula, T., Rapoza, B., Knight, W.A., and Boothroyd, G., 1996, Design for disassembly and the environment. *Annals of the CIRP*, **45**(1), 109–114.

Isaacs J.A. and Gupta, S.M., 1997, Economic consequences of increasing polymer content for the U.S. automobile recycling infrastructure. *Journal of Industrial Ecology*, 1(4), 19–33.

Jorden, W, 1991, *Konstruieren recyclinggerechter Produkte mit den neuen Richtlinie VDI 2243* (Design for disassembly with the new directive VDI 2243). In: Recycling: Eine Heurausforderung für den Konstrukteur (Recycling: a challenge for the designer). Düsseldorf: VDI. VDI-Berichte, vol. 906, p. 23–41 (in German).

Jovane, F., Alting, L., Armillotta, A., Eversheim, W., Feldmann, K., Seliger, G., and Roth, N, 1993, A key issue in product life cycle: disassembly. *Annals of the CIRP*, **42**(2), 651–658.

Jovane, F., Semeraro, Q., and Armillotta A., 1997, Computer-aided disassembly planning as a support to product redesign. In: Krause, F.L. and Seliger, G. (eds.), Life Cycle Networks. *Proceedings of 4th CIRP International Seminar on Life-Cycle Engineering*. London: Chapman and Hall. 388–399.

Jovane, F., Semeraro, Q., and Armillotta A., 1998, On the use of the profit rate function in disassembly process planning. *The Engineering Economist*, **43**(4), 309–330.

Knight, W.A., 1996, Software tools to evaluate cost benefits and environmental impact of designing for disassembly, recycling and the environment. *Technical Papers of the Design for the 21st Century*, Regional Technical Conference. Brookfield, CT: Society of Plastics Engineers (SPE). 27–38.

Knight, W.A. and Sodhi, M.S., 2000, Design for bulk recycling: analysis of materials separation. *Annals of the CIRP*, **49**(1), 83–86.

Kongar, E. and Gupta, S.M., 2002, A multi-criteria decision making approach for disassembly-to-order systems. *Journal of Electronics Manufacturing,* **11**(2), 171–183.

Kroll, E., 1996, Application of work-measurement analysis to product disassembly for recycling. *Concurrent Engineering*, **4**(2), 149–156.

Kroll, E., Beardsley, B., and Parulian, A., 1996, A methodology to evaluate ease of disassembly for product recycling. *IIE Transactions*, **28**, 837–845.

Kroll, E. and Carver, B.S., 1999, Disassembly analysis through time estimation and other metrics. *Robotics and Computer Integrated Manufacturing*, **15**, 191–200.

Kuo, T.C., Huang, S.H., and Zhang, H.C., 2001, Design for manufacture and design for 'X': concepts, applications, and perspectives. *Computers & Industrial Engineering*, **41**, 241–260.

Lambert, A.J.D. and Stoop, M.L.M., 2001, Processing of discarded household refrigerators: lessons from the Dutch example. *Journal of Cleaner Production*, **9**, 243–252.

Lee, B.H. and Ishii, K., 1997, Demanufacturing complexity metrics in design for recyclables. *Proceedings of IEEE Symposium on Electronics and the Environment*, 19–24.

Lee, S. and Shin, Y.G., 1990, Assembly planning based on subassembly extraction. *Proceedings of IEEE International Conference on Robotics and Automation*, 1606–1611.

McGlothlin, S. and Kroll, E., 1995, Systematic estimation of disassembly difficulties: application to computer monitors. *Proceedings of IEEE Conference on Electronics and the Environment*, 83–88.

Murayama, T., Kagawa, K., and Oba, F., 1999, Computer-aided redesign for improving recyclability. *Proceedings of 1st IEEE International Symposium on Environmentally Conscious Design and Inverse Manufacturing*, 746–751.

Navin-Chandra, D., 1994, The recovery problem in product design. *Journal of Engineering Design*, **5**(1), 65–86.

Reimer, B., Sodhi, M.S., and Knight, W.A., 2000, Optimizing electronics end-of-life disposal costs. *Proceedings of IEEE International Symposium on Electronics and the Environment*, 342–347.

Riller, P., 1991, *Wege zum recyclingfreundlichen Konstruktion von Elektrohaushaltgeräten* (Roads to the design for recycling of electric household appliances). In: Recycling: Eine Herausforderung für den Konstrukteur (Recycling: a challenge in product design). VDI-Berichte vol. 906. p. 291–321. Düsseldorf: Verein Deutscher Ingenieure (in German).

Salomonski, N. and Zussman, E., 1999, On-line predictive model for disassembly planning adaptation. *Robotics and Computer Integrated Manufacturing*, **15**, 211–220.

Samel, R., 2001, Materials in cars: options for change. *Materials Technology & Advanced Performance Materials*, **16**(1), 4–7.

Scholz-Reiter, B. and Scharke, H., 1997, Implementation and testing of a reactive disassembly planner. In: Krause, F.L. and Seliger, G. (eds.), Life cycle Networks, *Proceedings of 4th CIRP International Seminar on Life-Cycle Engineering*, 378–387.

Schuckert, M., 1993, *Die ganzheitliche Bilanzierung am Beispiel des Systems Automobil* (The gross energy balance in the example of the car). In: Kumulierte Energie- und Stoffbilanzen – ihre Bedeutung für Ökobilanzen (Gross energy and materials balances, their significance to LCA). VDI-Berichte vol. 1093. p. 57–71. Düsseldorf: Verein Deutscher Ingenieure (in German).

Shu, L.H. and Flowers, W.C., 1999, Application of a design-for-remanufacture framework to the selection of product life cycle fastening and joining methods. *Robotics and Computer Integrated Manufacturing*, **15**, 179–190.

Smith, D., Small, M., Dodds, R., Amagai, S., and Strong, T., 1995, Computer monitor recycling: a case study. *Proceedings of IEEE Conference on Clean Electronics Products and Technology* (CONCEPT), 124–128.

Sodhi, M.S. and Knight, W.A., 1998, Product design for disassembly and bulk recycling. *Annals of the CIRP*, **47**(1), 115–118.

Stoop, M.L.M. and Lambert, A.J.D., 1998, Processing of discarded refrigerators in the Netherlands. *Technovation*, **18**(2), 101–110.

Suarez, R. and Lee, S., 1997, Towards a standard cost measure of assembly operations. *Proceedings of IEEE International Conference on Robotics and Automation*, 620–625.

Tillman, A.M., Ekvall, T., Baumann, H., and Rydberg T., 1994, Choice of system boundaries in life cycle assessment. *Journal of Cleaner Production*, **2**(1), 21–29.

Tsoulfas, G.T., Pappis, C.P., and Minner, S., 2002, An environmental analysis of the reverse supply chain of SLI batteries. *Resources, Conservation and Recycling,* **36**, 135–154.

Veerakamolmal, P. and Gupta, S.M., 1998, Optimal analysis of lot size balancing for multi-products selective disassembly. *International Journal of Flexible Automation and Integrated Manufacturing*, **6**(3/4), 245–269.

Veerakamolmal, P. and Gupta, S.M., 1999, Analysis of design efficiency for the disassembly of modular electronic products. *Journal of Electronics Manufacturing*, **9**(1), 79–95.

Veerakamolmal, P. and Gupta, S.M., 2000, Design for disassembly, reuse and recycling. In: Goldberg, L. (ed.), *Green Electronics/Green Bottom Line: Environmentally Responsible Engineering*. Newnes: Butterworth-Heinemann, Chapter 5, p. 69–82.

Wang, M.H. and Johnson, M.R., 1995, Design for disassembly and recyclability: a concurrent engineering approach. *Concurrent Engineering*, **3**(2), 131–134.

Zandin, K.B., 1980, *MOST Work Measurement Systems*. New York: Marcel Dekker.

Zussman, E., Kriwet, A., and Seliger, G., 1994, Disassembly-oriented assessment methodology to support design for recycling. *Annals of the CIRP*, **43**(1), 9–14.

# *Part II*

## *Disassembly Sequencing*

# 4 Disassembly Network Features

## 4.1 INTRODUCTION

Chapter 2 and chapter 3 focused on the framework that influences the end-of life disassembly process. This and the following chapters will concentrate on the product itself and the potential disassembly sequences to take it apart. As was mentioned in subsection 1.5, two principal approaches of product modeling are discussed in this book; the mechanical approach and the hierarchical tree approach. An example of a hierarchical tree structure was provided in Figure 3.9. Hierarchical tree is related to the bill of materials (BOM). The disassembly sequence for obtaining a specific component is fixed in such a tree. There are, however, multiple disassembly sequences for reaching a predetermined final state. A still larger number of incomplete disassembly sequences is possible. Thus, the hierarchical tree approach, which is applicable for strongly modularized products, does not commonly represent all the possible disassembly sequences. The mechanical approach, on the other hand, permits a more comprehensive product representation that starts with the assembly drawing, or a virtual prototype, and further proceeds with summarizing all the relevant information. This usually commences with the construction of a connection diagram as described in subsection 3.2.9.

The disassembly theory, which evolved from assembly studies, has applications in three principal areas, including assembly planning, maintenance and repair planning, and end-of-life disassembly.

*Assembly planning* is often faced with the problem of deadlocks. *Deadlocks* are assembly sequences that fail to terminate in the complete product. A simple example is a container with a lid and some other components in it. If the lid is screwed on the container before placing the components in it, the components can no longer be inserted and the complete product will not be obtained. One way to plan for assembly and avoid deadlocks is to experiment by starting with the complete product and making an inventory of all the possible disassembly sequences, which results in all the possible assembly sequences when the process is reversed. This method is called *backward assembly planning*. The key assumption made here is that *assembly is reverse of disassembly.* This assumption, however, is an oversimplification of reality as it is only true under strict assumptions, which include the absence of forces and a rigid body approach.

*Maintenance and repair planning* usually involve both disassembly and (re)assembly operations. A typical problem involves finding a way to replace, maintain, or inspect a particular component such that a minimum number of other components are detached or disturbed.

*End-of-life disassembly* is an important and emerging area where disassembly theory is most applicable. In general, end-of life disassembly is not the reverse of assembly. While assembly is carried out to the completion of a product, end-of life disassembly is usually only carried out to a specific point (called disassembly depth) or only selective disassembly is performed depending on economic and environmental constraints.

In this chapter, we will discuss various methods for depicting disassembly sequences. We will derive expressions for the upper limit of size of the problem, considering different representations. The concept of topological constraint is discussed, and it is applied to various categories of weakly connected products. The concept of geometric constraint is introduced, and applied to a set of simple products. A graphical method is discussed for the determination of the complete set of disassembly sequences. Thus, a basic theory is obtained which provides for a starting point for automatically finding the optimum disassembly sequence.

## 4.2 DISASSEMBLY PROCESS REPRESENTATION

### 4.2.1 Disassembly Precedence Graph

A process can be broken down into a set of tasks. These tasks have to be performed in a certain order (e.g., task $C$ may not be performed prior to task $A$ and task $B$). This is expressed by precedence relationships. The set of precedence relationships is graphically represented in the form of a *precedence diagram*. Developed in the mid-1950s, such diagrams are commonly used in project planning techniques (such as critical path method (CPM) and project evaluation and review technique (PERT) for minimizing project time (Hillier and Lieberman, 2001). Precedence relationships are important in assembly line balancing problems and are used to exemplify the relationships between tasks (Prenting and Battaglin, 1964). Assembly line balancing problems are time-oriented as the basic objectives are to minimize cycle time and idle time. A way to accomplish this is to perform *parallel assemblies*, which essentially suggests preparing subassemblies and putting them together toward the goal of assembling the whole product.

The original assembly lines were manually operated and the resulting line balancing problems were solved using heuristic techniques. Most of these solutions resulted in reasonable but not necessarily optimum balancing. The introduction of robotic assembly shifted the solution methodology toward a technical approach, with the starting point of the solution based on the structure of the product to be assembled (Boothroyd et al., 1982). Boothroyd et al. realized that a single diagram could represent many different sequences and introduced assembly operations (instead of tasks) as basic units in the diagram. The diagram was organized in columns, with the *j*th column tasks to be carried out in the *j*th order of a sequence. Another example of an assembly precedence diagram was due to Fox and Kempf (1985). The detection of subassemblies remained as one of the most important objectives for maximizing parallel assembly. It is interesting to note that the names of the tasks were gradually replaced by the names of the components that had to be attached, thus corresponding to a component-oriented approach (see subsection 3.3.1).

Precedence diagrams are depicted as directed graphs in which nodes are operations and arcs are precedence relationships between operations. If an arc points from operation *A* towards operation *B*, it implies that operation *B* can only be performed after operation *A* has been performed (although not necessarily immediately after operation *A*).

If the precedence relationships are used in conjunction with the disassembly process, we could represent them in the form of a *disassembly precedence graph*. Just as in assembly line balancing problems, precedence relationships are also important in disassembly line balancing problems and are used to exemplify the relationships between disassembly tasks (Gungor and Gupta, 2002).

As an illustration of the disassembly precedence graph, consider a simple example where a box with a lid and some contents has to be disassembled. The following tasks are involved:

A. Place the box on the disassembly table
B. Open the lid
C. Remove the lid
D. Remove the contents
E. Remove the box from the disassembly table

The associated disassembly precedence graph is shown in Figure 4.1. From the figure, it is clear that there are two feasible disassembly sequences possible, viz., ABCDE and ABDCE. The two operations that must precede operation E have an AND relationship to each other: C AND D have to be performed prior to the performing of E.

Boothroyd et al. (1982) and Woo and Dutta (1991) confined the operations to the "attachment of a component" for assembly and the "detachment of a component" for disassembly, respectively. In these cases, the symbols that refer to the components are assigned to the respective operations. The next step involves abstraction in which there is neither any existence of forces nor is there any existence of fixtures, and components are assumed to be rigid bodies. Since no forces are considered, there is obviously no stability issue as there is no gravity! Similarly there is also no friction. The only way the configuration can change is if a component is detached (note that

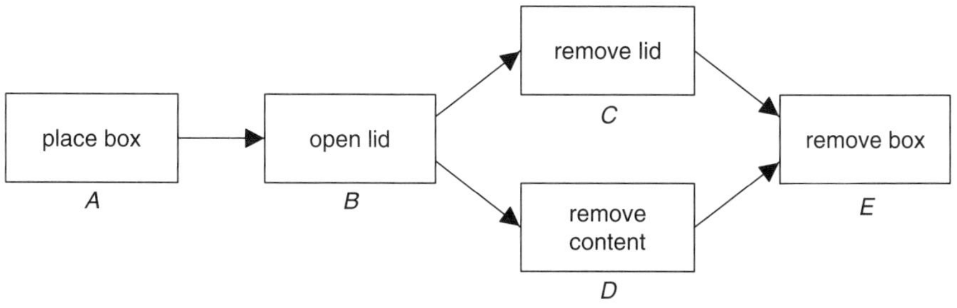


**FIGURE 4.1** Disassembly precedence graph for the simple example.

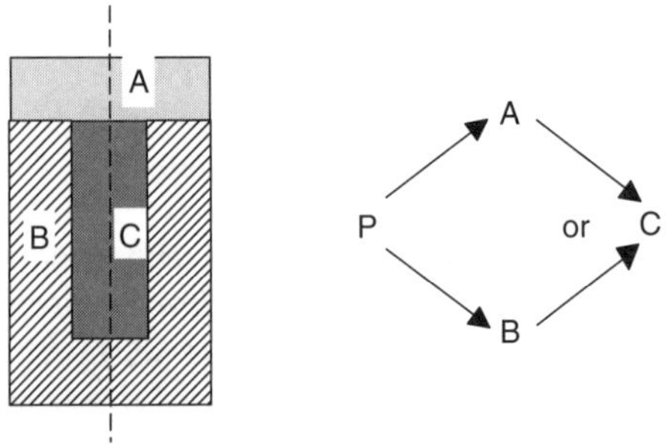


**FIGURE 4.2** Component-oriented disassembly precedence graph.

in such case the other components are left in place). Tools are not explicitly considered here. We assume *sequential disassembly*, which means that the components are removed one at a time. Any subassemblies or modules are temporarily considered as a single super-component. The basic idea is illustrated with the help of a simple example consisting of a cylindrical product that has three components: A, B, and C. The assembly drawing of the product and its disassembly precedence graph are depicted in Figure 4.2. The symbol P is assigned to the complete product.

We can start with the detachment of A, thus leaving us with BC, or start with the detachment of B, leaving us with AC. It is clear that C can only be detached once A OR B is detached. This is represented with an OR relationship in the graph. Thus, for this product, *four* complete sequences are possible, viz., ABC, ACB, BAC, and BCA. Although sequences such as BAC and BCA are virtually the same, for now, we distinguish between the component that is detached and the components that are left in place.

While the assumption of the absence of forces is seemingly unrealistic, by doing so we can nevertheless disentangle different causes of infeasibility and thus reduce the search space for the model. We can always add the presence of forces at a later stage of the study.

Even for a simple precedence graph the number of sequences can be high. Despite that a lot of work has been done in this field, the mathematics of disassembly precedence graph is only partly understood. One of the drawbacks of a disassembly precedence graph is that not all the disassembly sequences of a specific product are necessarily included in that graph. In general, several separate graphs are required to represent the complete set of disassembly sequences. This and other features of the disassembly precedence graph are elaborated in chapter 5.

### 4.2.2 Disassembly Tree

The disassembly tree is a popularly used representation of a disassembly process (see for example, Veerakamolmal and Gupta, 1999). However, it represents only a restricted subset of the disassembly sequences. A specific disassembly sequence is represented via an ordered disassembly tree. We choose the notation used by numerous

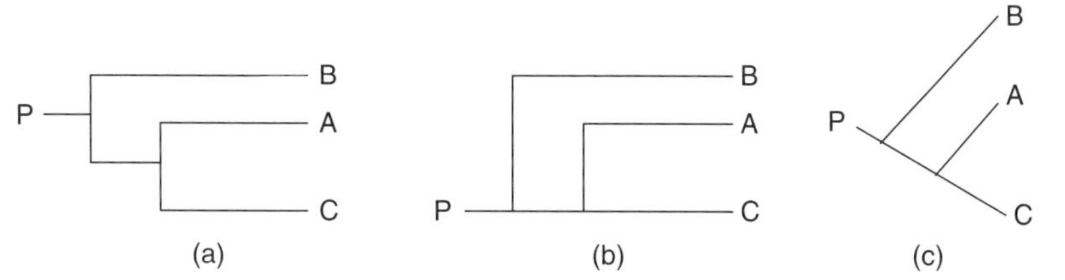


**FIGURE 4.3** (a) Disassembly tree; (b) Ordered disassembly tree; and (c) Fishbone tree for the product depicted in Figure 4.2.

authors, e.g., Henrioud (1989), Relange and Henrioud (2001), and Bratcu (2001). Another popular notation is known as fishbone notation (De Fazio and Whitney, 1987).

Examples of various types of disassembly trees are depicted in Figure 4.3 for the product presented in Figure 4.2. Note that the disassembly tree presented in Figure 4.3(a) can lead to two possible sequences, viz., BAC and BCA. The ordered disassembly tree presented in Figure 4.3(b) and the fishbone tree presented in Figure 4.3(c) represent the same one sequence, viz. BAC.

We will not go into further details of disassembly trees as more advanced representations have evolved that can represent all possible disassembly sequences in a single graph.

### 4.2.3 State Diagrams

The first study to include all possible disassembly sequences in a set of trees known as the *Bourjault's trees* was proposed by Bourjault (1984). Since the number of possible disassembly sequences increases exponentially with the number of components *N*, the number of these trees also increase dramatically with *N*. In chapter 5, we will demonstrate that Bourjault's reasoning intuitively results in a state diagram.

Bourjault's original work was connection-oriented, which assumes that a disassembly operation is the disestablishment of a connection. The first formulation of state diagrams was also connection-oriented. This is illustrated in Figure 4.4, which uses the product presented in Figure 4.2. In the connection diagram, the connections

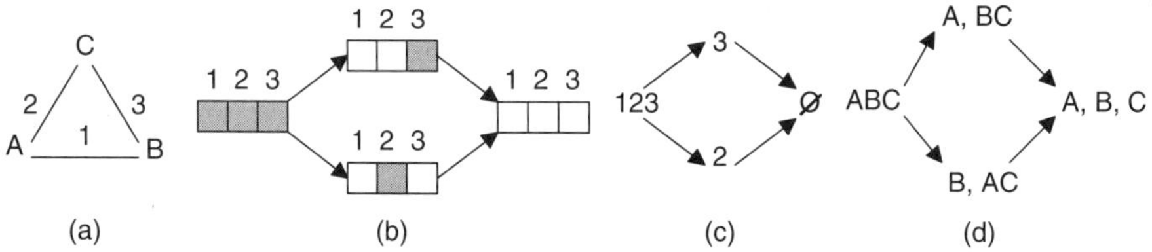


**FIGURE 4.4** Connection-oriented state diagrams for the product of Figure 4.2: (a) Connection diagram; (b) Notation due to De Fazio and Whitney (1987); (c) Connection-oriented state diagram; (d) Component-oriented state diagram.

are represented by the numbers 1 through 3. A *connection state* of a product is assigned to every feasible combination of established and disestablished connections. The state 123 means that all the connections in the product are established. Figure 4.4(a) presents the connection diagram, Figure 4.4(b) presents the notation originally used by De Fazio and Whitney (1987), and Figure 4.4(c) presents the *connection-oriented state diagram*. The connection-oriented state diagram is a directed graph with connection states as nodes and transitions between those states as arcs. Thus, in Figure 4.4(c), one starts with all the connections established, and eventually reaches to a state where no connection is established (a completely disassembled product). The symbol ∅ represents the state with no connection established. In this figure, all possible disassembly sequences are represented. Even though, the number of sequences in this example is modest, it is clear that in products with large $N$, this representation provides an advantage because of its compactness.

From this simple example, a major drawback of the connection-oriented approach is clear, viz., some connections can only be disestablished in combination with other connections. For instance, connection 1 cannot be disestablished without simultaneously also disestablishing connection 2 or connection 3. This may result in a complex set of constraints.

Alternatively, the *component-oriented approach* is advocated (see Figure 4.4(d)). This also results in a directed graph, with subassembly states as nodes and transitions between these states as arcs. In this approach, the product is considered a set of components, which is {A,B,C} for the product of Figure 4.2. The complete product is represented by ABC, and a *subassembly state* is a partition of this set. With the product defined as a 1-partition, a disassembly operation transforms it into a 2-partition, e.g., {A,{B,C}}. Similarly, the 3-partition for the example considered here is: {{A},{B},{C}}. All the subsets that result from the partitions are subassemblies or connected subsets of the product. The partition has to be geometrically feasible, which means that it can only be obtained from the product via a feasible disassembly process. We use A,BC as a short notation for {A,{B,C}}, and A,B,C as short for {{A},{B},{C}} (see Figure 4.4(d)). States with the same number of entities in the partition are in the same column or, alternatively in a vertical notation, in the same row. We will refer to the component-oriented state diagram briefly with the *state diagram*. It is also called a *diamond diagram* because of its characteristic shape if presented in the vertical notation. The component-oriented state diagram is due to Homem De Mello and Sanderson (1990, 1991), and Wolter (1992).

### 4.2.4 AND/OR Graphs

Instead of states as nodes, it is also possible to use subassemblies as nodes. In the subsequent section, we will notice that for the same value of $N$ (especially for large $N$), the number of subassemblies is far less than the number of states. Thus, if subassemblies are used as nodes, the reduced number leads to a relatively simpler graph. However, bear in mind that we are dealing with a hyper-graph where every disassembly operation has to be represented by a hyper-arc that points from the parent subassembly to the child subassemblies (AND relationship). Because a subassembly can typically be disassembled in several ways, it results in different hyper-arcs pointing

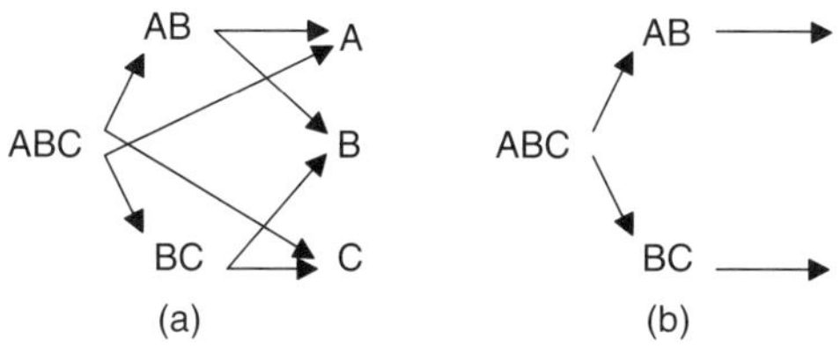


**FIGURE 4.5** The AND/OR graph; (a) Full notation; (b) Concise notation.

from the subassembly (OR relationship) (Homem De Mello and Sanderson, 1991; Moore et al., 2001). Several applications of AND/OR graphs in robot technology are described by Aleksander et al. (1983).

The AND/OR graph that is equivalent to the state diagram of Figure 4.4, is presented in Figure 4.5(a). The AND/OR graph appears to be very complex, which has led to its nickname "spaghetti diagram"! Note, however, that the components (say $S_1$ and $S_2$) that follow from applying the disassembly operation to a parent subset ($S$) have the following properties:

$$S \rightarrow S_1 + S_2; \quad S_1 \cap S_2 = \varnothing; \quad S_1 \cup S_2 = S \tag{4.1}$$

In the example presented in Figure 4.5(a), if $S$ = AB, then $S_1$ = A and $S_2$ = B. Similarly, if $S$ = BC, then $S_1$ = B and $S_2$ = C. This implies that, without any loss of information, only one of the two branches need to be included in the graph (as the other one follows unambiguously). The *concise AND/OR graph* equivalent of the AND/OR graph of Figure 4.5(a) is presented in Figure 4.5(b) (Lambert, 1999).

With the simplification included, the AND/OR graph does not appear to be as complicated. Thus, if a set of feasible subassemblies is available, the construction of the AND/OR graph is straightforward.

## 4.3 UNCONSTRAINED PRODUCTS

Before attempting to solve any disassembly problem, we would be interested in finding out the upper bound on the size of the problem. This can be obtained by considering a hypothetical product consisting of $N$ components and $K$ connections with the following assumptions:

1. Every component is connected to all other components. Therefore, every subset of components is connected and thus is a subassembly. In other words: no topological constraints are present. Such a product is said to be *strongly connected*. If a product is strongly connected, all the elements in the connectivity matrix are 1 (see Figure 3.6 for an example of a connectivity matrix).
2 Every parent subassembly can be disassembled in every possible combination of child subassemblies, such that Expression (4.1) is satisfied. In other words: no geometrical constraints are present.

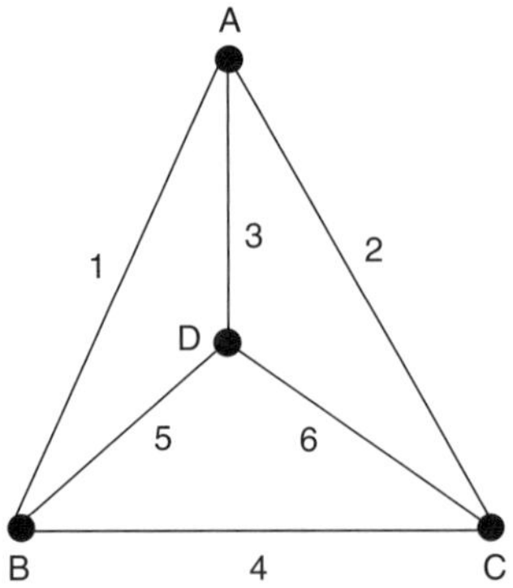


**FIGURE 4.6** Connection diagram for a strongly connected product with four components.

### 4.3.1 Components and Connections

For a strongly connected unconstrained product, the maximum value of $K$ is related to $N$ as follows:

$$K_{max} = \tfrac{1}{2} N(N-1) \tag{4.2}$$

This follows straight from the fact that each of the $N$ components is connected to $(N - 1)$ other components, which leads to $N(N - 1)$ connections. However, every connection is counted twice, thus the factor $\tfrac{1}{2}$. As an example, consider a product with four components, i.e., $N = 4$. Using Expression (4.2), we get $K_{max} = 6$ (see Figure 4.6).

### 4.3.2 Subassemblies, Connection States, and Subassembly States

Any subassembly of a product with $N$ components can be represented by a string of $N$ binary variables, each representing a different component. The value 1 is assigned to a variable if the corresponding component is present and the value 0, if it is not. There are $2^N$ such strings, each representing a subassembly except for the string with all zeroes. Therefore, the maximum number of different subassemblies $SUB_{max}$ is given by

$$SUB_{max} = 2^N - 1 = \sum_{k=1}^{N} \binom{N}{k} \tag{4.3}$$

where $\binom{N}{k}$ is the binomial coefficient (see appendices 4A and 4B).

The maximum number of different connection states $CST_{max}$ can be obtained in a similar way. Here, the connection states of a product with $K$ connections can be represented by a string of $K$ binary variables, each representing a different connection. The value 1 is assigned to a variable if the corresponding connection is present and the value 0, if it is not. There are $2^K$ such strings, each representing a connection state including the one with all zeros, which simply means that all the connections are absent or all the components are separated from each other (final state in complete disassembly). It thus follows that

$$CST_{max} = 2^K = \sum_{k=0}^{K} \binom{N}{k} \tag{4.4}$$

The determination of the maximum number of subassembly states $SST_{max}$ requires some combinatorial analysis. We know that a subassembly state is a partition of the set of components. A *set partition* of the set $\{S\}$ is a collection of disjoint subsets of $\{S\}$ whose union is $\{S\}$. The different subsets are called *blocks*. A partition with $k$ blocks is called a $k$-partition. The number of $k$-partitions $S(N,k)$ is given by the Stirling number[1] of the second kind $\left\{{N \atop k}\right\}$ (Wolter, 1992). For the maximum number of subassembly states, an addition over $k$ ranging from 1 through $N$ has to be performed, which results in the *Bell*[2] *number* $b_N$, which is given by

$$SST_{max} = b_N = \sum_{k=1}^{N} \left\{{N \atop k}\right\} = \sum_{k=1}^{N} \frac{1}{k!} \sum_{p=1}^{k} (-1)^{k-p} \binom{k}{p} p^N \tag{4.5}$$

Stirling numbers of the second kind and Bell numbers are discussed in appendices 4A and 4B.

## Example 1

Consider the connection diagram of Figure 4.6. Here $N = 4$. Therefore, using Expression (4.3), the maximum number of subassemblies $SUB_{max}$ equals 15. These subassemblies are

$k = 1$: A, B, C, D

$k = 2$: AB, AC, AD, BC, BD, CD

$k = 3$: ABC, ABD, ACD, BCD

$k = 4$: ABCD

The maximum number of connection states $CST_{max}$ using Expression (4.4), equals $2^6 = 64$.

The maximum number of subassembly states $SST_{max}$ using Expression (4.5), equals $b_4 = 15$. These are:

One of 1-partition: ABCD

Seven of 2-partitions: A,BCD; B,ACD; C,ABD; D,ABC; AB,CD; AC,BD; AD,BC

Six of 3-partitions: A,B,CD; A,C,BD; A,D,BC; B,C,AD; B,D,AC; C,D,AB

One of 4-partition: A,B,C,D.

Note that the sequel one, seven, six, one (or 1,7,6,1) is the fourth row in Table 4B.2, which represents the Stirling numbers of the second kind with $N = 4$.

---

[1] In honor of James Stirling (1692–1770).

[2] In honor of Eric Temple Bell (1883–1960).

### 4.3.3 Intermediate States and Nonmonotone Products

As is clear from Expression (4.5), the maximum number of subassembly states $SST_{max}$ increases sharply as $N$ increases. However, for $N > 2$, the maximum number of connection states $CST_{max}$ increases even faster than $SST_{max}$ as $N$ increases., see Expression (4.4). This means that a significant number of connection states does not correspond to any subassembly state and is therefore infeasible. However, because of internal degrees of freedom in a product or a subassembly, it is possible that a subassembly state corresponds to multiple connection states. We can define intermediate states that occur in the course of moving a component, (see Figure 4.7(a)(b)). In particular, this applies to *m*-movable products, which means that multiple translations or rotations have to be carried out for removing a component. The assembly of products with *m*-movable components is discussed by Schweikard and Schwarzer (1997). Internal degrees of freedom are also relevant in nonmonotone products (Wolter, 1992). These are products in which a set of components has to be moved prior to proceeding with disassembly. In real life, this frequently occurs, for example, via lock mechanisms or the opening of a cover, etc. Although intermediate states occur frequently, they rarely complicate the disassembly sequencing problem. We, therefore, will not consider this feature here. Similarly, because of its simplicity, we will confine ourselves to the subassembly states approach, even though the discussion of disassembly theory started by considering connection states (De Fazio and Whitney, 1987; Baldwin et al., 1991).

### 4.3.4 Cut-Sets and Disassembly Operations

We next proceed to determine the maximum number of cut-sets and disassembly operations for the unconstrained case.

#### 4.3.4.1 Cut-Sets

Cut-sets (see subsection 3.3) are actually 2-partitions of the set of components. They can also be considered as the minimum set of arcs of the connection diagram that, when removed, leave two disjoint subgraphs (Baldwin et al., 1991).

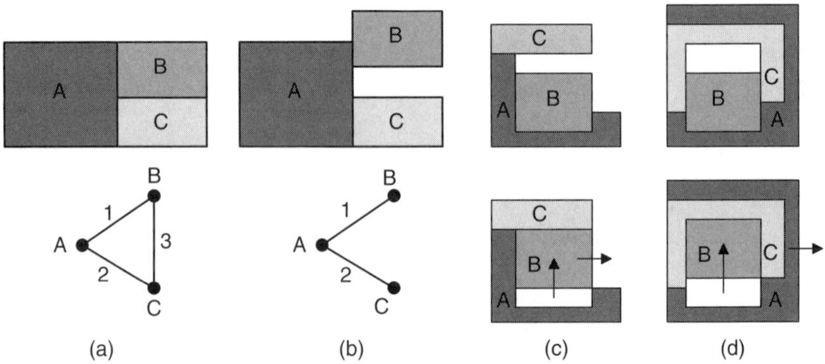


**FIGURE 4.7** Intermediate states. (a) and (b) subassembly state corresponding with multiple connection states; (c) 2-movable product; (d) nonmonotone product.

If a product consists of $N$ components, the maximum number of cut-sets in the product equals the Stirling number of the second kind with $k = 2$ (see appendices 4A and 4B). Consequently, the maximum number of cut-sets $CS_{max}$ is given by the following expression:

$$CS_{max} = 2^{N-1} - 1 \tag{4.6}$$

Note that, here we restrict ourselves to binary disassembly operations, which simply means that after each operation a $k$-partition transforms into a $k+1$-partition.

**Example 2**

For the product ABCD depicted in Figure 4.6, we have $N = 4$ which results in a maximum of 7 possible cut-sets. These correspond to the following state transitions:

ABCD → A,BCD; ABCD → B,ACD; ABCD → C,ABD; ABCD → D,ABC;

ABCD → AB,CD; ABCD → AC,BD; ABCD → AD,BC.

#### 4.3.4.2 Disassembly Operations in a State Diagram

In order to determine the maximum number of binary disassembly operations in a state diagram $SOP_{max}$, note that the expression for $CS_{max}$ in Expression (4.6) basically gives the maximum number of operations leading to different states (or 2-partitions) starting from the product node. For every subassembly in each new state, the maximum number of cut-sets has to be determined using the same formula (except $N$ that must be replaced by the number of components in the subassembly), and the results must be added. This procedure has to be repeated until the final state is obtained.

**Example 3**

The state A,BCD consists of a subassembly with the number of components $n = 1$, and a subassembly with $n = 3$. When substituted in Expression (4.6), this results in 0 and 3 cut-sets, respectively. Addition results in 3 cut-sets. Thus three operations point from state A,BCD toward the following states: A,B,CD, A,C,BD, and A,D,BC.

$SOP_{max}$ can be determined using the following formula (Wolter, 1992):

$$SOP_{max} = \sum_{k=0}^{N-2} \binom{N-k}{2} \left\{ \begin{matrix} N \\ N-k \end{matrix} \right\} \tag{4.7}$$

**Proof** Consider the reverse of disassembly problem, which is an unconstrained assembly problem. Start with the final state, which is an $N$-partition. There is $\left\{ \begin{smallmatrix} N \\ N \end{smallmatrix} \right\} = 1$ such partition. Assembly operations can merge any pair of subassemblies. There are

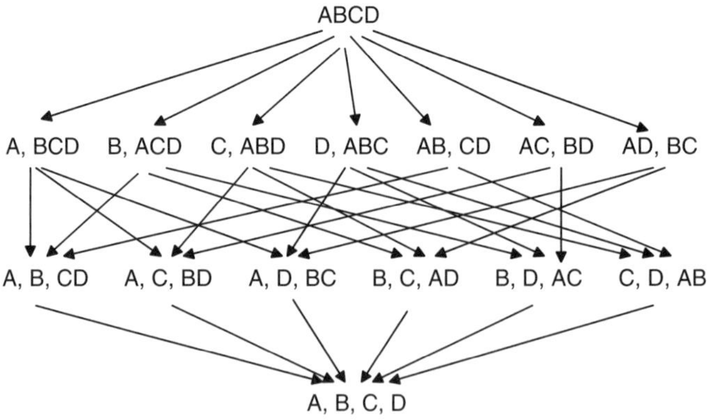


**FIGURE 4.8** State diagram for an unconstrained product with $N = 4$.

$\binom{N}{2}$ such pairs. This means that $\binom{N}{2}\left\{{N \atop N}\right\}$ arcs, each representing an assembly operation, emanate from this state. $(N - 1)$-partitions are considered next. Using analogous reasoning, it can be concluded that $\binom{N-1}{2}\left\{{N \atop N-1}\right\}$ assembly operations emanate from the available nodes. Continuing in a similar fashion, one finally ends up with $\binom{2}{2}\left\{{N \atop 2}\right\}$ assembly operations that transform the 2-partitions into the final product. Addition of all these terms results in Expression (4.7). It is obvious that, the same relationship is valid for disassembly operations, which can be established by reversing the directions of the arcs in the state diagram.

### Example 4

For the product ABCD depicted in Figure 4.6, we have $N = 4$. Using Expression (4.7), we can determine the maximum number of binary disassembly operations to be 31. These correspond to the following state transitions (see the corresponding state diagram in Figure 4.8):

ABCD → A,BCD; ABCD → B,ACD; ABCD → C,ABD; ABCD → D,ABC;

ABCD → AB,CD; ABCD → AC,BD; ABCD → AD,BC;

A,BCD → A,B,CD; A,BCD → A,C,BD; A,BCD → A,D,BC;

B,ACD → A,B,CD; B,ACD → B,C,AD; B,ACD → B,D,AC;

C,ABD → A,C,BD; C,ABD → B,C,AD; C,ABD → C,D,AB;

D,ABC → A,D,BC; D,ABC → B,D,AC; D,ABC → C,D,AB;

AB,CD → A,B,CD; AB,CD → C,D,AB;

AC,BD → A,C,BD; AC,BD → B,D,AC;

AD,BC → A,D,BC; AD,BC → B,C,AD;

A,B,CD → A,B,C,D; A,C,BD → A,B,C,D; A,D,BC → A,B,C,D;

B,C,AD → A,B,C,D; B,D,AC → A,B,C,D; C,D,AB → A,B,C,D;

An important property of a state diagram that follows directly from the previous proof is that the maximum number of arcs that point to a particular $p$-partition ($p \le N$) equals

$$\binom{p}{2} = \frac{1}{2}p(p-1) \tag{4.8}$$

### 4.3.4.3 Disassembly Operations in an AND/OR Graph

The maximum number of disassembly operations in an AND/OR graph $AOP_{\max}$ can be calculated using the following formula (Homem de Mello and Sanderson, 1991; Wolter, 1992; Gottipolu and Ghosh, 1997):

$$AOP_{\max} = \left\{ {N+1 \atop 3} \right\} = \frac{1}{2}(1+3^N) - 2^N \tag{4.9}$$

**Proof** The maximum number of hyperarcs that emanate from a subassembly equals the maximum number of cut-sets that can be applied to that subassembly. An addition has to be carried out on all the cut-sets that can be applied to all the subassemblies that are present in the AND/OR graph. For a product with $N$ components, there are $\{ {N \atop 2} \}$ cut-sets. Next, there are $\binom{N}{N-1}$ subassemblies with $N-1$ components, each of them has $\{ {N-1 \atop 2} \}$ cut-sets, and so on. When all the subassemblies in the AND/OR graph are considered, this adds up to $\sum_{k=0}^{N-2} \binom{N}{N-k} \{ {N-k \atop 2} \}$. This can be further simplified using the relationships (4A.5) and (4A.6c) as follows:

$$\sum_{k=0}^{N-2} \binom{N}{N-k} \left\{ {N-k \atop 2} \right\} = \left\{ {N+1 \atop 3} \right\} = \frac{1}{2}(1+3^N) - 2^N \tag{4.10}$$

This is the same as Expression (4.9).

**Example 5**

The AND/OR graph for the product ABCD, depicted in Figure 4.6 with $N = 4$, is presented in two layers in Figure 4.9(a) and Figure 4.9(b). To get the complete AND/OR graph, the second layer has to be superimposed on the first layer. In the first layer (Figure 4.9(a)), only sequential disassembly is shown, i.e., the smallest child subassembly has $n = 1$. In the second layer (Figure 4.9(b)), only parallel disassembly is shown, i.e., the smallest child subassembly has $n = 2$. Note that the first layer is shown in a concise notation, i.e., only one of the two subassemblies obtained after disassembly is shown (thus for example, when ABCD is disassembled into subassemblies ABC and D, only ABC is shown because it could be further disassembled and D is not shown because its release is intuitive). However, in the second layer, both child subassemblies must be shown as they both can be further disassembled.

Using Expression (4.9), we can determine that the maximum number of disassembly operations in the AND/OR graph is 25. Note that for $N = 4$, the maximum number of disassembly operations in an AND/OR graph is smaller than the maximum number of

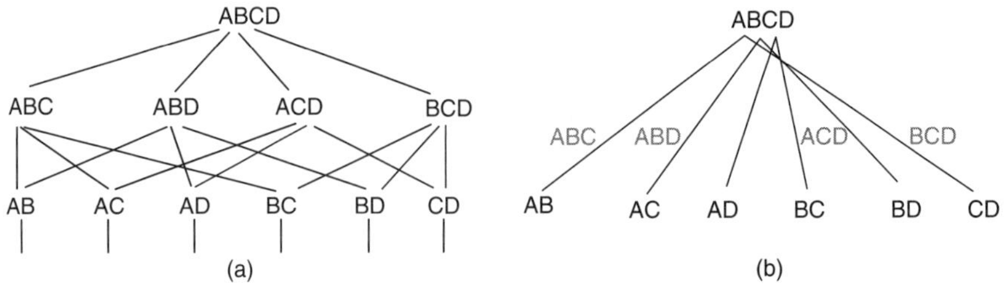


**FIGURE 4.9** AND/OR graph for the unconstrained product with $N = 4$, broken down in two layers: (a) first layer in concise notation and (b) second layer.

disassembly operations in a state diagram. The reason is that the state diagram distinguishes between the order in which parallel disassembly is carried out while the AND/OR graph is indifferent to this order.

**EXAMPLE 6**

For the product ABCD depicted in Figure 4.6, observe that the sequences:

$$\text{ABCD} \rightarrow \text{AB,CD} \rightarrow \text{A,B,CD} \rightarrow \text{A,B,C,D and}$$

$$\text{ABCD} \rightarrow \text{AB,CD} \rightarrow \text{C,D,AB} \rightarrow \text{A,B,C,D}$$

are represented by different trajectories in the state diagram of Figure 4.8, and with the same hyperarc in the AND/OR graph of Figure 4.9.

#### 4.3.4.4 Ternary Operations

Although we will confine ourselves to binary operations, some remarks on ternary and, in general, $m$-fold operations are in order. In state diagrams, these can be represented as a transition from a $p$-partition to a $(p + m - 1)$-partition. In AND/OR graphs, these can be represented with hyperarcs that point from a parent subassembly to $m$ child subassemblies. In practice, higher order operations are not encountered very often, although they are possible. An example of a non-contact coherent products which requires a ternary disassembly operation is presented in Figure 4.10 (Wolter, 1992). This product can only be disassembled via the operation:

ABC → A,B,C.

Although a method for the generation of assembly sequences with ternary operations has been proposed (Bonneville et al., 1995), this has never been further elaborated, even though this might be theoretically feasible.

### 4.3.5 Complete Disassembly Sequences

A *complete disassembly* sequence starts with the product and ends up with all the components separated from each other, which implies that all the connections in the product are disestablished.

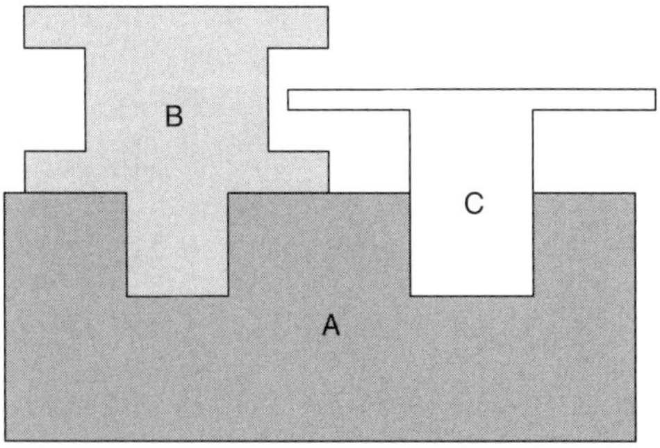


**FIGURE 4.10** A non-contact coherent product, which requires a ternary disassembly action for disassembly (Wolter, 1992).

#### 4.3.5.1 Complete Sequential Sequences

A complete sequential sequence is obtained using the *one-at-a-time approach*, because it is assumed that every disassembly operation disestablishes one single component only. The maximum number of complete sequential sequences $CSS_{max}$ can be calculated using the following expression (note that it is the same for state diagrams as well as AND/OR graphs).

$$CSS_{max} = \tfrac{1}{2} N! \tag{4.11}$$

**Proof** Starting with the product consisting of $N$ components, there are $N$ possible sequential disassembly operations, each leaving us with a component and a subassembly with N − 1 components. In the next step, for each of the N subassemblies, there are $N-1$ possible sequential disassembly operations, each leaving us with a component and a subassembly with $N-2$ components, and so on. This proceeds till subassemblies with 2 components are obtained. These can theoretically be disassembled in 2 ways. For instance, it is possible to detach either A or B from the subassembly AB. However, both operations can be considered to be identical, which is the reason for the factor $\frac{1}{2}$. If both operations are considered different from each other, for example, if one discerns between the static and the dynamic subassembly, the factor $\frac{1}{2}$ is omitted.

**Example 7**

For the product ABCD depicted in Figure 4.6, the 12 possible complete sequential sequences are as follows:

ABCD → ABC,D → AB,C,D → A,B,C,D or: ABCD → ABC → AB →

ABCD → ABC,D → AC,B,D → A,B,C,D or: ABCD → ABC → AC →

ABCD → ABC,D → BC,A,B → A,B,C,D or: ABCD → ABC → BC →

ABCD → ABD,C → AB,C,D → A,B,C,D or: ABCD → ABD → AB →

ABCD → ABD,C → AD,B,C → A,B,C,D or: ABCD → ABD → AD →

ABCD → ABD,C → BD,A,D → A,B,C,D or: ABCD → ABD → BD →

ABCD → ACD,B → AC,B,D → A,B,C,D or: ABCD → ACD → AC →

ABCD → ACD,B → AD,B,C → A,B,C,D or: ABCD → ACD → AD →

ABCD → ACD,B → CD,A,B → A,B,C,D or: ABCD → ACD → CD →

ABCD → BCD,A → BC,A,D → A,B,C,D or: ABCD → BCD → BC →

ABCD → BCD,A → BD,A,C → A,B,C,D or: ABCD → BCD → BD →

ABCD → BCD,A → CD,A,B → A,B,C,D or: ABCD → BCD → CD →

(Note that on the right, we have used the notation of the AND/OR graph).

In the state diagram of Figure 4.8, these sequences are represented by trajectories from the initial state ABCD to the final state A,B,C,D. Notice that the parallel states AB,CD; AC,BD; AD,BC, and the associated operations are not permitted in the sequential approach. Similarly, only the first layer in the AND/OR graph of Figure 4.9 is applicable.

### 4.3.5.2 Complete Sequences in the State Diagram

The maximum number of complete sequences in a state diagram $SCS_{\max}$ is given by the following expression (Wolter, 1992):

$$SCS_{\max} = \prod_{k=0}^{N-2} \binom{N-k}{2} = \frac{N!(N-1)!}{2^{N-1}} \tag{4.12}$$

**Proof** The number of arcs that enter a $p$-partition state of $N$ components, equals $\frac{1}{2}p(p-1)$, (see Expression (4.8)). Consider the reverse of disassembly process (which is actually the assembly process). From the final state $\frac{1}{2}N(N-1)$ assembly arcs depart to the $N-1$-partitions. From any of the $(N-1)$-partitions, $\frac{1}{2}(N-1)(N-2)$ arcs depart to $(N-2)$-partitions, and so on, till the 2-partition is reached, from which only one arc departs. This results in the following product:

$$\left(\tfrac{1}{2}N(N-1)\right)\cdot\left(\tfrac{1}{2}(N-1)(N-2)\right)\cdot\dots\cdot\left(\tfrac{1}{2}\cdot 2\cdot 1\right) = \left(\tfrac{1}{2}\right)^{N-1}\cdot N!\cdot(N-1)!$$

which is the right hand side of expression (4.12). Note that the following recursive relationship can be easily derived from Expression (4.12):

$$SCS_{\max}(N+1) = \tfrac{1}{2}N\cdot(N+1)\cdot SCS_{\max}(N) \tag{4.13}$$

**Example 8**

In the state diagram of Figure 4.8, the number of complete disassembly sequences (calculated using Expression (4.12) is equal to 18. The 12 completely sequential sequences have been previously listed in example 7. The additional 6 sequences deal with parallel states. They are as follows:

ABCD → AB,CD → A,B,CD → A,B,C,D

ABCD → AB,CD → C,D,AB → A,B,C,D

ABCD → AC,BD → A,C,BD → A,B,C,D

ABCD → AC,BD → B,D,AC → A,B,C,D

ABCD → AD,BC → A,D,BC → A,B,C,D

ABCD → AD,BC → B,C,AD → A,B,C,D

#### 4.3.5.3 Complete Sequences in the AND/OR Graph

The maximum number of complete sequences in an AND/OR graph $ACS_{max}$ is given by the following expression (Wolter, 1992):

$$ACS_{max} = \frac{(2N-2)!}{2^{N-1}(N-1)!} \qquad (N>1) \tag{4.14}$$

**Proof** The proof makes use of a recursive relationship that is found as follows. Let the maximum number of complete sequences of an assembly that consists of $N$ components be $ACS_{max}(N)$. A product with $N$ components can be decomposed into one subassembly with $(N-p)$ components and another subassembly with $p$ components in $\binom{N}{p}$ ways. The number of possible disassembly operations on the resulting subassemblies can also be determined in a similar way. Continuing this way leads to the following recursive relationship:

$$ACS_{max}(N) = \frac{1}{2}\sum_{p=1}^{N-1}\binom{N}{p} ACS_{max}(p)\cdot ACS_{max}(N-p) \tag{4.15}$$

The factor 1/2 is introduced to avoid double counting. In addition, $ACS_{max}$ (1) should be put equal to 1 to obtain the correct result.

Although nontrivial, one can nevertheless show that Expression (4.15) can be rewritten as follows:

$$ACS_{max}(N) = (2N-3)ACS_{max}(N-1), \quad N > 2 \text{ and } ACS_{max}(2) = 1 \tag{4.16}$$

Note that the Expression (4.14) can also be written as a double factorial as follows:

$$ACS_{max}(N) = (2N-3)!! = 1\cdot 3\cdot 5\cdot\dots\cdot(2N-3) \qquad (N>1) \tag{4.17}$$

A close examination reveals that Expression (4.16) and Expression (4.17) are identical.

**Example 9**

In the AND/OR graph of Figure 4.9 there are $ACS_{max}(4) = 15$ complete disassembly sequences. The 12 completely sequential sequences have been previously listed in example 7. The additional 3 sequences are as follows:

$$\text{ABCD} \rightarrow \text{AB} + \text{CD} \rightarrow$$

$$\text{ABCD} \rightarrow \text{AC} + \text{BD} \rightarrow$$

$$\text{ABCD} \rightarrow \text{AD} + \text{BC} \rightarrow$$

When compared to the analogous 6 sequences of the state diagram listed in example 8, we observe that pairs of two are combined into one in the AND/OR graph, because it is indifferent to the order in which two parallel subassemblies are separated.

### 4.3.6 Incomplete Disassembly Sequences

As previously mentioned disassembly is often carried out to a certain depth only (rather than complete disassembly) except, of course, when used to study assembly. This implies that disassembly sequences are often incomplete. It is therefore clear that the set of all complete sequences is a subset of the set of all incomplete sequences (Lambert, 1997). In this subsection we explore the maximum possible number of incomplete sequences in an unconstrained product.

#### 4.3.6.1 Incomplete Sequential Sequences

The maximum number of incomplete sequential sequences $ISS_{max}$ can be calculated using the following expression:

$$ISS_{max} = N!\sum_{p=1}^{N}\frac{1}{p!} - \tfrac{1}{2}N! \qquad (4.18)$$

**Proof** Starting with the complete product, there are two choices: leave it intact (i.e. incomplete sequence) or sequentially decompose it into two subassemblies ($N$ possible ways) with a set of one and a set of $N - 1$ components, respectively. If sequentially decomposed, each of the $N - 1$ subassemblies (each with a set of $N - 1$ components) can either be left intact or sequentially decompose it into two subassemblies ($N - 1$ possible ways) with a set of one and a set of $N - 2$ components, respectively. Continuing to proceed along these lines, we eventually reach the point where all the components are disjoint. Thus, the number of ways to accomplish all this can be written as follows:

$$1 + N + N(N-1) + N(N-1)(N-2) + \cdots + N!$$

This can be rewritten as

$$N!\sum_{p=1}^{N}\frac{1}{p!}$$

Modifying this to offset double counting leads us to Expression (4.18).

**Example 10**

In the product ABCD, $N = 4$. Using Expression (4.18), we determine that the maximum number of incomplete sequential sequences is 29, of which 12 (complete sequential sequences) were listed in example 7. We list the additional 17 incomplete sequences as follows:

ABCD

ABCD → ABC; ABCD → ABD; ABCD → ACD; ABCD → BCD;

ABCD → ABC → AB; ABCD → ABC → AC; ABCD → ABC → BC;

ABCD → ABD → AB; ABCD → ABD → AD; ABCD → ABD → BD;

ABCD → ACD → AC; ABCD → ACD → AD; ABCD → ACD → CD;

ABCD → BCD → BC; ABCD → BCD → BD; ABCD → BCD → CD

### 4.3.6.2 Incomplete Sequences in the State Diagram

For the maximum number of incomplete sequences in a state diagram $SIS_{max}$ there is no closed form expression. It can, however, be calculated using the following expression:

$$SIS_{max} = \left[\left[\left[\left[\cdots\left[\left[\left[\left[\binom{N}{2} + \left\{ N \atop N-1 \right\}\right]\cdot\binom{N-1}{2}\right] + \left\{ N \atop N-2 \right\}\right]\cdot\binom{N-3}{2}\right] + \cdots\right] + \left\{ N \atop 2 \right\}\right]\cdot\binom{2}{2}\right] + \left\{ N \atop 1 \right\}\right] \tag{4.19}$$

At first glance, this expression seems rather complicated. However, its structure follows straight from the reasoning that was advanced in the derivation of Expression (4.12). The Stirling numbers of the second kind are added to represent the additional sequences that stop at the corresponding subassembly states. Omitting these leads to Expression (4.12).

**Example 11**

Using Expression (4.19), we can determine that the number of incomplete sequences in the state diagram of Figure 4.8 as follows:

$$SIS_{max}(4) = \left[\left[\left[\left[\left[\binom{4}{2} + \left\{{4 \atop 3}\right\}\right] \cdot \binom{3}{2}\right] + \left\{{4 \atop 2}\right\}\right] \cdot \binom{2}{2}\right] + \left\{{4 \atop 1}\right\}\right] = 44$$

We have already listed 29 incomplete sequential sequences in example 10. The additional 15 incomplete parallel sequences (of which the last six are actually complete parallel sequences) are as follows:

ABCD $\rightarrow$ AB,CD; ABCD $\rightarrow$ AC,BD; ABCD $\rightarrow$ AD,BC;

ABCD $\rightarrow$ AB,CD $\rightarrow$ AB; ABCD $\rightarrow$ AB,CD $\rightarrow$ CD;

ABCD $\rightarrow$ AC,BD $\rightarrow$ AC; ABCD $\rightarrow$ AC,BD $\rightarrow$ BD;

ABCD $\rightarrow$ AD,BC $\rightarrow$ AD; ABCD $\rightarrow$ AD,BC $\rightarrow$ BC;

ABCD $\rightarrow$ AB,CD $\rightarrow$ AB $\rightarrow$; ABCD $\rightarrow$ AB,CD $\rightarrow$ CD $\rightarrow$;

ABCD $\rightarrow$ AC,BD $\rightarrow$ AC $\rightarrow$; ABCD $\rightarrow$ AC,BD $\rightarrow$ BD $\rightarrow$;

ABCD $\rightarrow$ AD,BC $\rightarrow$ AD $\rightarrow$; ABCD $\rightarrow$ AD,BC $\rightarrow$ BC $\rightarrow$

### 4.3.6.3 Incomplete Sequences in the AND/OR Graph

In AND/OR graphs, the maximum number of incomplete sequences $AIS_{max}$ can be found using the following recursive relationship (originally found by Lambert (1997) via a graphical method):

$$AIS_{max}(N) = \frac{1}{2}\sum_{p=1}^{N-1}\left[\binom{N}{p} \cdot AIS_{max}(p) \cdot AIS_{max}(N-p)\right] + 1 \tag{4.20}$$

This relationship follows straight from the recursive relationship that was used in the proof of Expression (4.14). However, one has to increase the number at any subassembly that is encountered by 1 as this refers to the extra possibility of "no further action" that is permitted in incomplete sequences. Therefore, following an inductive approach and for convenience, replacing $AIS_{max}(N)$ with $M_N$, we start with $M_1 = 1$. Next, we observe that $M_2 = \frac{1}{2}\binom{2}{1}M_1 \cdot M_1 + 1 = 2$. This is because both the sequences AB and AB $\rightarrow$ A + B are permitted. Along those lines, we arrive at: $M_3 = \frac{1}{2}[\binom{3}{1}M_1 \cdot M_2 + \binom{3}{2}M_2 \cdot M_1] + 1$, which results in 7 possible sequences, and so on.

**Example 12**

Using Expression (4.20), we determine that the maximum number of incomplete sequences in the AND/OR graph of Figure 4.9 is 41, of which 29 are incomplete sequential sequences already listed in example 10. We list the additional 12 incomplete parallel sequences (of which the last three are actually complete parallel sequences) as follows:

$$\text{ABCD} \rightarrow \text{AB,CD; ABCD} \rightarrow \text{AC,BD; ABCD} \rightarrow \text{AD,BC;}$$

$$\text{ABCD} \rightarrow \text{AB,CD} \rightarrow \text{AB; ABCD} \rightarrow \text{AB,CD} \rightarrow \text{CD;}$$

$$\text{ABCD} \rightarrow \text{AC,BD} \rightarrow \text{AC; ABCD} \rightarrow \text{AC,BD} \rightarrow \text{BD;}$$

$$\text{ABCD} \rightarrow \text{AD,BC} \rightarrow \text{AD; ABCD} \rightarrow \text{AD,BC} \rightarrow \text{BC;}$$

$$\text{ABCD} \rightarrow \text{AB,CD} \rightarrow\text{; ABCD} \rightarrow \text{AC,BD} \rightarrow\text{; ABCD} \rightarrow \text{AD,BC} \rightarrow$$

Note that here only three complete parallel sequences are listed instead of the six listed in the state diagram (example 11). This is because there are three pairs of sequences where the AND/OR graph is indifferent to the order in which two parallel subassemblies are separated.

### 4.3.7 Alternative Approaches

It is clear that the "number of disassembly sequences" in a product depends on the way a distinction is made between different sequences. For example, depending on whether a distinction is made between parallel sequences or not, the number of disassembly sequences can be different. This is why we saw a difference between the number of sequences in a state diagram and an AND/OR graph. An additional distinction between the order in which the components are disassembled (also known as *asymmetric* approach) may cause further differences. For example, a subassembly AB can be disassembled in two different ways: one with A as a static component and B as a dynamic component that is removed, and vice versa (Ben-Arieh and Kramer, 1994).

#### Example 13

To illustrate the difference between the order in which components are assembled, consider four components A, B, C, and D. One may combine them in the following way:

Fix A, assemble B, assemble C, assemble D.

This is denoted as (((AB)C)D).

Alternatively, one may combine them in the following way:

Fix A, fix B, assemble C to B, assemble BC to A, assemble D to ABC.

This is denoted by ((A(BC))D).

Proceeding along the same lines one can obtain the maximum number of complete sequences in a strongly connected product with $N$ components in a state diagram using an asymmetric approach for an assembly (and hence disassembly) of the product as follows:

$$CS_{\max}(N) = Q(N-1) = \frac{(2N-2)!}{(N-1)!} \tag{4.21}$$

where $Q(N)$ are quadruple factorial numbers given by

$$Q(N) = \frac{(2N)!}{(N)!} \tag{4.22}$$

The relationship of Expression (4.21) with the number of sequences in the state diagram $ACS_{max}$ given in Expression (4.14) is evident; multiplication by $2^{N-1}$ reveals Expression (4.21). The multiplication with this factor can be made plausible by considering $N = 2$, in which the asymmetric approach distinguishes between (AB) and (BA), which causes the multiplication by 2.

With this, we come to the end of determining the number of various features in a strongly connected, unconstrained product. All the numbers for these features have been calculated for $N = 1$ through 11 and summarized in appendix 4C.

## 4.4 TOPOLOGICALLY CONSTRAINED PRODUCTS

From the discussion so far, it is clear that for unconstrained products, the number of sequences increases exponentially with the number of components $N$. This means that this and associated problems tend to be NP-complete (Tovey, 2002). In chapter 7, we will show that under some circumstances, the problem of searching for the optimum disassembly sequence can be solved with linear programming techniques, thus circumventing NP-completeness. Furthermore, in real life situations, the size of the problem is considerably reduced compared to the unconstrained products with the same number of components because of a variety of constraints that limit the number of subassemblies and the number of operations. There are three different types of constraints that introduce this limitation, viz., topological constrains, geometric constraints, and technical constraints. In this section, we will focus on topological constraints.

So far, we have only discussed strongly connected products. In these products, every subset of components is connected. In other words, in a strongly connected product every subset of components is a subassembly. Real life products, however, are rarely strongly connected. In fact, in most cases, the number of connections is far less than the maximum possible except, of course, for products with very small $N$. We introduce $\alpha$ as the *index of complexity,* which is defined as the average number of connections per component:

$$\alpha = \frac{2K}{N} \tag{4.23}$$

The maximum value of $\alpha$ is $N-1$, which follows directly from Expression (4.2).

The minimum number of connections can be found as follows. In an arbitrary graph, if one can make a loop, it is always possible to remove one connection without leaving the graph disjoint. If one keeps removing a connection without leaving the graph disjoint, one ends up with $N - 1$ connections, which is the minimum number of connections needed to keep the product connected. We illustrate this in Figure 4.11 using a strongly connected product of Figure 4.6. There are many other ways of

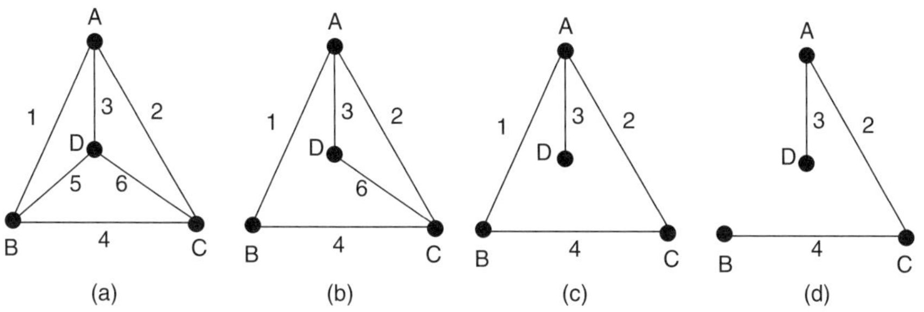


**FIGURE 4.11** Reduction of the number of connections.

reducing the number of connections, but all those result in a connected graph with three connections. If we remove an additional connection, we will be left with a disjoint graph.

### Example 14

Consider the connection diagram in Figure 4.12(a), which represents an automatic transmission with $N = 11$ (De Fazio and Whitney, 1987). A diagram with the minimum number of connections is shown in Figure 4.12(b). To prove that $N - 1$ is indeed the minimum number of connections we proceed as follows. Consider an assembly with $N = 1$. It has 0 connections. If one adds a component ($N = 2$), at least one connection has to be established. Every additional component adds one extra connection to the connection diagram for keeping the network coherent. By this inductive reasoning, it is clear that $N - 1$ is the minimum number of connections required.

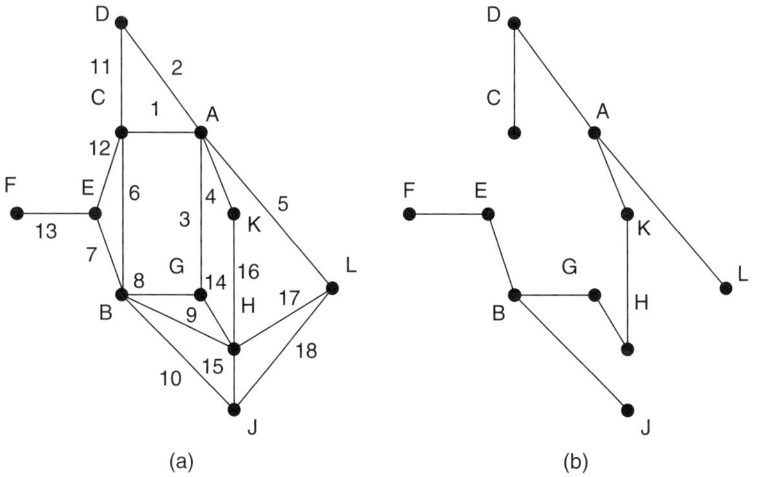


**FIGURE 4.12** Determination of a connection diagram with a minimum number of connections.

A product with the minimum possible number of connections is called *weakly connected* (Homem De Mello and Sanderson, 1991). In weakly connected products, the index of complexity $\alpha = 2 - \frac{2}{N}$ (see Expression (4.23)). Therefore in weakly connected products, $\alpha \to 2$ as $N \to \infty$. In practice, $\alpha$ is somewhere between two and four (Whitney, 2001). The value of $\alpha$ in Figure 4.12(a) is 3.27 and in Figure 4.12(b) is 1.82.

A component with only one connection is called a *leaf*. Every other component is called a *root*. In the hierarchical tree approach, which is a special case of such a product representation, the concepts of root and leaf have a still more distinct significance.

If a product is not strongly connected, there always exist at least one subset of components that is not connected. Therefore, these subsets do not correspond to a subassembly, because a subassembly is assumed to be connected. For example, the subset BD in Figure 4.11(b) is not a subassembly because B and D are not connected. This makes some states in the state diagram and some subassemblies in the AND/OR graph infeasible because of the topological constraints. Thus, the state diagram for an unconstrained product of Figure 4.8 has to be pruned because the states AC,BD and A,C,BD are infeasible. Consequently, the following operations are infeasible:

$$\text{ABCD} \to \text{AC,BD; A,BCD} \to \text{A,C,BD; C,ABD} \to \text{A,C,BD;}$$

$$\text{AC,BD} \to \text{A,C,BD; AC,BD} \to \text{B,D,AC; A,C,BD} \to \text{A,B,C,D.}$$

Similarly, in the AND/OR graph of Figure 4.9, subassembly BD is infeasible. Consequently, the following operations are infeasible:

$$\text{ABCD} \to \text{AC,BD; ABD} \to \text{BD; BCD} \to \text{BD; BD} \to$$

## 4.5 WEAKLY CONNECTED PRODUCTS

### 4.5.1 INTRODUCTION

As was mentioned before, a product with the minimum possible number of connections is called weakly connected. There are two principal reasons for considering weakly connected products.

1. It was observed in the previous section that in real life products, the index of complexity is generally between two and four, which is closer to weakly connected products than to strongly connected products.
2. The hierarchical tree representation (see subsection 3.3.4) is a special case of a weakly connected product, because tree structures are often weakly connected.

In a strongly connected product, the structure of the connection diagram follows straight from $N$. However, in a weakly connected product, different structures are possible with the same value of $N$. This is because the following characteristics can be different:

1. The maximum number of connections to a component $k_{max}$
2. The number of leafs $N_{leaf}$ which are peripheral components that have only one connection with the rest of the product

Two extreme configurations can be defined for $N > 2$ (Wolter, 1992):

1. The *string configuration* (Figure 4.13(a)), with $k_{max} = 2$ and $N_{leaf} = 2$, which has both minimum $k_{max}$ and $N_{leaf}$
2. The *star configuration* (Figure 4.13(b)), with $k_{max} = N - 1$, and $N_{leaf} = N - 1$, which are the maximum possible values

The *tree configuration*, which frequently occurs in practice, is the third kind of configuration, with its properties in between the string and the star configurations (Figure 4.13). The hierarchical tree of Figure 4.13(c) is the connection diagram, the type of which has been used to represent the organization of such varied products as a TV set or computers (Krikke et al., 1998, 1999; Veerakamolmal and Gupta, 1998, 1999; Lambert and Gupta, 2002; and Lambert, 2002). In the tree configuration of Figure 4.13(c), we observe that $k_{max} = 5$ and $N_{leaf} = 6$. For any tree configuration, the following relationship holds:

$$k_{max} \leq N_{leaf} \leq N - 1 \tag{4.24}$$

**Proof** We start with a string configuration, with $k_{max} = N_{leaf} = 2$, and with $N > 4$. This configuration is transformed by detaching a leaf and attaching it to a component that is not a leaf and, consequently, a *root*. This transformation increases both $k_{max}$ and $N_{leaf}$ by 1. We repeat this operation with another leaf. Attaching it to a leaf neither increases $k_{max}$ nor $N_{leaf}$. Attaching it to a root increases $N_{leaf}$ by 1, and may increase $k_{max}$ by 1 but only if it is the root with maximum number of connections. In other words, it is not possible to increase $k_{max}$ without increasing $N_{leaf}$. Therefore the relationship given in Expression (4.24) is true.

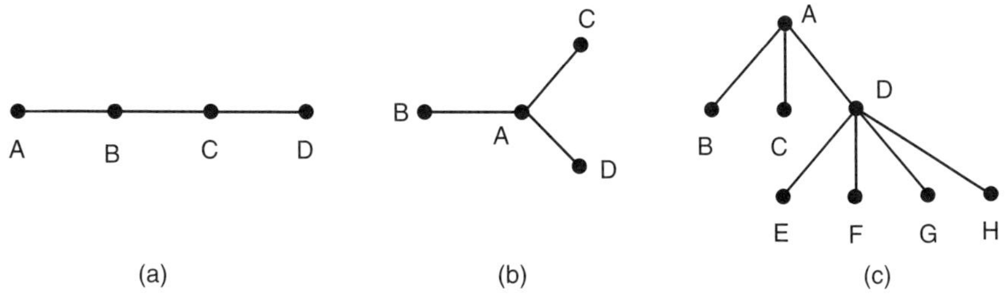


**FIGURE 4.13** Weakly connected products. (a) String configuration; (b) Star configuration; (c) Tree configuration.

In the following subsections the upper bounds on the number of basic features in the disassembly of string and star configurations are obtained. All the numbers for these features have been calculated for $N = 1$ through 11 and summarized in appendix 4D.

### 4.5.2 Connections, Subassemblies, Connection States, and Subassembly States

Consider an unconstrained weakly connected product with $N$ components. This implies that, apart from topological constraints, no additional geometric or technical constraints are present.

The maximum number of connections in a weakly connected product is given by the following relationship and is independent of its specific structure:

$$K_{\text{weak}} = N - 1 \tag{4.25}$$

The maximum number of subassemblies depends on the configuration of the product.

A *string* configuration with $N$ components has the following possible subassemblies: $N$ subassemblies with 1 component, $N - 1$ subassemblies with 2 components and so on, up to 1 subassembly with $N$ components. The maximum number of subassemblies $SUB_{\text{string}}$, is found via the addition of a geometric series as follows:

$$SUB_{\text{string}} = \sum_{k=0}^{N-1} (N-k) = \frac{1}{2} N \cdot (N+1) \tag{4.26}$$

A *star* configuration has $N$ subassemblies with 1 component. All the other subassemblies have to include the central component, e.g., A in Figure 4.13(b). This restricts the number of subassemblies with $k$ components to $\binom{N-1}{k-1}$, and thus the maximum number of subassemblies $SUB_{\text{star}}$, is given by the following relationship:

$$SUB_{\text{star}} = N + \sum_{k=2}^{N-1} \binom{N-1}{k-1} = 2^{N-1} + (N-1) \tag{4.27}$$

An interesting property of a weakly connected product is that every possible subassembly is also weakly connected.

**Proof** A subassembly can be created via a sequence of cut-sets. Since, in the course of this transformation, no loop is created, the resulting subassemblies are weakly connected.

#### Example 15

The string configuration of Figure 4.13(a) has the following 10 subassemblies:

A, B, C, D, AB, BC, CD, ABC, BCD, ABCD

The star configuration of Figure 4.13(b) has the following 11 subassemblies:

A, B, C, D, AB, AC, AD, ABC, ABD, ACD, ABCD

It is evident from Expression (4.26) and Expression (4.27) that with increasing value of $N$, the computational complexity (Tovey, 2002) of the number of subassemblies of a string configuration is $O(N^2)$ and that of a star configuration is $O(2^N)$. This implies that the number of subassemblies in a star configuration increases at a faster rate than the number of subassemblies in a string configuration with increasing value of $N$. Moreover, for a given $N$, the number of subassemblies in a tree configuration is somewhere between the number of subassemblies in a string and a star configuration.

### Example 16

A string configuration with $N=8$ has 36 subassemblies while a star configuration with $N=8$ has 135 subassemblies. It can be shown that the tree configuration of Figure 4.13(c) (also with $N=8$) has 81 subassemblies (which is between 36 and 135).

The number of connection states is equal for both the string and the star configurations and can be obtained by inserting Expression (4.25) into Expression (4.4) as follows:

$$CST_{\text{weak}} = 2^{N-1} \tag{4.28}$$

Since a connection state corresponds to a subassembly state and vice versa, the number of subassembly states is given as follows:

$$SST_{\text{weak}} = 2^{N-1} \tag{4.29}$$

### Example 17

The string configuration of Figure 4.13(a) with $N = 4$ has the following eight states:

1-partition: ABCD
2-partitions: A,BCD; AB,CD; ABC,D
3-partitions: A,B,CD; A,BC,D; AB,C,D
4-partitions: A,B,C,D

The star configuration of Figure 4.13(b) with $N = 4$ has the following eight states:

1-partition: ABCD
2-partitions: B,ACD; C,ABD; D,ABC
3-partitions: AB,C,D; AC,B,D; AD,B,C
4-partitions: A,B,C,D

Note that a tree configuration with $N = 4$ is a special case and is equivalent to a star configuration.

### 4.5.3 Cut-Sets and Disassembly Operations

#### 4.5.3.1 Cut-Sets

In a weakly connected product, a cut-set is equivalent to the detachment of only one connection. Therefore, for string, star, and tree configurations, the following expression for the maximum number of cut-sets holds:

$$CS_{\text{weak}} = N - 1 \tag{4.30}$$

#### 4.5.3.2 Disassembly Operations in a State Diagram

The maximum number of disassembly operations in a state diagram of a product with string configuration $SOP_{\text{string}}$ is determined as follows, which is partly analogous to the reasoning in subsection 4.3.4.2:

We observe that there are $\binom{N-1}{k-1}$ $k$-partitions of a string configuration, because there are $N-1$ connections, from which $k-1$ are disestablished. Each of these $k$-partitions can be transformed into a $(k+1)$-partition in $(N-k)$ ways, as there are still $(N-k)$ established connections left. If we proceed from $k=1$, which is the original product or initial state, through $k=N$, which is the final state, the number of operations adds to

$$\sum_{k=1}^{N-1}\binom{N-1}{N-k}\cdot(N-k) = \sum_{k=1}^{N-1}\frac{(N-1)!}{(N-k)!(k-1)!}\cdot(N-k)$$

$$= (N-1)\sum_{k=1}^{N-1}\frac{(N-2)!}{(N-k-1)!(k-1)!}$$

This transforms, via the substitution: $k^* = k - 1$, into

$$(N-1)\sum_{k^*=0}^{N-2}\frac{(N-2)!}{(N-k^*-2)!k^*!} = (N-1)\sum_{k^*=0}^{N-2}\binom{N-2}{k^*} = 2^{N-2}(N-1)$$

For the last step we used (4A.2b).

For the star configuration and other configurations of weakly connected products, although topologically different, the same reasoning holds. Thus the maximum number of disassembly operations for any weakly connected product in a state diagram is given by

$$SOP_{\text{weak}} = 2^{N-2}(N-1) \tag{4.31}$$

**Example 18**

The state diagram of the string and star configuration from Figure 4.13, with $N = 4$, is depicted in Figure 4.14(a) and Figure 4.14(c) respectively.

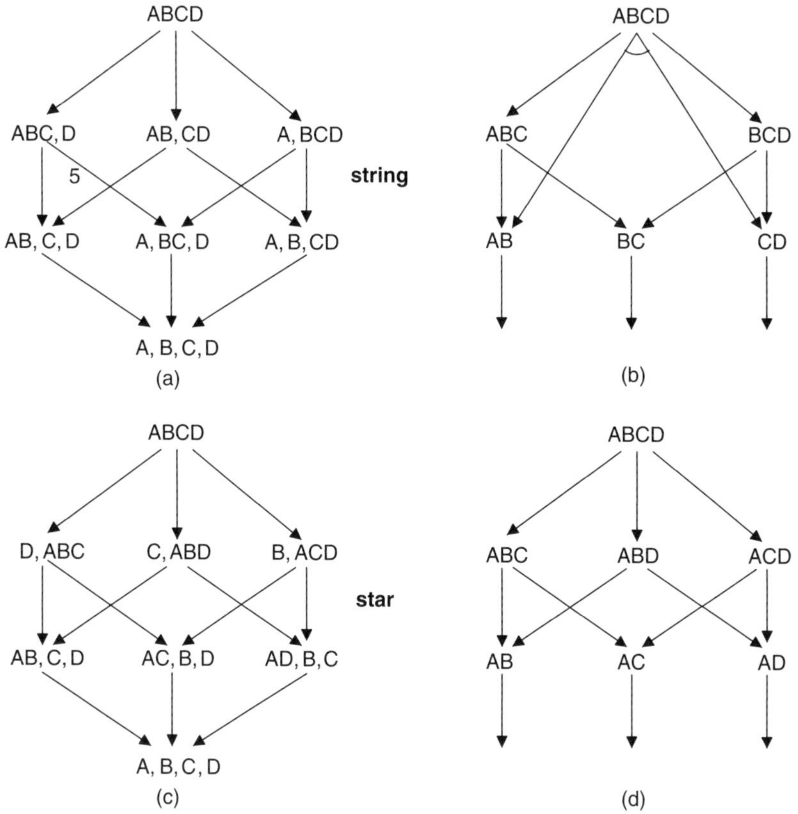


**FIGURE 4.14** (a) State diagram of the string configuration; (b) AND/OR graph of the string configuration; (c) state diagram of the star configuration; (d) AND/OR graph of the star configuration.

Two properties are evident here: (1) The figures here are much simpler than the ones for unconstrained products. This is because many states and subassemblies are infeasible here; and (2) Parallel disassembly is present in string configurations, but is absent in star configurations.

### 4.5.3.3 Disassembly Operations in an AND/OR Graph

The AND/OR graph discriminates between the string configuration and the star configuration in the number of disassembly operations. Therefore, we consider these configurations separately.

In the string configuration there are $N-1$ hyperarcs departing from every subassembly with $N$ components (see Expression (4.30)). There is one subassembly with $N$ components; there are two subassemblies with $N-1$ components and so on. This results in a maximum number of disassembly operations for the string configuration $AOP_{\text{string}}$ as follows:

$$AOP_{\text{string}} = \sum_{k=1}^{N-1} k(N-k) = \tfrac{1}{6} N(N+1)(N-1) \tag{4.32}$$

In the star configuration, the number of disassembly operations in the AND/OR graph $AOP_{\text{star}}$ equals that in the corresponding state diagram, which is given by Expression 4.31, because no parallel disassembly is possible. Consequently, compare Expression 4.31:

$$AOP_{\text{star}} = 2^{N-2}(N-1) \tag{4.33}$$

### 4.5.4 Complete Sequential Sequences

A sequential sequence corresponds to the serial detachment of a leaf. In the string configuration, one has two leafs and thus two possibilities for every operation, till only two components are left, thus

$$CSS_{\text{string}} = 2^{N-2} \tag{4.34}$$

In the star configuration, one starts with $N - 1$ leafs, next $N - 2$, and so on till two leafs are left. Thus

$$CSS_{\text{star}} = (N - 1)! \tag{4.35}$$

**Example 19**

There are following four complete sequential sequences for the string configuration of Figure 4.13(a):

ABCD → ABC,D → AB,C,D → A,B,C,D

ABCD → ABC,D → A,BC,D → A,B,C,D

ABCD → A,BCD → A,B,CD → A,B,C,D

ABCD → A,BCD → A,BC,D → A,B,C,D

There are following six complete sequential sequences for the star configuration of Figure 4.13(b):

ABCD → ABC,D → AB,C,D → A,B,C,D

ABCD → ABC,D → AC,B,D → A,B,C,D

ABCD → ABD,C → AB,C,D → A,B,C,D

ABCD → ABD,C → AD,B,C → A,B,C,D

ABCD → ACD,B → AC,B,D → A,B,C,D

ABCD → ACD,B → AD,B,C → A,B,C,D

### 4.5.5 Complete Sequences in the State Diagram

The state diagrams for both the string and the star configurations have a similar structure, thus a similar reasoning can be followed to determine the number of complete sequences, see Figure 4.14(a) and Figure 4.14(c). One starts with the 1-partition, from which $N-1$ arcs depart, each pointing to a 2-partition. From every 2-partition, $N-2$ arcs depart, each pointing to a 3-partition, and so on until the $N-2$-partitions are reached, from each of which 1 arc departs. This results in the following:

$$SCS_{\text{weak}} = (N-1)! \tag{4.36}$$

Note that this is not an astonishing result for the star configuration as only sequential disassembly is possible here and thus the same result as Expression 4.35 is obtained. For the string configuration, however, the number of complete sequences exceeds the number of complete sequential sequences, because additional parallel sequences are permitted in the case of complete sequences.

### 4.5.6 Complete Sequences in the AND/OR Graph

The number of complete sequences in the AND/OR graph is different for the string configuration and the star configuration.

Let the maximum number of complete sequences of a product with string configuration consisting of $N$ components be $ACS_{\text{string}}$. For convenience, let us represent this number as $P_N$. The product can be separated in an $N-p$ component subassembly and a $p$ component subassembly in two ways. This leaves us with the recursive relationship:

$$P_N = \sum_{p=1}^{N-1} P_p \cdot P_{N-p} \tag{4.37}$$

From this, it follows that (note that $P_1 = 1$):

$$ACS_{\text{string}} = P_N = \frac{\binom{2N}{N}}{N+1} = \frac{(2N-2)!}{N!(N-1)!} \tag{4.38}$$

The proof is nontrivial and needs similar theory as for the proof of Expression 4.14. Note that $ACS_{\text{string}}$ has a relationship with the *Catalan*[3] *numbers* $C(N)$, viz. $P_N = C(N)$. Catalan numbers represent the number of ways to insert $N$ pairs of parentheses in a word of $N+1$ letters. It is evident that these also represent the number of sequences that can be obtained.

[3] In honor of Eugène Charles Catalan (1814–1894).

**Example 20**

The number of complete sequences in the AND/OR graph of Figure 4.14(b) is $P_4 = C(4) = 5$, which can be written as follows (Ben-Arieh and Kramer, 1994):

ABCD → ABC + D → AB + C + D → A + B + C + D **or** (((AB)C)D),

ABCD → ABC + D → BC + A + D → A + B + C + D **or** ((A(BC)D),

ABCD → AB + CD → A + B + C + D **or** ((AB)(CD))

ABCD → BCD + A → BC + A + D → A + B + C + D **or** (A((BC)D))

ABCD → BCD + A → CD + A + B → A + B + C + D **or** (A(B(CD)))

For the star configuration, the number of complete sequences in the AND/OR graph equals that of the state diagram as follows:

$$ACS_{\text{star}} = (N-1)! \tag{4.39}$$

**Example 21**

Since $N = 4$, the numbers of complete sequences in the state diagrams in Figure 4.14(a) and Figure 4.14(c) and the AND/OR graph in Figure 4.14 (d) are each equal to $(N - 1)! = 6$.

### 4.5.7 Incomplete Disassembly Sequences

#### 4.5.7.1 Incomplete Sequential Sequences

Let the number of incomplete sequential sequences in a product with string configuration consisting of $N$ components be $ISS_{\text{string}}$. For convenience, let us represent this number as $P_N$. The following recursive relationship holds:

$$ISS_{\text{string}} = P_N = 1 + 2P_{N-1} \quad (N > 2) \tag{4.40}$$

with $P_1 = 1$, $P_2 = 2$. Consequently,

$$ISS_{\text{string}} = 3 \cdot 2^{N-2} - 1 \quad (N > 2) \tag{4.41}$$

**Proof** Consider a product ABCD (with $N = 4$) (like the one shown in Figure 4.13(a)) that can be sequentially disassembled. The number of incomplete sequential sequences $P_4$ is the sum of doing nothing (one possibility), starting with the detachment of A ($P_3$ possibilities) and starting with the detachment of D ($P_3$ possibilities). Therefore, $P_4 = 1 + 2P_3$. By inductive reasoning, it will clearly lead to Expression (4.40).

For the star configuration, we follow an analogous reasoning. Here, however, the following recursive relationship holds:

$$ISS_{\text{star}} = P_N = 1 + (N-1)\cdot P_{N-1} \tag{4.42}$$

or

$$ISS_{\text{star}} = (N-1)!\sum_{k=0}^{N-1}\frac{1}{k!} \tag{4.43}$$

#### 4.5.7.2 Incomplete Sequences in the State Diagram

Since the state diagram of string configuration is similar to the state diagram of star configuration, the disassembly of both the string and star configurations is always sequential. Therefore, the number of incomplete sequential sequences is also equivalent. Thus, from Expression (4.43):

$$SIS_{\text{weak}} = ISS_{\text{star}} \tag{4.44}$$

#### 4.5.7.3 Incomplete Sequences in the AND/OR Graph

For the string configuration, the reasoning is similar to that for the complete sequences. However, the recursive relation (4.37) has to be modified as follows:

$$P_N = \sum_{p=1}^{N-1} P_p \cdot P_{N-p} + 1 \quad \text{with } P_1 = 1,\ P_2 = 2 \tag{4.45}$$

This is known as the *binomial transform of Catalan numbers*. A linear recursive relationship is as follows:

$$(N+2)\cdot P_{N+2} = (6N+4)\cdot P_{N+1} - 5N\cdot P_N \tag{4.46}$$

Thus, the number of incomplete sequences in an AND/OR graph is as follows:

$$AIS_{\text{string}} = P_N \tag{4.47}$$

And for the star configuration, the number of incomplete sequences in both the state diagram and the AND/OR graph are equal. Thus

$$AIS_{\text{star}} = ISS_{\text{star}} \tag{4.48}$$

### 4.5.8 Tree Configurations

Products with pure string or star configurations are not that common in practice. However, a large number of products have tree structures, an example of which is given in Figure 4.13(c).

Consider an assembly with $N$ components. We investigate the number of subassemblies with $k$ components ($k \leq N$) and relate this to the number of subsets with $k$ components, which is given by $\binom{N}{k}$. Since a weakly connected product is topologically constrained, the number of subassemblies, which have to be connected, is less than the number of subsets (which are not necessarily connected) by a certain factor. According to subsection 4.5.2, there are $N - k + 1$ subassemblies with $k$ components in a string configuration, and there are $\binom{N-1}{k-1}$ subassemblies with $k$ components in a star configuration, provided $k > 1$. Therefore, the fraction of subset that is a subassembly, in a string configuration is $\frac{(N-k+1)!\cdot k!}{N!}$, and in a star configuration is $\frac{k}{N}$.

We depicted this in Figure 4.15 for $N = 8$. In this figure, we have also inserted the share of subassemblies for the hierarchical tree of Figure 4.13(c). We can observe that both the string and the star configuration are limiting cases for the weakly connected products. General structures occupy an intermediate position between

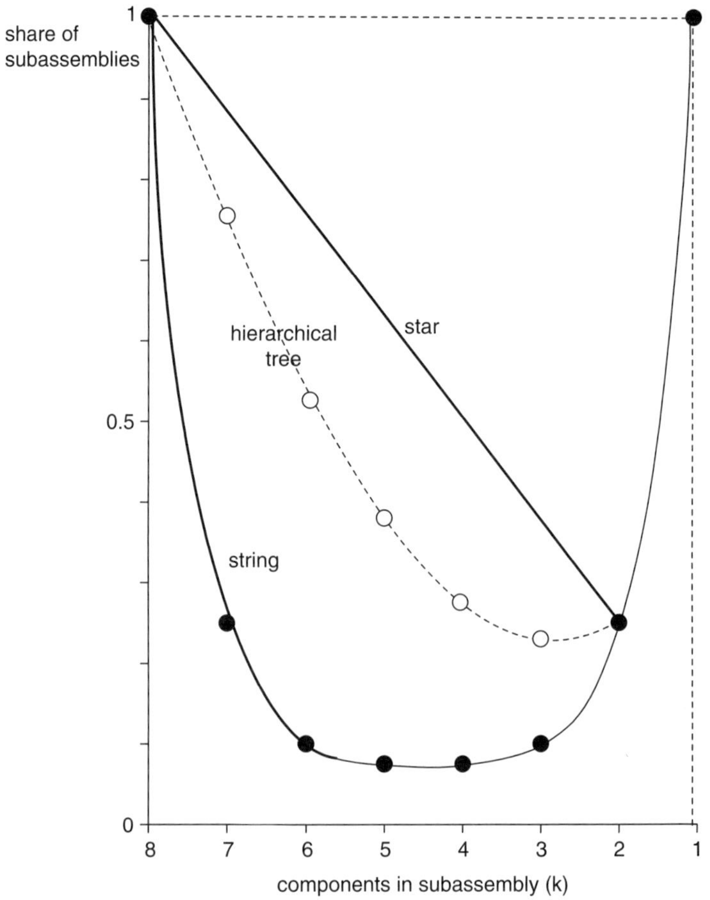


**FIGURE 4.15.** Number of subassemblies with $k$ components in a weakly connected product (with $N = 8$), related to the number of subsets.

these extremes (Lambert, 2002). Evidently, further research is needed for different manifestations of tree structures.

## 4.6 GEOMETRICALLY CONSTRAINED PRODUCTS

### 4.6.1 SUBASSEMBLY SELECTION IN TOPOLOGICALLY CONSTRAINED PRODUCTS

It is clear from the discussions in the preceding subsections that the numbers of subassemblies, disassembly operations and disassembly sequences are a function of topological constraints in weakly connected products. For strongly connected products, there is an upper bound on the number of various features in disassembly sequencing. Real life products are rarely strongly connected and therefore also have topological constraints. Even though formulae for real life products are not available, there are nevertheless systematic methods available for obtaining a list of possible subassemblies. In such methods the connectivity matrix (see subsection 3.2.9) plays an important role. The connectivity matrix based on the connection diagram of Figure 4.12(a) is shown in Figure 4.16(a). For convenience, the information in the lower triangle of the

| | A | B | C | D | E | F | G | H | J | K | L |
|---|---|---|---|---|---|---|---|---|---|---|---|
| A | | | | | | | | | | | |
| B | 0 | | | | | | | | | | |
| C | 1 | 1 | | | | | | | | | |
| D | 1 | 0 | 1 | | | | | | | | |
| E | 0 | 1 | 1 | 0 | | | | | | | |
| F | 0 | 0 | 0 | 0 | 1 | | | | | | |
| G | 1 | 1 | 0 | 0 | 0 | 0 | | | | | |
| H | 0 | 1 | 0 | 0 | 0 | 0 | 1 | | | | |
| J | 0 | 1 | 0 | 0 | 0 | 0 | 0 | 1 | | | |
| K | 1 | 0 | 0 | 0 | 0 | 0 | 0 | 1 | 0 | | |
| L | 1 | 0 | 0 | 0 | 0 | 0 | 0 | 1 | 1 | 0 | |

(a)

| | A | B | C | D | E | F | G | H | J | K | L |
|---|---|---|---|---|---|---|---|---|---|---|---|
| A | - | 0 | 1 | 1 | 0 | 0 | 1 | 0 | 0 | 1 | 1 |
| B | 0 | - | 1 | 0 | 1 | 0 | 1 | 1 | 1 | 0 | 0 |
| C | 1 | 1 | - | 1 | 1 | 0 | 0 | 0 | 0 | 0 | 0 |
| D | 1 | 0 | 1 | - | 0 | 0 | 0 | 0 | 0 | 0 | 0 |
| E | 0 | 1 | 1 | 0 | - | 1 | 0 | 0 | 0 | 0 | 0 |
| F | 0 | 0 | 0 | 0 | 1 | - | 0 | 0 | 0 | 0 | 0 |
| G | 1 | 1 | 0 | 0 | 0 | 0 | - | 1 | 0 | 0 | 0 |
| H | 0 | 1 | 0 | 0 | 0 | 0 | 1 | - | 1 | 1 | 1 |
| J | 0 | 1 | 0 | 0 | 0 | 0 | 0 | 1 | - | 0 | 1 |
| K | 1 | 0 | 0 | 0 | 0 | 0 | 0 | 1 | 0 | - | 0 |
| L | 1 | 0 | 0 | 0 | 0 | 0 | 0 | 1 | 1 | 0 | - |

(b)

**FIGURE 4.16.** Connectivity matrices for product of Figure 4.12(a): (a) lower triangular version and (b) full matrix version.

connectivity matrix can be duplicated in the upper triangle by taking its transpose, see Figure 4.16(b). Note that there are 11 rows and 11 columns in the connectivity matrix because $N = 11$. We further observe that $K = 18$, which is also the number of nonzero elements in the triangular connectivity matrix. The index of complexity $\alpha = 3.27$ (see Expression 4.23).

*A priori selection:* We will briefly describe the procedure for listing all the subassemblies of the product of Figure 4.12(a) using the full matrix version of connectivity matrix in Figure 4.16.

Start by listing all components. Next, list all subassemblies that *contain A*. Look at the nonzero elements in row A (first row), which are C, D, G, K, and L. Using A and these elements, list all the subassemblies with two components, viz., AC, AD, AG, AK, and AL, all the subassemblies with three components, viz., ACD, ACG, ACK, ACL, ADG, etc. through AKL, all the subassemblies with four components, viz., ACDG, ACDK, ACDL, ACGK, ACGL, ACKL, ADGK, ADGL, ADKL, AGKL, and so on, till the subassembly ACDGKL is obtained. Next proceed with the two-component subassembly AC and observe column C. Using the non-zero elements, the subassemblies ACB (or: ABC), ACD, and ACE are obtained. Repeat the procedure with the remaining two component subassemblies till ALJ is obtained. Similarly, next proceed with the three-component subassembly, then with four-component subassembly and so on. Delete duplicate subassemblies from the list.

Next delete the row and the column of the matrix that correspond to A. Repeat the above procedure with this smaller matrix, starting with B. This way, all the subassemblies that contain B and not A are listed. Subsequently, delete the row and the column that correspond to B and start with C and so on.

*A posteriori selection:* It is also possible to obtain the list of potential subassemblies by considering geometric constraints. The connectivity matrix can be used to test whether a potential subassembly is connected or not.

Obviously, the *a priori* listing of the topologically feasible subassemblies becomes virtually impracticable for products with an increasing number of components.

### 4.6.2 Geometric Constraints

Although geometric constraints will be discussed in detail in the next chapter, we nevertheless would like to make a few remarks here. Geometric constraints explain the impracticality of specific disassembly operations, namely those that are obstructed by the presence of other components. This is in contrast with the description in the preceding subsections, where geometric properties of the product were not taken into account. We will illustrate this additional feature with the help of Figure 4.17.

This product, which is assumed to be 2-dimensional, has a strongly connected structure with the connection diagram given in Figure 4.6. A careful examination of Figure 4.17 reveals that due to the geometry of the product, some subassemblies cannot be obtained via disassembly. For example, the subassemblies ABC, AB, AC, and BC cannot be obtained because all of these require the removal of D and the

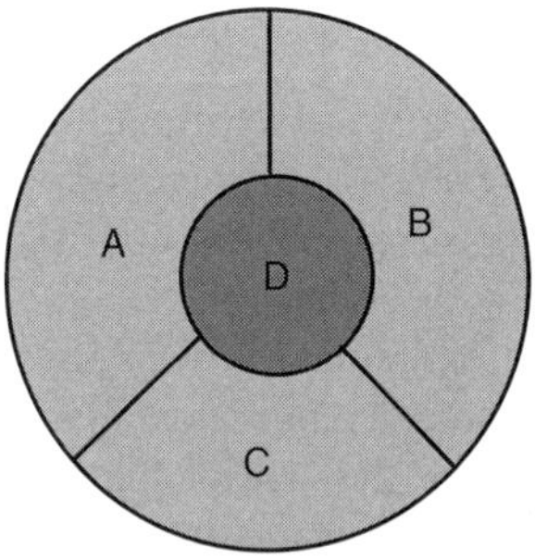


**FIGURE 4.17** Example of a topologically unconstrained but geometrically constrained product.

removal of D can only proceed if two of the peripheral components are removed first (because of obstruction, see Figure 4.17).

A product may be topologically constrained, geometrically constrained, or both topologically and geometrically constrained. In the product of Figure 4.18(a), the potential subassemblies AC and BD are topologically infeasible. In the product of Figure 4.18(b), the subassemblies AB, AC, BC, and ABC, are geometrically infeasible. In the product of Figure 4.18(c), the potential subassemblies AC and BD are topologically infeasible, and any subassembly that contains component B (except for the complete product) is geometrically infeasible. Note that even though the products in Figure 4.18(a) and Figure 4.18(c) have exactly the same connection diagram, they, nevertheless, have a completely different set of feasible subassemblies because of geometric constraints.

It is thus clear that in addition to the connection diagram, further information on the product is required to accommodate the geometric constraints. We will consider this in more detail in the next chapter.

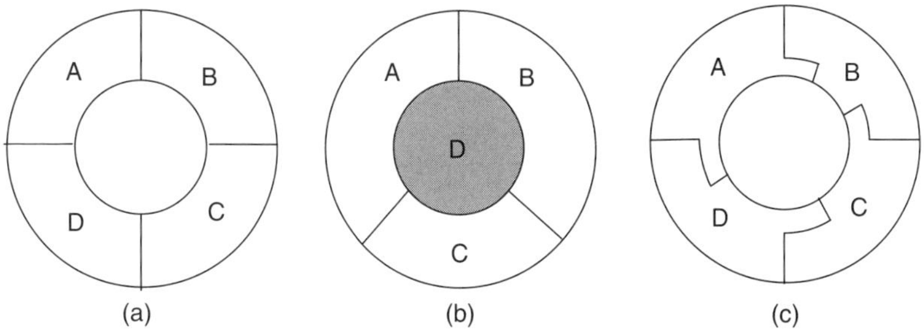


**FIGURE 4.18.** Products that are; (a) topologically constrained; (b) geometrically constrained; (c) both topologically and geometrically constrained.

### 4.6.3 Graphical Methods

In several earlier subsections we focused on strongly connected unconstrained products and derived closed form expressions or recursive relationships for the number of states, subassemblies, actions, and disassembly sequences for products with $N$ components. We noted that the size of the problem of finding all possible disassembly sequences and searching for optimum solutions dramatically increased with the increase in the value of $N$. In practice, however, a large number of products are not strongly connected thus reducing the size of the problem. Additional constraints (such as topological, geometric, and technical) further reduce the size of the problem. This results in disassembly graphs that may contain a significantly reduced number of nodes and arcs. The actual number of sequences can be derived from the corresponding disassembly graphs using graphical methods. We will elaborate on these methods for both the state diagram and the AND/OR graph in the next couple of subsections.

#### 4.6.3.1 State Diagrams

Consider the state diagram given in Figure 4.8. If subassemblies AB and AD are infeasible, the state diagram has to be modified. The modified state diagram is shown in Figure 4.19.

*Complete sequences:* For determining the number of complete sequences $SCS$ we start at the final state and proceed upstream. We put the value 1 at the final state, as it corresponds to a single complete sequence. Next, a value is transferred via the entering arcs upstream to the states that are at the level immediately above it, which contains the $(N-1)$-partitions. The value of the final state is given to these entering arcs and the added values of these arcs are given to states from where they emanate. Proceed the same way to the next level upstream, which contains the $(N-2)$-partitions. A similar procedure to transfer the value to the arcs and states is carried out. Note that in Figure 4.19(a), there are three outgoing arcs from the state A,BCD and there are two outgoing arcs from the state B,ACD. The departing arcs of a state are related to each other via (exclusive) OR-relationships. Since we are determining the number of sequences, we add the numbers that are assigned to the arcs. The result of this addition is given to the corresponding node and written in bold. Proceeding upward this way leads us to the initial state, revealing the number 10 given to that state. This means that there are 10 complete sequences for this case, which is a significant reduction in value compared to a strongly connected product for which $SCS_{max}$ is equal to 18 (see Table 4C.2).

*Incomplete sequences:* Here also we proceed using a similar reasoning. However, when a node is encountered upstream, a one is added to the total of the values of the arcs emanating from the node. This accounts for the additional "zero operation" that corresponds to ending the sequence at that particular node. From Figure 4.19(b) we discover that this results in 26 sequences. This is a significant reduction in value compared to a strongly connected product for which $SIS_{max}$, is equal to 44 (see Table 4C.2).

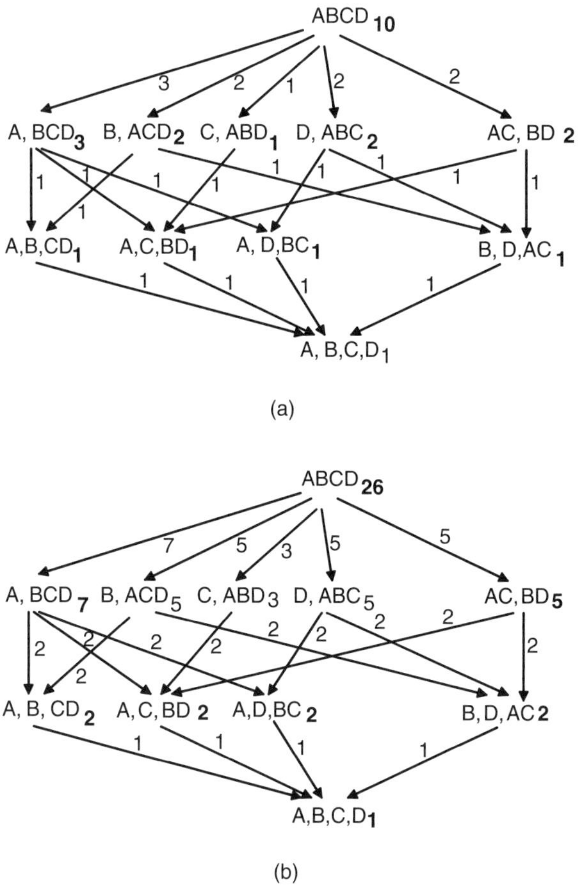


**FIGURE 4.19** Graphical method for obtaining the number of disassembly sequences in a state diagram. (a) complete sequences; (b) incomplete sequences.

### 4.6.3.2 AND/OR Graphs

Consider the state diagram given in Figure 4.19 (which is a modified diagram of Figure 4.8 with subassemblies AB and AD removed). The corresponding AND/OR graph is given in Figure 4.20 (Lambert, 1997). Note that as opposed to the state diagram where the elimination of subassemblies AB and AD propagated itself to multiple states, in an AND/OR graph this is only limited to the subassemblies involved. Here also, we follow reasoning similar to that of the state diagram. However, an additional complication is introduced by the AND relations that results from parallel disassembly operations, such as the following operation in Figure 4.20:

$$ABCD \rightarrow AB + CD$$

The values at the branches of the resulting hyperarc have to be multiplied with each other. In the specific case of Figure 4.20, the parallel disassembly operation listed

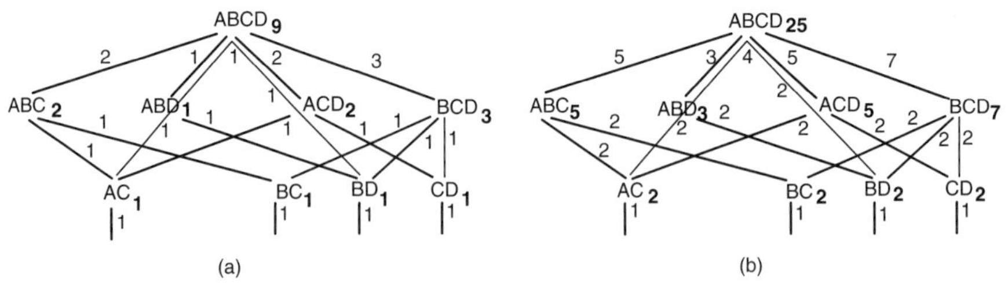


**FIGURE 4.20** Graphical method for obtaining the number of disassembly sequences in an AND/OR graph. (a) Complete sequences; (b) incomplete sequences.

earlier is the only one for this case. Note that, if one of the branches points to a single component (and therefore not depicted in AND/OR graphs) the corresponding value is 1, which, of course, has no effect on the result of the multiplication.

*Complete sequences:* For the case of complete sequences, both branches of the hyperarc that represent the parallel operations have a value 1 assigned to them. This means that this operation contributes to only one complete sequence. Thus, there are 9 complete sequences for this case compared to a strongly connected product for which $ACS_{max} = 15$, which is a reduction of 6 sequences.

*Incomplete sequences:* Here also we proceed using a similar reasoning. When determining the number of incomplete sequences, a 1 is added to the total of the values of the arcs emanating from the node. This accounts for the additional "zero operation" that corresponds to ending the sequence at that particular node. In dealing with the parallel operation, for the case of Figure 4.20(b), we notice that the value that is assigned to both branches of the corresponding hyperarc equals 2, which results in $2 \times 2 = 4$ different sequences that start at the parallel disassembly action, namely:

$$\mathrm{ABCD} \rightarrow \mathrm{AB} + \mathrm{CD} \rightarrow \mathrm{A} + \mathrm{B} + \mathrm{C} + \mathrm{D}$$

$$\mathrm{ABCD} \rightarrow \mathrm{AB} + \mathrm{CD} \rightarrow \mathrm{A} + \mathrm{B} + \mathrm{CD}$$

$$\mathrm{ABCD} \rightarrow \mathrm{AB} + \mathrm{CD} \rightarrow \mathrm{AB} + \mathrm{C} + \mathrm{D}$$

$$\mathrm{ABCD} \rightarrow \mathrm{AB} + \mathrm{CD}$$

We observe that in this case 25 incomplete sequences are possible. This is a reduction of 16 compared to a strongly connected product for which $AIS_{max} = 41$.

The reduction of the number of sequences that is derived from the AND/OR graph, compared with the number of sequences derived from the equivalent state diagram, is evident from the presence of a parallel disassembly operation. In the AND/OR graph there is no distinction in the ordering of disassembling AC and BD.

## 4.7 CONCLUSION

In this chapter we compiled the theoretical maximum of a number of basic features in disassembly sequencing. This included the results that have been published in the literature, together with some additional discrete mathematics and coverage of incomplete sequences, useful in the domain of end-of life disassembly and repair. We focused on topological constraints and compared it to the complexity of real world products. Related to this, specific cases of weakly connected products were discussed. These cases showed a drastic reduction in the numbers of basic features. We also discovered that geometric constraints could further reduce the size of the search space. Since optimization can be carried out via linear programming methods (see chapter 7), the size of the search space does not reflect the size of the problem, which is actually determined by the number of subassemblies (or subassembly states), and by the number of feasible operations.

## APPENDIX 4A: BINOMIAL COEFFICIENTS, STIRLING NUMBERS OF THE SECOND KIND, BELL NUMBERS, AND CATALAN NUMBERS

Various properties of binomial coefficients and Stirling numbers are given in Abramowitz and Stegun (1968). Principal properties and literature on the Bell numbers is given in Sloane (2003).

### 4A.1 Binomial Coefficients

Binominal coefficients are represented as follows:

$$\binom{N}{k} = \binom{N}{N-k} = \frac{N!}{k!(N-k)!} \tag{4A.1}$$

A relevant property of binomial coefficients is as follows (Gradshteyn and Ryzhik, 1981):

$$\sum_{k=0}^{N} \binom{N}{k} r^k = (r+1)^N \tag{4A.2a}$$

from which it follows, that

$$\sum_{k=0}^{N} \binom{N}{k} = 2^N \tag{4A.2b}$$

$$\sum_{k=0}^{N} \binom{N}{k} 2^k = 3^N \text{ etc.} \tag{4A.2c}$$

Binomial coefficients represent the number of ways for selecting $k$ elements out of $N$.

## 4A.2 Stirling Numbers of the Second Kind

Stirling numbers of the second kind can be expressed in a closed form as follows:

$$\left\{ \begin{matrix} N \\ k \end{matrix} \right\} = \frac{1}{k!} \sum_{p=1}^{k} (-1)^{k-p} \binom{k}{p} p^N \tag{4A.3}$$

The following two recursive relationships are relevant:

$$\left\{ \begin{matrix} N \\ k \end{matrix} \right\} = k \left\{ \begin{matrix} N-1 \\ k \end{matrix} \right\} + \left\{ \begin{matrix} N-1 \\ k-1 \end{matrix} \right\} \tag{4A.4}$$

and

$$\binom{k}{r} \left\{ \begin{matrix} N \\ k \end{matrix} \right\} = \sum_{m=k-r}^{N-r} \binom{N}{m} \left\{ \begin{matrix} N-m \\ r \end{matrix} \right\} \left\{ \begin{matrix} m \\ k-r \end{matrix} \right\} \tag{4A.5}$$

For special numbers, the following relationships are relevant:

$$\left\{ \begin{matrix} N \\ 1 \end{matrix} \right\} = \left\{ \begin{matrix} N \\ N \end{matrix} \right\} = 1 \tag{4A.6a}$$

$$\left\{ \begin{matrix} N \\ 2 \end{matrix} \right\} = 2^{N-1} - 1 \tag{4A.6b}$$

$$\left\{ \begin{matrix} N \\ 3 \end{matrix} \right\} = \frac{1}{2}(1 + 3^{N-1}) - 2^{N-1} \tag{4A.6c}$$

The Stirling numbers of the second kind represent the number of $k$-partitions of $N$ elements.

## 4A.3 Bell Numbers

There is no closed formula for deriving the Bell numbers, but they can be found using the following recursive relationship:

$$b_{N+1} = \sum_{k=0}^{N} \binom{N}{k} \cdot b_k, \qquad \text{where } b_0 = 1 \tag{4A.7}$$

Bell numbers give the number of partitions of a set with $N$ elements.

## 4A.4 Catalan Numbers (Also Known as Segner Numbers)

Catalan numbers are given by the following relationship:

$$C(N) = \frac{\binom{2N}{N}}{N+1} = \frac{(2N-2)!}{N!(N-1)!} \tag{4A.8}$$

Among other things, the Catalan numbers provide the number of ways to insert $N$ pairs of parentheses in a word of $N+1$ letters.

## APPENDIX 4B: TABLES OF BINOMIAL COEFFICIENTS, STIRLING NUMBERS OF THE SECOND KIND, AND BELL NUMBERS

**TABLE 4B.1**
**Binomial Coefficients**

| $k\rightarrow$ $N\downarrow$ | 1 | 2 | 3 | 4 | 5 | 6 | 7 | 8 | 9 | 10 | 11 |
|---|---|---|---|---|---|---|---|---|---|---|---|
| 1 | 1 | . | . | . | . | . | . | . | . | . | . |
| 2 | 2 | 1 | . | . | . | . | . | . | . | . | . |
| 3 | 3 | 3 | 1 | . | . | . | . | . | . | . | . |
| 4 | 4 | 6 | 4 | 1 | . | . | . | . | . | . | . |
| 5 | 5 | 10 | 10 | 5 | 1 | . | . | . | . | . | . |
| 6 | 6 | 15 | 20 | 15 | 6 | 1 | . | . | . | . | . |
| 7 | 7 | 21 | 35 | 35 | 21 | 7 | 1 | . | . | . | . |
| 8 | 8 | 28 | 56 | 70 | 56 | 28 | 8 | 1 | . | . | . |
| 9 | 9 | 36 | 84 | 126 | 126 | 84 | 36 | 9 | 1 | . | . |
| 10 | 10 | 45 | 120 | 210 | 252 | 210 | 120 | 45 | 10 | 1 | . |
| 11 | 11 | 55 | 165 | 330 | 462 | 462 | 330 | 165 | 55 | 11 | 1 |

**TABLE 4B.2**
**Stirling Numbers of the Second Kind**

| $k\rightarrow$ $N\downarrow$ | 1 | 2 | 3 | 4 | 5 | 6 | 7 | 8 | 9 | 10 | 11 |
|---|---|---|---|---|---|---|---|---|---|---|---|
| 1 | 1 | . | . | . | . | . | . | . | . | . | . |
| 2 | 1 | 1 | . | . | . | . | . | . | . | . | . |
| 3 | 1 | 3 | 1 | . | . | . | . | . | . | . | . |
| 4 | 1 | 7 | 6 | 1 | . | . | . | . | . | . | . |
| 5 | 1 | 15 | 25 | 10 | 1 | . | . | . | . | . | . |
| 6 | 1 | 31 | 90 | 65 | 15 | 1 | . | . | . | . | . |
| 7 | 1 | 63 | 301 | 350 | 140 | 21 | 1 | . | . | . | . |
| 8 | 1 | 127 | 966 | 1,701 | 1,050 | 266 | 28 | 1 | . | . | . |
| 9 | 1 | 255 | 3,025 | 7,770 | 6,951 | 2,646 | 462 | 36 | 1 | . | . |
| 10 | 1 | 511 | 9,330 | 34,105 | 42,525 | 22,827 | 5,880 | 750 | 45 | 1 | . |
| 11 | 1 | 1,023 | 28,501 | 145,750 | 246,730 | 179,487 | 63,987 | 11,880 | 1,155 | 55 | 1 |

**TABLE 4B.3**
**Bell Numbers**

| $N$ | $b_N$ | $N$ | $b_N$ |
|---|---|---|---|
| 0 | 1 | 11 | 678,570 |
| 1 | 1 | 12 | 4,213,597 |
| 2 | 2 | 13 | 27,644,437 |
| 3 | 5 | 14 | 190,899,322 |
| 4 | 15 | 15 | 1,382,958,545 |
| 5 | 52 | 16 | 10,480,142,147 |
| 6 | 203 | 17 | 82,864,869,804 |
| 7 | 877 | 18 | 682,076,806,159 |
| 8 | 4,140 | 19 | 5,832,742,205,057 |
| 9 | 21,147 | 20 | 51,724,158,235,372 |
| 10 | 115,975 | 21 | 474,869,816,156,751 |
| | | 22 | 4,506,715,738,447,323 |

## APPENDIX 4C: TABLES OF MAXIMUM NUMBER OF FEATURES FOR UNCONSTRAINED PRODUCTS

**TABLE 4C.1**
**Maximum Number of Features in Disassembly Graphs**

| | Connections | Sub-Assemblies | Connection States | Subassembly States | Cut-Sets | Operations State Diagram | Operations AND/OR |
|---|---|---|---|---|---|---|---|
| $N$ | $K_{max}$ (4.2) | $SUB_{max}$ (4.3) | $CST_{max}$ (4.4) | $SST_{max}$ (4.5) | $CS_{max}$ (4.6) | $SOP_{max}$ (4.7) | $AOP_{max}$ (4.9) |
| 1 | — | 1 | — | 1 | — | — | — |
| 2 | 1 | 3 | 2 | 2 | 1 | 1 | 1 |
| 3 | 3 | 7 | 8 | 5 | 3 | 6 | 6 |
| 4 | 6 | 15 | 64 | 15 | 7 | 31 | 25 |
| 5 | 10 | 31 | 1,024 | 52 | 15 | 160 | 90 |
| 6 | 15 | 63 | 32,768 | 203 | 31 | 856 | 301 |
| 7 | 21 | 127 | 2,097,152 | 877 | 63 | 4,802 | 966 |
| 8 | 28 | 255 | 268,435,456 | 4,140 | 127 | 28,337 | 3,025 |
| 9 | 36 | 511 | $\sim 69 \cdot 10^9$ | 21,147 | 255 | 175,896 | 9,330 |
| 10 | 45 | 1,023 | $\sim 35 \cdot 10^{12}$ | 115,975 | 511 | 1,146,931 | 28,501 |
| 11 | 55 | 2,047 | $\sim 36 \cdot 10^{15}$ | 678,570 | 1,023 | 7,841,108 | 86,526 |

**TABLE 4C.2**
**Maximum Number of Sequences in Disassembly Graphs**

| $N$ | Sequential Complete $CSS_{max}$ (4.11) | State Complete $SCS_{max}$ (4.13) | AND/OR Complete $ACS_{max}$ (4.16) | Sequential Incomplete $ISS_{max}$ (4.18) | State Incomplete $SIS_{max}$ (4.19) | AND/OR Incomplete $AIS_{max}$ (4.20) | Asymmetric State Complete $CS_{max}$(4.21) |
|---|---|---|---|---|---|---|---|
| 1 | — | — | — | 1 | 1 | 1 | 1 |
| 2 | 1 | 1 | 1 | 2 | 2 | 2 | 2 |
| 3 | 3 | 3 | 3 | 7 | 7 | 7 | 12 |
| 4 | 12 | 18 | 15 | 29 | 44 | 41 | 120 |
| 5 | 60 | 180 | 105 | 146 | 451 | 346 | 1,680 |
| 6 | 360 | 2,700 | 945 | 877 | 6,872 | 3,797 | 30,240 |
| 7 | 2,520 | 56,700 | 10,395 | 6,140 | 145,867 | 51,157 | 665,280 |
| 8 | 20,160 | 1,587,600 | 135,135 | 49,121 | 4,116,044 | 816,356 | 17,297,280 |
| 9 | 181,440 | 57,153,600 | 2,027,025 | 442,090 | 149,047,171 | 15,050,581 | 518,918,400 |
| 10 | 1,814,400 | 2,571,912,000 | 34,459,425 | 4,420,901 | 6,737,849,792 | 314,726,117 | 17,643,225,600 |
| 11 | 19,958,400 | 141,455,160,000 | 654,729,075 | 48,629,912 | 371,943,215,227 | 7,359,554,636 | 670,442,572,800 |

The names of the series (Sloane, 2003)

$K_{max}$ Triangular numbers
$CST_{max}$ Graphs on $N$ labeled nodes
$SST_{max}$ Bell numbers
$SOP_{max}$ Driving-point impedances of an $n$-terminal network
$AOP_{max}$ Stirling numbers of second kind
$ACS_{max}$ Double factorial numbers
$ISS_{max}$ *Unknown series*
$SIS_{max}$ *Unknown series*
$AIS_{max}$ Planted binary phylogenetic trees with $N$ labels
$CS_{max}$ Quadruple factorial numbers

## APPENDIX 4D: TABLES OF MAXIMUM NUMBER OF FEATURES FOR STRING AND STAR CONFIGURATIONS

**TABLE 4D.1**
**Maximum Number of Features in Disassembly Graphs**

| | Subassemblies | | States | Operations State Diagram | Operation AND/OR Graph |
|---|---|---|---|---|---|
| $N$ | $SUB_{string}$ (4.26) | $SUB_{star}$ (4.27) | $CST_{weak}$; $SST_{weak}$ (4.28); (4.29) | $SOP_{weak}$ (4.31) | $AOP_{string}$ (4.32) |
| 1 | 1 | 1 | 1 | — | — |
| 2 | 3 | 3 | 2 | 1 | 1 |
| 3 | 6 | 6 | 4 | 4 | 4 |
| 4 | 10 | 11 | 8 | 12 | 10 |
| 5 | 15 | 20 | 16 | 32 | 20 |
| 6 | 21 | 37 | 32 | 80 | 35 |
| 7 | 28 | 70 | 64 | 192 | 56 |
| 8 | 36 | 135 | 128 | 448 | 84 |
| 9 | 45 | 264 | 256 | 1,024 | 120 |
| 10 | 55 | 521 | 512 | 2,304 | 165 |
| 11 | 66 | 1,034 | 1,024 | 5,120 | 220 |

**TABLE 4D.2**
**Maximum Number of Sequences in Disassembly Graphs**

| | Sequential Complete | | AND/OR Complete | Sequential Incomplete | | AND/OR Incomplete |
|---|---|---|---|---|---|---|
| $N$ | $CSS_{string}$ (4.34) | $CSS_{star}$ (4.35) | $ACS_{string}$ (4.38) | $ISS_{string}$ (4.41) | $ISS_{star}$ (4.43) | $AIS_{string}$ (4.47) |
| 1 | 1 | 1 | 1 | 1 | 1 | 1 |
| 2 | 1 | 1 | 2 | 2 | 2 | 2 |
| 3 | 2 | 2 | 5 | 5 | 5 | 5 |
| 4 | 4 | 6 | 14 | 11 | 16 | 15 |
| 5 | 8 | 24 | 42 | 23 | 65 | 51 |
| 6 | 16 | 120 | 132 | 47 | 326 | 188 |
| 7 | 32 | 720 | 429 | 95 | 1,957 | 731 |
| 8 | 64 | 5,040 | 1,430 | 191 | 13,700 | 2,950 |
| 9 | 128 | 40,320 | 4,862 | 383 | 109,601 | 12,235 |
| 10 | 256 | 362,880 | 16,796 | 767 | 986,410 | 51,822 |
| 11 | 512 | 3,628,800 | 58,786 | 1,535 | 9,864,101 | 223,191 |

The names of the series (Sloane, 2003):

$SUB_{string}$ Triangular numbers
$AOP_{string}$ Tetrahedral (or pyramidal) numbers
$CSS_{star}$ Factorial numbers

$ACS_{\text{string}}$ Catalan numbers
$ISS_{\text{star}}$ Total number of arrangements of a set with $N$ elements
$AIS_{\text{string}}$ Binomial transform of Catalan numbers

Unlisted series, which are either trivial or are similar to others:

$K_{\text{weak}}$ $N-1$, not listed, see (4.25)
$CS_{\text{weak}}$ $N-1$, not listed, see (4.30)
$AOP_{\text{star}}$ Equals $SOP_{\text{weak}}$, see (4.33)
$SCS_{\text{weak}}$ Equals $CSS_{\text{star}}$, see (4.36)
$ACS_{\text{star}}$ Equals $CSS_{\text{star}}$, see (4.39)
$SIS_{\text{weak}}$ Equals $ISS_{\text{star}}$, see (4.44)
$AIS_{\text{star}}$ Equals $ISS_{\text{star}}$, see (4.48)

## REFERENCES

Abramowitz, M. and Stegun, I.A., 1968, *Handbook of Mathematical Functions*, New York: Dover Publications.

Aleksander, I., Farreny, H. and Ghallab, M., 1983, *Decision and Intelligence*. In P. Coiffet (ed), Robot technology series (London, UK: Kogan Page), **6**, chapters 6 and 7

Baldwin, D.F., Abell, T.E., Lui, M.M., De Fazio, T.L., and Whitney, D.E., 1991, An integrated computer aid for generating and evaluating assembly sequences for mechanical products. *IEEE Transactions on Robotics and Automation*, **7**, 78–94.

Ben-Arieh, D. and Kramer, B., 1994, Computer-aided process planning for assembly: generation of assembly operations sequence. *International Journal of Production Research*, **32**(3), 643–656.

Bonneville, F., Henrioud, J.M. and Bourjault, A., 1995, Generation of assembly sequences with ternary operations. *Proceedings of 1995 IEEE International Symposium on Assembly and Task Planning*, 245–249.

Boothroyd, G., Poli, C., and Murch, L.E., 1982, Automatic Assembly. In G. Boothroyd (ed.), *Manufacturing and Materials Processing*, Volume **6**. New York: Marcel Dekker. p. 280–283.

Bourjault, A., 1984, *Contribution à une approche méthodologique de l'assemblage automatisé: elaboration automatique des séquences opératoires*, (Contribution to a systematic approach of automatic assembly: automatic determination of operation sequences). Ph.D. Thesis, Besançon, France: Université de Franche-Comté, (in French).

Bratcu, A., 2001, *Détermination systématique des graphes de précédence et équilibrage des lignes d'assemblage* (Systematic determination of precedence graphs and assembly line balancing). Ph.D Thesis, Besançon, France: Université de Franche-Comté, (in French).

De Fazio, T.L. and Whitney, D.E., 1987, Simplified generation of all mechanical assembly sequences. *IEEE Journal of Robotics and Automation*, **RA-3**(6), 640–658.

Fox, B.R. and Kempf, K.G., 1985, Opportunistic scheduling for robotic assembly. *Proceedings of 1985 IEEE International Conference on Robotics and Automation*, 880–889.

Gottipolu, R.B. and Ghosh, K., 1997, Representation and selection of assembly sequences in computer-aided process planning. *International Journal of Production Research*, **35**(12), 3447–3465.

Gradshteyn, I.S. and Ryzhik, I.M., 1981, *Table of Integrals, Series, and Products*, New York: Academic Press.

Gungor, A. and Gupta, S.M., 2002, Disassembly line in product recovery. *International Journal of Production Research*, **40**(11), 2569–2589.

Henrioud, J.M., 1989, *Contribution à la conceptualisation de l'assemblage automatisée* (Contribution to the theory of automatic assembly). Ph.D. Thesis, Besançon, France: Université de Franche-Comté, (in French).

Hillier, F.S. and Lieberman, G.J., 2001, *Introduction to Operations Research*, 7th edition, New York: McGraw-Hill.

Homem De Mello, L.S. and Sanderson, A.C., 1990, AND/OR graph representation of assembly plans. *IEEE Transactions on Robotics and Automation*, **6**(2), 188–189.

Homem De Mello, L.S. and Sanderson, A.C., 1991, A correct and complete algorithm for the generation of mechanical assembly sequences. *IEEE Transactions on Robotics and Automation*, **7**(2), 228–240.

Krikke, H.R., Van Harten, A. and Schuur, P.C., 1998, On a medium term product recovery and disposal strategy for durable assembly products. *International Journal of Production Research*, **36**(1), 111–139.

Krikke, H.R., Van Harten, A. and Schuur, P.C., 1999, Business case Roteb: recovery strategies for monitors. *Computers and Industrial Engineering*, **36**(4), 739–757.

Lambert, A.J.D., 1997, Optimal disassembly of complex products. *International Journal of Production Research*, **35**(9), 2509–2523.

Lambert, A.J.D., 1999, Linear programming in disassembly/clustering sequence generation. *Computers and Industrial Engineering*, **36**, 723–738.

Lambert, A.J.D. and Gupta, S.M., 2002, Demand-driven disassembly optimisation for electronic consumer goods. *Journal of Electronics Manufacturing,* **11**(2), 121–135.

Lambert, A.J.D., 2002, Determining optimum disassembly sequences in electronic equipment. *Computers and Industrial Engineering,* **43**, 553–575.

Moore, K.E., Gungor, A., and Gupta, S.M., 2001, Petri net approach to disassembly process planning for products with complex AND/OR precedence relationships. *European Journal of Operations Research*, **135**, 428–449.

Prenting, T.O. and Battaglin, R.M., 1964, The precedence diagram: a tool for analysis in assembly line balancing. *The Journal of Industrial Engineering*, **15**(4), 208–213.

Relange, L. and Henrioud, J.M., 2001, Systematic determination of assembly state transition diagrams. *Proceedings of 4th IEEE International Symposium on Assembly and Task Planning*. 55–60.

Schweikard, A. and Schwarzer, F., 1997, General translational assembly planning. *Proceedings of the 1997 IEEE International Conference on Robotics and Automation*, 612–619.

Sloane, N.J.A., 2003, *The On-line Encyclopedia of Integer Sequences.* AT&T Research. Available on Internet via: http://www.research.att.com/~njas/sequences

Tovey, C.A., 2002, Tutorial on Computational Complexity. *Interfaces*, **32**(3), 30–61.

Veerakamolmal, P. and Gupta, S.M., 1998, Optimal analysis of lot size balancing for multi-products selective disassembly. *International Journal of Flexible Automation and Integrated Manufacturing*, **6**(3/4), 245–269.

Veerakamolmal, P. and Gupta, S.M., 1999, Analysis of design efficiency for the disassembly of modular electronic products. *Journal of Electronics Manufacturing*, **9**(1), 79–95.

Whitney, D.E., 2001, Assembly in the 21st Century. Oral presentation at the *International Symposium on Assembly and Task Planning.*

Wolter, J.D., 1992, A combinatorial analysis of enumerative data structures for assembly planning. *Journal of Design and Manufacturing*, **2**(2), 93–104.

Woo, T.C. and Dutta, D., 1991, Automatic disassembly and total ordering in three dimensions. *Journal of Engineering for Industry*, **113**, 207–213.

# 5 Geometrical Constraints and Precedence Relationships

## 5.1 INTRODUCTION

In the previous chapter we learned that the topological constraints significantly limit the number of possible disassembly sequences in a typical product. Constraints that further limit the number of possible sequences include geometrical and technical constraints. In this chapter we focus on geometrical constraints.

When a disassembly operation of separating a parent subassembly into two subassemblies takes place, it is inherently assumed that the subassembly that is detached is not obstructed and that a collision free path to infinity exists for the detachment to take place. This is referred to as the *detachability* condition. A related condition is the *movability* condition, which implies that a child subassembly can be moved an infinitesimal distance with respect to the remainder of the parent subassembly. Another condition, known as the *reachability* condition, requires that the subassembly that has to be detached should be accessible by an appropriate tool. Finally, the *stability* condition is necessary to make sure that the subassemblies do not fall apart spontaneously.

Reachability is tool dependent. Detachability requires the presence of some free, and thus an accessible surface on the subassembly that has to be removed. This enables at least some tool, such as a sucking disk, to be attached to the subassembly for detaching it. Stability is related to forces. A subassembly cannot fall apart spontaneously if no forces are present. Movability as an additional condition is relevant to nonmonotone products (see Figure 4.7). Movability is a weaker condition than detachability, because motion is a prerequisite for detachment. Stability and movability will be concisely discussed in chapter 6. A comprehensive study of this topic is beyond the scope of this book.

There are two approaches to detachability, namely, a general approach and a restricted approach. The *general approach* is to determine by human inspection whether or not two subassemblies are detachable. The process is supported by computer software, which is based on rules. An appropriate application of these rules restricts the amount of human inspections that is required for the construction of the AND/OR graph or the state diagram, that represent all the possible disassembly sequences. We will elaborate on this approach in this chapter.

The *restricted approach* is confined to 1-movable product. Apart from this, there is a restricted set of directions of motion (usually orthogonal). The information on movability and detachability can be condensed in interference matrices, which can be easily derived. Thus a fully automatic generation of the AND/OR graph and the disassembly sequences becomes possible. This circumvents the need for sophisticated software that simulates realistic motion but consumes a large amount of CPU time and is less flexible. The restricted approach will be addressed in chapter 6.

We make the following assumptions with respect to the general approach:

1. Forces are absent.
2. The components are rigid bodies, thus not deformable.

With these assumptions, both internal forces such as friction and elasticity and external forces such as gravity are excluded. Apart from this, the properties of the components and the connections between them are fully determined by the dimensions, the positions, and the orientations of the components. An important consequence of this assumption is that the disassembly processes are reversible, which implies that assembly can be considered as reversed disassembly under these assumptions. Therefore, inclusion of geometric constraints in disassembly studies offers an indispensable way to model the end-of life processing, maintenance or repair, and assembly optimization. This is also one of the characteristics that propelled the development of disassembly theory.

We will describe this theory but will also modify it, which will result in a widely applicable approach based on subassemblies rather than on connections. Standard theory starts with precedence relationships, which stem from task planning, as discussed in subsection 4.2.1. We will show that precedence relationships can be dealt with in the most systematic way if a subassembly-oriented rather than a connection-oriented approach is followed. It will be demonstrated that precedence relationships can be converted to selection rules, which apply to subassemblies rather than to operations, thus selecting those subassemblies that can be obtained by disassembly operations only. Once the feasible subassemblies are known, the nodes of the AND/OR graph are defined and, subsequently, the disassembly sequences can be determined. This method will be applied to several products.

One of the final sections of this chapter is devoted to disassembly precedence graphs. This instrument has not only been valuable as a precursor for the use of state diagrams and AND/OR graphs in disassembly, but it is frequently applied by various authors in disassembly studies. Apart from this, it has interesting applications in disassembly to order systems.

Once the AND/OR graph is subjected to geometrical constraints there may still be subassemblies and operations present that are unfavorable from a technical point of view. An important characteristic of dealing with *technical constraints* is that these may require a greater level of detail on the product's structure and physical properties, thus trivializing the assumptions on the absence of forces and rigidity of components. Note that the technical constraints do not simply depend on the product structure, but also on its surroundings such as the tools available and managerial factors such as costs. An important aspect of technical constraints is that they are normally different for disassembly and assembly. Thus, for example, when disassembly is used in assembly studies, technical constraints must be introduced only after reversing the disassembly graph that was earlier used in the disassembly study incorporating only the topological and geometrical constraints. However, some exceptions are occasionally possible (for example, at times some components are deformable such as electric cables).

In addition to the three types of constraints discussed so far—topological, geometric, and technical—*soft constraints* could also play a role here. These constraints are not as strict and stem from process control, scheduling, and other management issues. Soft constraints might be in the domain of complexity reduction, reliability, economy, environment, legislation, safety, logistics, ergonomics, ease, market demand, etc. Soft constraints are used for the preliminary selection of viable options from the set of options that is determined by the hard constraints. This is in fact a kind of a posteriori pruning procedure such as the ones described in the next subsection. After this, a procedure for determining the optimum disassembly (or assembly) sequence can be carried out; which is the subject of chapter 7.

Prior to the optimization procedure, we have to gauge an overview of all the possible disassembly sequences. Many authors have used this methodology. Despite various preferences articulated by different authors, we will agree with Homem De Mello and Sanderson (1990, 1991) that the AND/OR graph is an appropriate tool for representing disassembly sequences. Furthermore, due to a simplified and layered notation, it provides a clear depiction of all the possible disassembly operations and sequences. The combinatorial explosion that is observed by many authors and demonstrated in chapter 4, can be mitigated in many cases from practice where, due to topological and geometrical constraints, the number of possible sequences will be of the order of $O(N^2)$ rather than $O(e^N)$, where $N$ is the number of components. We have already noticed in the last chapter that even in the case of unconstrained products, the number of nodes and arcs in AND/OR graphs is far less than in the state diagrams (see appendix 4C).

## 5.2 EARLIER RESEARCH ON PRECEDENCE RELATIONSHIPS

### 5.2.1 Connection-Oriented Precedence Relationships

Bourjault (1984) was first to make use of precedence relationships in disassembly theory. We will discuss his way of thinking in a somewhat modified form, noticing that Bourjault's original work was based on a connection approach. This is further discussed in section 5.3. Bourjault's theory was aimed at automatically determining disassembly sequences, for the use in decision support systems.

Huang and Lee (1989, 1990, 1991) introduced the algebra for precedence relationships based on the predicates "must precede" and "no later than." Other authors have also worked on similar ideas. Homem De Mello and Sanderson (1990, 1991) formulated an algorithm for the determination of all possible disassembly sequences. De Fazio and Whitney (1987) and Baldwin et al. (1991) formulated connection-oriented precedence relationships according to the 2-type formalism that resulted in a distinction between "must precede" and "not until" operators. Some valuable rules, such as the *subset rule* and the *superset rule*, were also formulated. These rules enabled the formulation of an algorithm for determining a state diagram in a semi-automatic manner, by querying the user and avoiding redundancy. The authors demonstrated that using this method, the minimum number of queries was far less than the exponentially increasing number of queries that were needed to implement Bourjault's original method. The set of precedence relationships was automatically generated based on the answers of the queries. In contrast with Homem De Mello and Sanderson, these authors chose to use state diagrams rather than AND/OR graphs.

Waarts et al. (1992) described a semiautomatic sequence planner. Because the size of the problem using this approach increases considerably with the increase in *N*, a heuristic was proposed. The heuristic involved (1) the introduction of a predefined modular structure that enabled the treatment of a module that is composed of multiple components as a single super-component and (2) a manual pruning of the state diagram by excluding infeasible states and transitions based on both geometrical and technical considerations. In this approach, pruning proceeds in a rather intuitive manner. Authors such as Wilson and Rit (1990), and Wilson (1995) also designed software tools that returned a minimum number of queries. Each answer resulted in an additional precedence relationship or in a new possibility for disassembly.

A strictly formal logical framework that deals with connection-oriented precedence relationships is due to Seow and Devanathan (1994). Although interesting from a mathematical point of view, this formalism will not be discussed in this book, because its purpose can also be fulfilled by using more transparent methods discussed in this chapter. Finally, we mention the work of Erdos et al. (2001) who worked with connection-oriented precedence relationships aimed at automatic generation of disassembly AND/OR graphs.

### 5.2.2 Subassembly-Oriented Precedence Relationships

All the studies discussed above were based on connection-oriented approaches. Several other authors advocate the subassembly-oriented approach, in which the detachment of a subassembly rather than the disestablishment of a connection is considered the basic disassembly operation.

Wolter (1989) took a subassembly-oriented approach to precedence relationships, which he called "sequencing assertions" and "subassembly assertions." Lin and Chang (1993) presented a hybrid set of precedence relationships that referred to both the disestablishment of connections and the detachment of components as operators. Rajan and Nof (1996) also took a subassembly-oriented approach. With a method that is based on constraints in the direction of motion, the authors automatically derived a set of precedence relationships for each component in the assembly. Using Boolean algebra, reduction in the precedence relationships was obtained. This method will be discussed in chapter 6. Although elegant, there are two disadvantages of the method. First, the method is not well suited for complex 3-dimensional structures and second, the method does not exclude the generation of redundant precedence constraints. Further theory on subassembly-oriented precedence relationships is due to Lambert (2000, 2001, 2002ab). This approach makes the logical description simpler as it avoids the 2-type precedence relationships that were described in the preceding subsection.

We must remark here that the disassembly precedence graphs (Figure 4.2) are commonly based on subassembly-oriented approach (or component-oriented approach if components instead of subassemblies are involved) rather than on connection-oriented approach. Although we will continue to mention some aspects of the connection-oriented approach, we will mostly confine ourselves to the more transparent subassembly-oriented approach.

## 5.3 BOURJAULT'S METHOD

### 5.3.1 Introduction

We start by describing Bourjault's (1984) work, partly because the original work was reported in the French language (for an overview in the English language, see Bourjault, 1987) and partly because of its influence on subsequent research. We will follow the author's approach and see how to naturally arrive at state diagrams. Bourjault's work was mainly inspired by assembly problems where detecting possible assembly sequences without deadlocks was the principal objective. He noticed that certain freedom existed in the ordering of assembly operations, provided the constraints were not violated.

### 5.3.2 Primitive Tables

Consider a ballpoint pen that was extensively examined by Bourjault and has become an elementary test example for many researchers (see Figure 5.1).

The ballpoint pen consists of six components: body (A), head (B), cartridge (C), ink (D), button (E), and cap (F). The assembly process starts with these six disjoint components. It we start, for example, by attaching E and B to A, it is obvious that complete assembly cannot take place because of obstruction, as C and D can never be placed into their positions, because a deadlock results. In other words, a geometrical constraint obstructs further assembly. In order to avoid deadlocks, Bourjault studied the disassembly process, which he considered as reverse of assembly (however, we will see later that this assumption is only valid if the constraints are purely topological or geometric). This indeed implies that the assembly consists of rigid components and neither internal nor external forces play a role. A second fruitful idea was the introduction of formal reasoning, including graph theory, set theory, and the formulation of logical relationships. This included the introduction of the connection diagram (see subsection 3.2.9). The appearance of formal reasoning was stimulated by the replacement of human labor with robots. It was further motivated by the desire

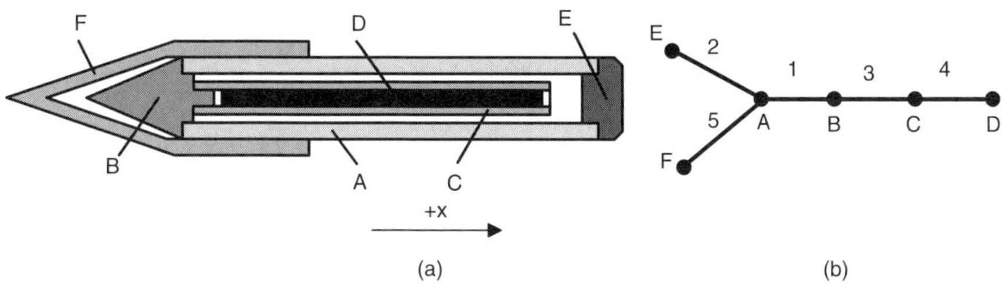


**FIGURE 5.1** Simple axially symmetric example: Bourjault's ballpoint. (a) assembly; (b) connection diagram. (Source: Bourjault, 1984.)

to automate the assembly planning and sequencing process, which coincided with the research on decision support systems (very popular during those years).

Let us take a closer look at the connection diagram in Figure 5.1(b). It has six nodes and five arcs (i.e., $N = 6$ and $K = 5$), which represent the components and connections, respectively. The ballpoint is weakly connected (see section 4.5). Therefore, from Expression 4.28 and Expression 4.29, there are 32 connection states and 32 subassembly states. In Bourjault's original work, a vector represented the connection states. For example, the subassembly ABF would be represented by $1\underline{234}5$, which indicated that only the connections 1 and 5 were established. For convenience, we will replace this with 15, thus only listing the connections that are established. For computers, a binary notation such as 10001 for the subassembly ABF would be even better, as it could be used for products with arbitrary $K$.

In general, this allows many possible sequences for assembly. Before selecting a preferable sequence, an inventory of all feasible sequences needs to be made. However, if we start with a single component and try to combine it with others in an attempt to reach the complete product, we may find that many of these potential sequences result in a deadlock. The desire to avoid deadlocks motivated Bourjault to use disassembly to list all feasible sequences.

Disassembly operations consist of the disestablishment of one or more connections, which is the same as the cut-set approach (see subsections 3.3.1 and 4.3.4). In the ballpoint example considered here (Figure 5.1), however, the cut-set is equivalent to the disestablishment of only a single connection as the ballpoint is weakly connected.

We define an operator $R_k$ that disestablishes connection $k$. We proceed with the search for all possible precursors of every disassembly operation. Precursors are those connection states on which $R_k$ can be applied. Bourjault used a table, which he called a *primitive table*. The columns of the table correspond to the operators $R_k$ and the rows are called *levels*, which correspond to the order in which operators are applied. The cells in the table are filled with precursors. In the ballpoint example, only the operators $R_2$ or $R_5$ can be applied to the initial state 12345, which appears in the corresponding columns of the first level (see Table 5.1). Obviously, the operator $R_j$ can only be applied to a connection state that has the connection $j$ established. As opposed to Bourjault's original approach,

**TABLE 5.1**
**Primitive Table for the Ballpoint of Figure 5.1 (Excluding Technical Constraints)**

| Operator | $R_1$ | $R_2$ | $R_3$ | $R_4$ | $R_5$ |
|---|---|---|---|---|---|
| Level | | | | | |
| 1 | | 12345 | | | 12345 |
| 2 | 1234 | 1234 | 1345 | 1345 | 1345 |
| 3 | 134 | 234 | 134 or 135 or 234 | 134 or 145 or 234 | 135 or 145 |
| 4 | 13 or 14 | 23 or 24 | 13 or 23 or 34 | 14 or 24 or 34 | 15 |
| 5 | 1 | 2 | 3 | 4 | |

we restrict ourselves to *geometric* constraints only. Bourjault also included some technical constraints, which interfered with the reversibility of assembly and disassembly. We will apply technical constraints separately.

It is clear from Figure 5.1(a) that, if connection 5 is established, the operator $R_1$ cannot be applied because of a geometric constraint. We will denote this condition as $R_5 \rightarrow R_1$. Here, the symbol $\rightarrow$ means "must precede," which corresponds to the notation in disassembly precedence graphs. Because only two operators can be applied on the assembly at the first level, only two different connection states, viz., 1234 and 1345 will appear in level 2. We have to investigate which operator can be applied to each of these two states. This will determine the level 2 elements. It can be seen from these that only connection states 134, 135, and 234 are possible at the 3rd level. Once again we have to investigate which operators can be applied to these states to determine the level 3 elements. Proceeding along these lines, we obtain the primitive table (Table 5.1).

Since this primitive table is only restricted to geometric constraints, the sequences derived from here are reversible and can be applied to both assembly and disassembly.

In Bourjault's original presentation, a clear distinction between geometric and technical constraints was not made. An example of a technical constraint for the *assembly* sequence of the ballpoint is that the ink can be put in the cartridge only if the head is applied to the cartridge first. This, therefore, implies that connection 4 can only be established after connection 3 is established. This is a technical condition as the ink is not a rigid body, but rather deformable and subjected to forces such as adhesion. Since we are studying the assembly process via a disassembly process, we have to revert this condition. This leaves us with a somewhat artificial constraint, which says that connection 3 can only be disestablished if connection 4 is disestablished first, i.e., $R_4 \rightarrow R_3$.

Due to this additional constraint, the connection state 1345 disappears from the cell that corresponds to the operator $R_3$ at the second level. Other levels are also similarly affected. A connection-oriented precedence relationship $R_i \rightarrow R_j$ can thus be reformulated in a connection-oriented selection rule, stating that all the connection states with $i$ established and $j$ disestablished are infeasible. This can be concisely denoted by

$$i \text{ not } j \tag{5.1}$$

Table 5.2(a) depicts the modified primitive table, with the assembly technical constraint included. Reverting this table results in the feasible assembly sequences.

We have already mentioned that, in general, the technical constraints are not reversible. If the objective is *disassembly* instead of assembly in the ballpoint example, it is clear that the ink can only be removed from the cartridge if the head is removed first. This results in the precedence relationship $R_3 \rightarrow R_4$. This means that those connection states with 3 established and 4 not established should be removed from Table 5.1. By doing this, we end up with Table 5.2(b).

### 5.3.3 Bourjault's Tree

Next, we illustrate how assembly and disassembly trees are constructed from the primitive tables.

**TABLE 5.2(a)**
**Primitive Table for the Ballpoint of Figure 5.1 (Reverse Assembly)**

| Operator | $R_1$ | $R_2$ | $R_3$ | $R_4$ | $R_5$ |
|---|---|---|---|---|---|
| Level | | | | | |
| 1 | | 12345 | | | 12345 |
| 2 | 1234 | 1234 | | 1345 | 1345 |
| 3 | 134 | 234 | 135 | 134 or 234 | 135 |
| 4 | 13 | 23 | 13 or 23 | 34 | 15 |
| 5 | 1 | 2 | 3 | | |

**TABLE 5.2(b)**
**Primitive Table for the Ballpoint of Figure 5.1 (Disassembly)**

| Operator | $R_1$ | $R_2$ | $R_3$ | $R_4$ | $R_5$ |
|---|---|---|---|---|---|
| Level | | | | | |
| 1 | | 12345 | | | 12345 |
| 2 | 1234 | 1234 | 1345 | | 1345 |
| 3 | 134 | 234 | 134 or 234 | 145 | 145 |
| 4 | 14 | 24 | 34 | 14 or 24 | 15 |
| 5 | 1 | 2 | | 4 | |

The *assembly tree* is derived from Table 5.2(a) by starting from the highest-level (5th level) and moving up to level 1. We note that at the 5th level, we can start by establishing connection 1, 2, or 3. If we have connection 1 established, we proceed with the next operation, which can be found in level 4 of the table. This might result in either 13 or 15. Subsequently, we move upward in the table, and depict the resulting connection states as in the graph of Figure 5.2.

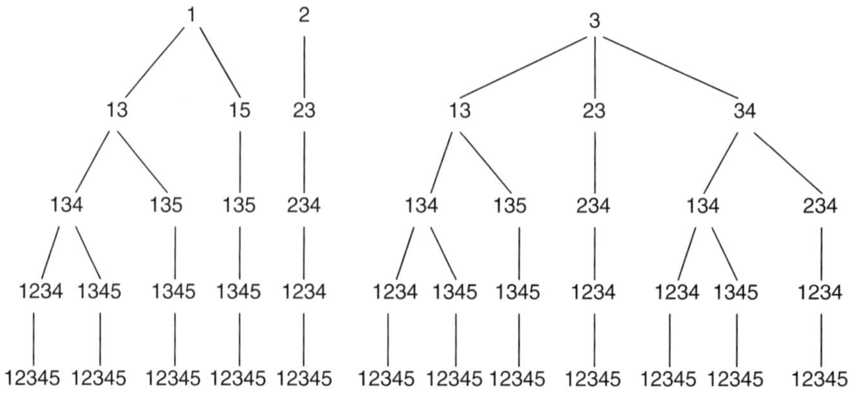


**FIGURE 5.2** Assembly tree or Bourjault's tree for the ballpoint, derived from Table 5.2(a).

This type of representation is called *Bourjault's tree* (note that a slightly different notation from Bourjault's original presentation is used here). It is a directed graph (although arrows are not shown) where one proceeds in the downward direction. Each assembly sequence is represented here by a linear subgraph. A node corresponds to a connection state and an arc to an assembly operation.

As is clear from Figure 5.2, there are 12 assembly sequences possible. We also note that several subsequences in the tree appear multiple times. The connection state 13, e.g., appears twice. Obviously, the complete subtree that is associated with it, also appears twice. This feature will be exploited later for reducing the tree.

The *disassembly tree* is derived from Table 5.2(b) by starting from level 1 and moving toward highest level (5th level) for we are starting here with the complete product. At level 1, two different operators can be applied to the product, which results in two branches. Proceeding along similar lines as in the assembly case, we obtain the tree that is depicted in Figure 5.3(a). The connection state ∅ represents an empty state (i.e, no connections are established), which corresponds to the set of disjoint components.

### 5.3.4 Reduction of Bourjault's Tree

Note that in drawing the disassembly tree of Figure 5.3(a), if a connection state appears two or more times, we have only shown the subtree the first time and encircled it the subsequent times while dropping the corresponding subtree. This results in a *reduced tree* that still contains the complete information of the original version of Bourjault's tree. For depicting the complete tree, the corresponding subtrees could simply be copied at the encircled states.

Once the reduced tree is obtained, merging the identical connection states can further modify it. This actually results in the connection-state diagram of Figure 5.3(b). This type of notation was first applied by De Fazio and Whitney (1987) (see subsection 4.2.3).

From the connection diagram in Figure 5.1(b) it follows that every feasible connection state corresponds to a subassembly-state. This is depicted in the state diagram of Figure 5.3(c), which is topologically identical to Figure 5.3(b), but with the nodes representing subassembly states. The connection state 234, for instance, corresponds to the subassembly state AE,BCD,F.

### 5.3.5 Number of Queries

We observe that, even in the simple example of the ballpoint, searching for all the possible precursors of a given operation is cumbersome and could lead to errors. With $K$ connections, we have a theoretical maximum of $2^K$ connection states. For every state, we have to answer the question whether or not one of the $n$ different operators can be applied to the state. Since on an average only half of the connections are established in a given state, it results in a theoretical maximum of

$$K \cdot 2^{K-1} \tag{5.2}$$

queries that have to be answered for constructing the primitive table. A typical example of such a query is: can connection 2 be disestablished from the state 1235?

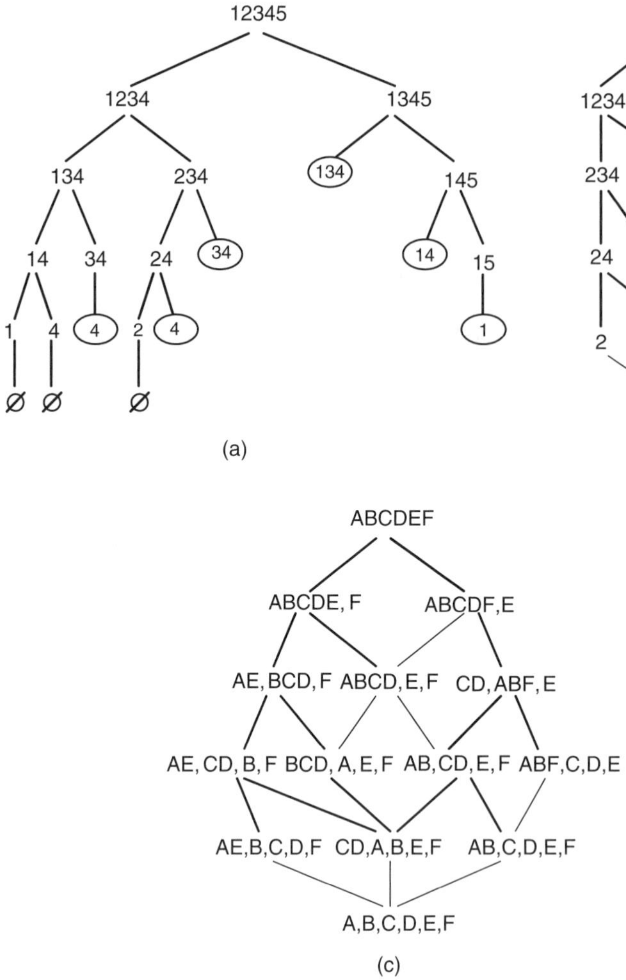

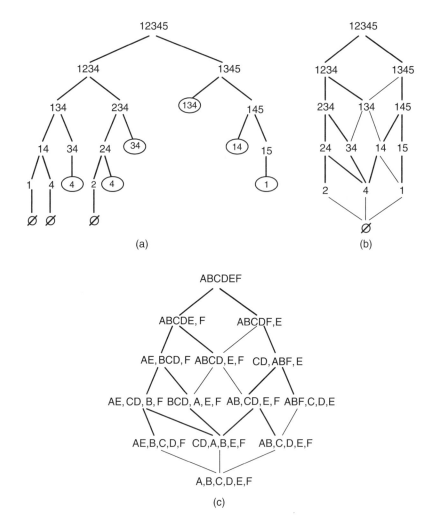


**FIGURE 5.3** Disassembly of Bourjault's ballpoint. (a) Reduced tree; (b) connection-state diagram; (c) subassembly-state diagram.

Although a simple "yes" or "no" is sufficient for each query, this still results in a time consuming task, which becomes virtually unmanageable with increasing value of $K$. Fortunately, there are methods that strongly reduce this expense. In the sequel, we will discuss the tools that are available to considerably reduce the problem size.

## 5.4 THE CUT-SET METHOD

### 5.4.1 Connection-Oriented Approach: Superset and Subset Rules

Although Bourjault's method was a breakthrough, it was impracticable for most of the real life assemblies. We discussed a number of modifications in the preceding subsection. De Fazio and Whitney (1987) dealt with the exploding number of queries and reduced the maximum number of queries to 2*K*. Using their approach, for every connection, one has to list the following:

1. The connections that have to be disestablished prior to disestablishing a particular connection.
2. The connections that can only be disestablished if the connection under review was disestablished beforehand.

However, answering each individual query appears complex. Even a simple case such as that of Figure 5.1 cannot be appropriately studied in this way. This becomes even more intricate if one or more meshes are present in the connection diagram, because this causes the simultaneous disestablishment of multiple connections for some specific cut-sets. This is known as the *mesh-closure rule* or *loop-closure rule* (Baldwin et al., 1991). This rule states that if there are meshes in the connection diagram, two simultaneous disestablishments of connections have to be performed for every mesh.

### Example 1

The mesh-closure rule can be demonstrated using the assembly of Figure 5.4. There are three simple meshes in this example: ABD, ACD, BCD. The apparent fourth mesh ABC is a composite of other three and is not counted as a separate mesh. The presence of three meshes causes the presence of three additional connections apart from the $N - 1$ connections that should have been present if the product were weakly connected. Starting with the initial product, one can only proceed with the simultaneous disestablishment of (1 and 2 and 6), or (1 and 3 and 5), or (2 and 3 and 4) corresponding to the removal of A, B, and C, respectively.

### Example 2

The removal of B from the initial assembly in Figure 5.4 is related to two meshes, viz., ABD and BCD. Therefore, three simultaneous disestablishments have to take place. Once B is removed, the removal of A is related to only one mesh, viz., ACD and thus two simultaneous disestablishments have to take place, i.e., those of 2 and 6.

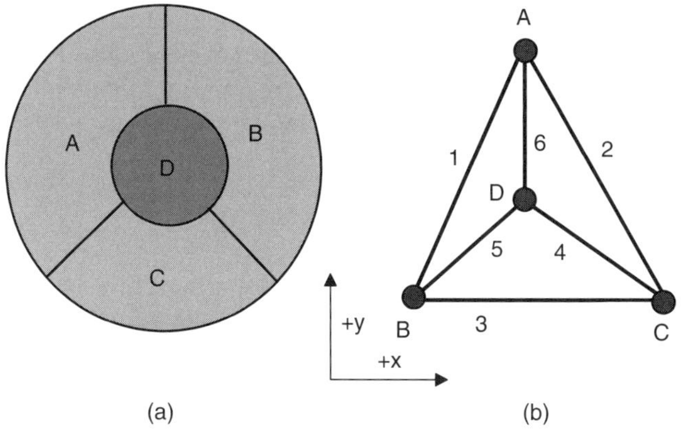


**FIGURE 5.4** Simple two-dimensional example: (a) assembly; (b) connection diagram.

Due to the complexity in answering the $2K$ queries, Baldwin et al. (1991) found alternative methods to reduce the number of queries and limit the queries to those that could be answered with simple "yes" or "no." For this purpose, two further rules were formulated.

The first of these is the *superset rule*. This rule states that if a particular component cannot be removed from a given subassembly, the addition of more components to that subassembly does not change this situation. The second rule is the *subset rule*. This rule states that if a particular child subassembly can be detached from a given subassembly, it can also be removed from a subassembly that consists of a subset of the components of the original subassembly, provided it contains all the components of the particular child subassembly.

**EXAMPLE 3**

Component D in Figure 5.4 cannot be removed from ABD because of geometrical reasons. Therefore, it can also not be removed from ABCD, which is a superset of ABD. Component A in Figure 5.4 can be removed from ABCD. Therefore, it can also be removed from subset subassemblies ABD or AB or AC or AD or ACD.

A modified cut-set method queries every possible cut-set, feasible or not. A cut-set is feasible if:

1. Both resulting subassemblies are connected.
2. Detachment of one of the resulting subassemblies is possible.

The maximum number of cut-sets that need to be investigated in an unconstrained product depicted in the form of a state diagram and an AND/OR graph, are given by the Expression 4.7 and Expression 4.9, respectively.

### 5.4.2 SUBASSEMBLY-ORIENTED APPROACH

Before proceeding along these lines, we will depart from the connection approach and shift to the subassembly approach, which is slightly different from the method that was presented by Baldwin et al. (1991). This does not influence the number of queries, but it facilitates responding to the queries and, above all, greatly simplifies deriving the expressions for precedence relationships. Note that we prefer to use the AND/OR graph instead of the state diagram.

**EXAMPLE 4**

Consider the ballpoint example depicted in Figure 5.1. We systematically carry out a search for all cut-sets, starting with those that can be applied to the complete product. We note that, if no restrictions were present, the product depicted in the form of a state diagram or an AND/OR graph, would potentially have 856 or 301 operations to be investigated respectively. Every subassembly that is encountered is subjected to *a posteriori* automatic check on topological feasibility, using the method that was presented in subsection 4.6.1. We proceed as follows. We start with the complete

**TABLE 5.3(a)**
**Sequential Cut-Sets of ABCDEF**

| Cut-Set | Topological Check | Geometrical Check | Reason for Geometrical Rejection | New Feasible Cut-Set |
|---|---|---|---|---|
| A,BCDEF | reject | — | — | — |
| B,ACDEF | reject | — | — | — |
| C,ABDEF | reject | — | — | — |
| D,ABCEF | select | reject | BE not D | — |
| E,ABCDF | select | select | — | E,ABCDF |
| F,ABCDE | select | select | — | F,ABCDE |

product ABCDEF and list all the sequential cut-sets and apply the topology check (Table 5.3(a)).

We observe that the first three cut-sets in Table 5.3(a) are rejected due to topological infeasibility (see Figure 5.1(b)). The fourth cut-set D,ABCEF is topologically feasible and is therefore selected and subjected to the following query:

*Is the cut-set ABCDEF → D,ABCEF geometrically feasible?*

Clearly, the answer to this query is NO (see Figure 5.1(a)). This results in the rejection of the cut-set D,ABCEF because we see that D cannot be removed due to the presence of both B and E. To translate this in the "connection" language is not always transparent and is therefore prone to errors. Fortunately, we can use a subassembly-oriented *precedence relationship* as follows:

$$\text{Detachment of B or E} \rightarrow \text{Detachment of D} \tag{5.3}$$

The arrow → means "must precede." The "or" operator expresses that at least one of the two components (B, E) must be detached prior to the detachment of D.

Relationships such as Expression 5.3 can simply be transformed into a *selection rule*. It follows straight from Expression 5.3 that there cannot exist any subassembly that includes both B and E but not D. This selection rule can be written as follows:

$$\text{BE not D} \tag{5.4}$$

A set of such selection rules can be applied to quickly weed out cut-sets based on geometric constraints. This is a consequence of the superset rule. It implies that not only subassembly BE is geometrically infeasible, but all other subassemblies such as ABE, BCE, BEF, ABCE that include B and E but do not include D are also geometrically infeasible regardless of their topological properties.

**TABLE 5.3(b)**
**Parallel Cut-Sets of ABCDEF**

| Cut-Set | Topological Check | Geometrical Check | Reason for Geometrical Rejection | New Feasible Cut-Set |
|---|---|---|---|---|
| AB,CDEF | reject | — | — | — |
| AC,BDEF | reject | — | — | — |
| AD,BCEF | reject | — | — | — |
| AE,BCDF | reject | — | — | — |
| AF,BCDE | reject | — | — | — |
| BC,ADEF | reject | — | — | — |
| BD,ACEF | reject | — | — | — |
| BE,ACDF | reject | — | — | — |
| BF,ACDE | reject | — | — | — |
| CD,ABEF | select | reject | BE not D | — |
| CE,ABDF | reject | — | — | — |
| CF,ABDE | reject | — | — | — |
| DE,ABCF | reject | — | — | — |
| DF,ABCE | reject | — | — | — |
| EF,ABCD | reject | — | — | — |
| ABC,DEF | reject | — | — | — |
| ABD,CEF | reject | — | — | — |
| ABE,CDF | reject | — | — | — |
| ABF,CDE | reject | — | — | — |
| ACD,BEF | reject | — | — | — |
| ACE,BDF | reject | — | — | — |
| ACF,BDE | reject | — | — | — |
| ADE,BCF | reject | — | — | — |
| ADF,BCE | reject | — | — | — |
| AEF,BCD | select | reject | AF not B | — |

Next on the list in Table 5.3(a) is the cut-set E,ABCDF. Since this is topologically feasible, a query similar to the one in the above cut-set returns an answer YES. That means this is geometrically feasible as well. Analogously, the cut-set F,ABCDE is also both topologically and geometrically feasible.

Note that, due to the subset rule, many cut-sets such as E,AB, E,AC,...; E,ABC, ... E,ABCD; ... E,BCDF are also feasible because E,ABCDF is feasible. This way we can determine the feasibility for a large number of cut-sets automatically without any human intervention!

Next, we check all the parallel cut-sets of ABCDEF (see Table 5.3(b)). It results in no feasible cut-sets.

We note the following. Many cut-sets are *a priori* rejected due to topological constraints. Some of these, such as AD,BCEF would have also been rejected because of selection rule BE not D. The cut-set CD,ABEF, although topologically feasible, is rejected because of selection rule BE not D. The only other query that had to be checked for geometrical feasibility was AEF,BCD. This cut-set is infeasible (see Figure 5.1(a)), which results in an additional selection rule, i.e., AF not B.

So far, we have two selection rules (viz., BE not D; and AF not B), a list of two feasible cut-sets (viz., E,ABCDF; and F,ABCDE) and thus, the subassemblies ABCDE, ABCDF, E, and F.

Next, the cut-sets of the feasible subassemblies with five components (viz., ABCDE and ABCDF) are investigated (see Table 5.3(c)). We notice again that many of the cut-sets are topologically infeasible because at least one of the two resulting

**TABLE 5.3(c)**
**Cut-Sets for Subassemblies with Five Components**

| Cut-Set | Topological Check | Geometrical Check | Reason for Geometrical Rejection | New Feasible Cut-Set |
|---|---|---|---|---|
| A,BCDE | reject | — | — | — |
| B,ACDE | reject | — | — | — |
| C,ABDE | reject | — | — | — |
| D,ABCE | select | reject | BE not D | — |
| E,ABCD | select | select[a] | — | — |
| AB,CDE | reject | — | — | — |
| AC,BDE | reject | — | — | — |
| AD,BCE | reject | — | — | — |
| AE,BCD | select | select | — | AE,BCD |
| BC,ADE | reject | — | — | — |
| BD,ACE | reject | — | — | — |
| BE,ACD | reject | — | — | — |
| CD,ABE | select | reject | BE not D | — |
| CE,ABD | reject | — | — | — |
| DE,ABC | reject | — | — | — |
| A,BCDF | reject | — | — | — |
| B,ACDF | reject | — | — | — |
| C,ABDF | reject | — | — | — |
| D,ABCF | select | select | — | D,ABCF |
| F,ABCD | select | select[a] | — | — |
| AB,CDF | reject | — | — | — |
| AC,BDF | reject | — | — | — |
| AD,BCF | reject | — | — | — |
| AF,BCD | select | reject | AF not B | — |
| BC,ADF | reject | — | — | — |
| BD,ACF | reject | — | — | — |
| BF,ACD | reject | — | — | — |
| CD,ABF | select | select | — | CD,ABF |
| CF.,ABD | reject | — | — | — |
| DF,ABC | reject | — | — | — |

[a] Selected because of the previously selected cut-set and the subset rule.

**TABLE 5.3(d)**
**Cut-Sets for Subassemblies with Four Components**

| Cut-Set | Topological Check | Geometrical Check | Reason for Geometrical Rejection | New Feasible Cut-Set |
|---|---|---|---|---|
| A,BCD | select | select[a] | — | — |
| B,ACD | reject | — | — | — |
| C,ABD | reject | — | — | — |
| D,ABC | select | select[a] | — | — |
| AB,CD | select | select[a] | — | — |
| AC,BD | reject | — | — | — |
| AD,BC | reject | — | — | — |
| A,BCF | reject | — | — | — |
| B,ACF | reject | — | — | — |
| C,ABF | select | select[a] | — | — |
| F,ABC | select | select[a] | — | — |
| AB,CF. | reject | — | — | — |
| AC,BF | reject | — | — | — |
| AF,BC | select | reject | AF not B | — |

[a] Selected because of the previously selected cut-set and the subset rule.

subassemblies is disjoint. The first topologically feasible cut-set that is encountered is D,ABCE. However, it is rejected due to the selection rule BE not D. The next topologically feasible cut-set is E,ABCD. This is selected due to the existence of the feasible cut-set E,ABCDF and the subset rule, and so on.

Therefore, so far, we have two selection rules (viz., BE not D; and AF not B), a list of five feasible cut-sets (viz., E,ABCDF, F,ABCDE, AE,BCD, D,ABCF, and CD,ABF) and thus, the subassemblies: ABCDE, ABCDF, ABCD, ABCF, ABF, BCD, AE, CD, D, E, and F.

Next, the cut-sets of the feasible subassemblies with four components (viz., ABCD and ABCF) are investigated (see Table 5.3(d)).

We note from Table 5.3(d) that there are six cut-sets that are topologically feasible. Five of these are geometrically feasible and are selected due to the existence of a previously feasible cut-set and the subset rule. Finally, AF,BC is topologically feasible but is rejected due to the selection rule AF not B. The list of subassemblies is further supplemented with ABC, AB, A, and C.

Therefore, so far, we have two selection rules (viz., BE not D; and AF not B), a list of five feasible cut-sets (viz., E,ABCDF; F,ABCDE; AE,BCD; D,ABCF; and CD,ABF) and, the subassemblies: ABCDE; ABCDF; ABCD; ABCF; ABC; ABF; BCD, AB; AE; CD; A; C; D; E; and F.

The results of the cut-set analysis of the subassemblies with three components (viz., ABC, ABF, and BCD) are listed in Table 5.3(e).

We note from Table 5.3(e) that there are six cut-sets that are topologically feasible. Five of these are geometrically feasible and are selected due to the existence

**TABLE 5.3(e)**
**Cut-Sets for Subassemblies with Three Components**

| Cut-Set | Topological Check | Geometrical Check | Reason for Geometrical Rejection | New Feasible Cut-Set |
|---|---|---|---|---|
| A,BC | select | select[a] | — | — |
| B,AC | reject | — | — | — |
| C,AB | select | select[a] | — | — |
| A,BF | reject | — | — | — |
| B,AF | select | reject | AF not B | — |
| F,AB | select | select[a] | — | — |
| B,CD | select | select[a] | — | — |
| C,BD | reject | — | — | — |
| D,BC | select | select[a] | — | — |

[a] Selected due to the previously selected cut-set and the subset rule.

of a previously feasible cut-set and the subset rule. Finally, B,AF is topologically feasible but is rejected due to the selection rule AF not B. The list of subassemblies is further supplemented with BC; and B.

Therefore, we have a list of two selection rules (viz., BE not D, and AF not B), a list of five feasible cut-sets (viz., E,ABCDF; F,ABCDE; AE,BCD; D,ABCF; and CD,ABF) and, a list of subassemblies (including the complete product) as follows: ABCDEF; ABCDE; ABCDF; ABCD; ABCF; ABC; ABF; BCD; AB; AE; BC, CD; A; B; C; D; E; and F.

Note that only seven queries were required here to check geometrical feasibility. Five of them are returned with a YES answer while two returned with a NO answer (which led to two selection rules). Therefore, from 301 possible cut-sets, only 84 had to be checked, of which only 7 had to be checked via human intervention and the results for others were automatically derived using the subset and the superset rules. We will demonstrate in chapter 6 that in some special cases with a restricted number of directions of motion, the problem could be solved without any human intervention.

### 5.4.3 Technical Constraints

In the previous subsection, the technical constraints were not addressed. This situation corresponds to the disassembly AND/OR graph that is shown in Figure 5.5(a). However, technical constraints can be easily included *a posteriori.* A technical constraint for disassembling the ballpoint that was discussed in subsection 5.3.2 can be stated as follows. The removal of the ink D from cartridge C is not possible before the head B is removed. This means that a subassembly that includes B and C but does not include D, should be rejected. This results in the following selection rule:

$$\text{BC not D} \tag{5.5}$$

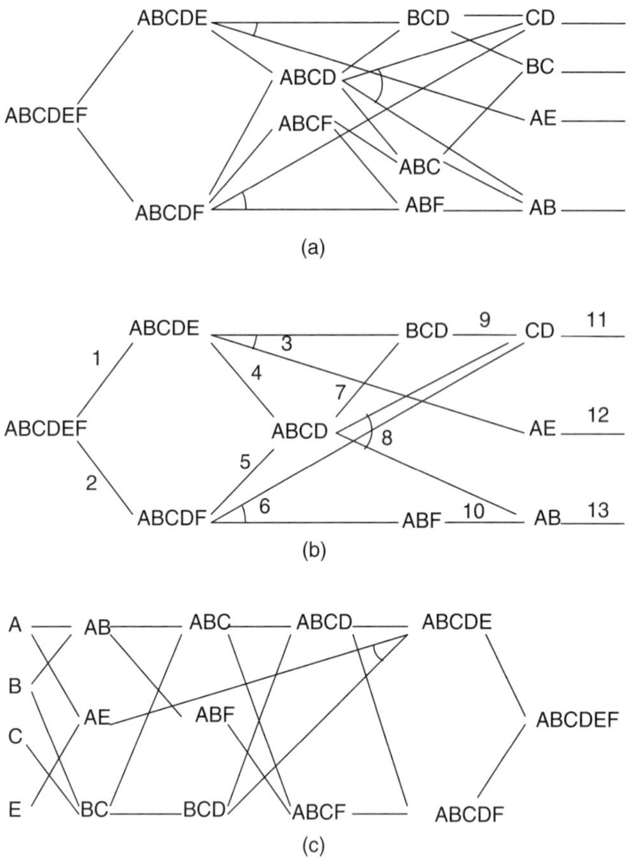


**FIGURE 5.5** AND/OR graphs for the ballpoint of Figure 5.1. (a) Disassembly graph with topological and geometrical constraints only; (b) disassembly graph with topological, geometrical, and technical constraints (note that the numbers refer to the enumeration of the operations); (c) assembly graph with topological, geometrical and technical constraints.

This can be added to the list of selection rules, and applied to the already available list of subassemblies (see example 4). As a result, ABCF, ABC, and BC have to be removed. Thus, the new list of subassemblies (that includes the topological, geometric, and technical constraints) is as follows: ABCDEF, ABCDE, ABCDF, ABCD, ABF, BCD, AB, AE, CD, A, B, C, D, E, and F. The corresponding AND/OR graph is shown in Figure 5.5(b). Note that the corresponding subassembly-state diagram is shown in Figure 5.3(c).

If we were to consider assembly instead of disassembly, we would have a different technical constraint. In this case, the head B has to be mounted on cartridge C before the ink D can be placed in it. This means that every subassembly that includes C and D, but not B, should be excluded from the list, which results in the selection rule:

$$\text{CD not B} \tag{5.6}$$

Using this, we see that only subassembly CD is excluded from the original list of subassemblies. Thus, the new list of subassemblies (for the assembly case) is as follows: ABCDEF, ABCDE, ABCDF, ABCD, ABCF, ABC, ABF, BCD, AB, AE, BC, A, B, C, D, E, and F. The resulting AND/OR graph is shown in Figure 5.5(c).

### 5.4.4 Redundancy

Theoretically, one can derive many selection rules. The derivation of all possible rules will become extremely laborious for complicated products. Even in the simple case of the ballpoint example (Figure 5.1), the following 10 selection rules can be formulated:

AF not B, BE, not A, BE not C, BE not D, CF not B, DF not B, EF not A, EF not B, EF not C, and EF not D.

However, we used only two of them (viz., AF not B and BE not D). The others are redundant. The number of selection rules tends to increase exponentially with the increase in *N*. The set of possible directions of motion is another factor that adds to the complexity of deriving all the possible selection rules.

The choice of a selection rule based on a negative response of a query is not unequivocal. Intuitively, we attempt to choose the one that is as discerning as possible. This way, the number of required selection rules would be kept at a minimum.

### 5.4.5 The Sequential Method

Although the number of cut-sets to be investigated can be considerably reduced by using a systematic approach, it could nevertheless remain a large number. We can further reduce the number of cut-sets to be investigated by limiting the search to sequential disassembly. The main drawback of the sequential method is that it cannot handle *m-disassemblable* products. *m*-disassemblability refers to the necessity of performing at least one parallel disassembly operation that results in a smallest child subassembly of *m* components, in order to completely disassemble a product or obtain a particular subassembly. An example of a 2-disassemblable product (2-dimensional) is given in Figure 5.6. Here, it is necessary to perform three parallel disassembly

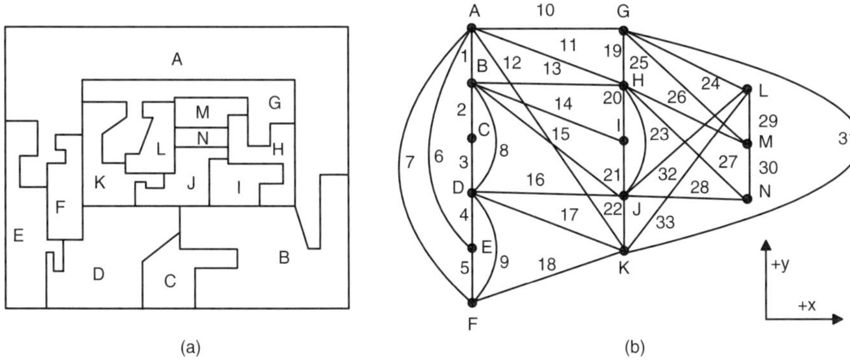


**FIGURE 5.6** Moderately complex two-dimensional example: (a) assembly; (b) connection diagram. (Source: Chen et al., 1997.)

operations to completely disassemble the product. These result in the smallest child subassemblies BC, AF (or GK), and HI, respectively. We will further discuss this type of disassembly in the next chapter.

If a product is *m*-disassemblable, the value of *m* is generally modest, such as in the case of Figure 5.6 (where $m = 2$). Once the product is *m*-disassembled, each subassembly can be considered a module. This module can be treated as a product with much fewer components, which can in most cases, be treated as the usual 1-disassemblable product. If not, it could be further broken down into smaller modules and so on. Many products, however, are 1-disassemblable and therefore are directly suited for the sequential method.

### Example 5

Consider the ballpoint example again (Figure 5.1). This is a 1-disassemblable product. We repeat the procedure of example 4; this time using sequential cut-sets only.

The cut-sets in Table 5.3(a) are all sequential and thus this table remains unchanged, which results in the selection rule BE not D and the feasible cut-sets E,ABCDF and F,ABCDE. The investigation of the cut-sets in Table 5.3(b) is skipped because none of them is sequential. This implies that the selection rule AF not B is not discovered at this time. We proceed with a modification of Table 5.3(c) so as to only include cut-sets for sequential subassemblies with five components, viz., ABCDE and ABCDF (see Table 5.4(a)).

We notice that one query is returned, which results in the addition of D,ABCF to the list of feasible cut-sets. The list of subassemblies with four components here is the same as in example 4. We therefore proceed with a modification of Table 5.3(d) so as

**TABLE 5.4(a)**
**Cut-Sets for Subassemblies with Five Components, Sequential Method**

| Cut-Set | Topological Check | Geometrical Check | Reason for Geometrical Rejection | New Feasible Cut-Set |
|---|---|---|---|---|
| A,BCDE | reject | — | — | — |
| B,ACDE | reject | — | — | — |
| C,ABDE | reject | — | — | — |
| D,ABCE | select | reject | BE not D | — |
| E,ABCD | select | select[a] | — | — |
| A,BCDF | reject | — | — | — |
| B,ACDF | reject | — | — | — |
| C,ABDF | reject | — | — | — |
| D,ABCF | select | select | — | D,ABCF |
| F,ABCD | select | select[a] | — | — |

[a] Selected due to the previously selected cut-set and the subset rule.

**TABLE 5.4(b)**
**Cut-Sets for Subassemblies with Four Components, Sequential Method**

| Cut-Set | Topological Check | Geometrical Check | Reason for Geometrical Rejection | New Feasible Cut-Set |
|---|---|---|---|---|
| A,BCD | select | select | — | A,BCD |
| B,ACD | reject | — | — | — |
| C,ABD | reject | — | — | — |
| D,ABC | select | select[a] | — | — |
| A,BCF | reject | — | — | — |
| B,ACF | reject | — | — | — |
| C,ABF | select | select | — | C,ABF |
| F,ABC | select | select[a] | — | — |

[a] Selected because of the previously selected cut-set and the subset rule.

to only include cut-sets for sequential subassemblies with four components, viz., ABCD and ABCF (see Table 5.4(b)).

Note that two new feasible cut-sets are added here, viz., A,BCD and C,ABF. We therefore proceed with a modification of Table 5.3(e) so as to only include cut-sets for sequential subassemblies with three components, viz., ABC, ABF and BCD (see Table 5.4(c)).

It is interesting to note that the selection rule AF not B does not appear until this phase. Apart from this, an additional feasible cut-set B,CD also appears here.

**TABLE 5.4(c)**
**Cut-Sets for Subassemblies with Three Components, Sequential Method**

| Cut-Set | Topological Check | Geometrical Check | Reason for Geometrical Rejection | New Feasible Cut-Set |
|---|---|---|---|---|
| A,BC | select | select[a] | — | — |
| B,AC | reject | — | — | — |
| C,AB | select | select[a] | — | — |
| A,BF | reject | — | — | — |
| B,AF | select | reject | AF not B | — |
| F,AB | select | select[a] | — | — |
| B,CD | select | select | — | B,CD |
| C,BD | reject | — | — | — |
| D,BC | select | select[a] | — | — |

[a] Selected because of the previously selected cut-set and the subset rule.

Note that the number of queries is now eight. This is because we do not have the benefit of parallel disassembly, which restricts the application of the subset rule. The number of cut-sets that need to be investigated has been reduced from 84 to 33. More importantly, the number of cut-sets investigated does not increase exponentially with $N$, but will rather increase as $O(N^2)$.

A weakness of the sequential method is that, as of now, no method is available that can check the *completeness* of the set of subassemblies that is found in this way. In the sequel, we will see some examples in which one or more feasible subassemblies fail. We will also discuss some methods for partly circumventing this problem. A second but less serious shortcoming is that the parallel disassembly operations need to be included. However, once the full list of subassemblies is established, a list of potential parallel operations can be found automatically via a systematic search for triplets $S_1$, $S_2$, and $S_3$ of subassemblies in this list, which are considered sets of components. The union of $S_2$ and $S_3$ should be equal to $S_1$, and the intersection of $S_2$ and $S_3$ should equal the empty set $\emptyset$. We will call such a triplet a *complementary triplet*. In the ballpoint example (Figure 5.1) with no technical constraints included, this search results in the following parallel operations:

$$\text{ABCDE} \rightarrow \text{AE,BCD}$$

$$\text{ABCDF} \rightarrow \text{CD,ABF}$$

$$\text{ABCD} \rightarrow \text{AB,CD}$$

*A posteriori* check of the first two operations confirms their feasibility. The feasibility of the third operation follows from the second one, via the subset rule.

### 5.4.6 Reverse Method

For products with a large number of components $N$ where $m$-disassemblability cannot be avoided, we will have to carry out a search for many more cut-sets. Although several of these may be rejected after an automatic selection on topological constraints is carried out, the required CPU time for checking all possibilities would nevertheless increase exponentially as $N$ increases. However, if the value of $m$ were known, we could confine ourselves to those cut-sets, in which the smallest child subassembly has a maximum of $m$ components. This way, the size of the problem would increase polynomially instead of exponentially, namely, as $O(N^m)$. For the ballpoint example (Figure 5.1), with $m = 2$, the number of cut-sets is reduced by only ten (see Table 5.3(b)). Of course, for products with larger values of $N$, the problem size would have reduced even more.

We could further confine the CPU time by making use of the so-called reverse method. Using this method, one simply starts with the 2-subassemblies, checks these for geometric feasibility, merges these, checks the resulting 3-subassemblies, merges these with the 2-subassemblies, etc. The typical number of checks that has to be carried out has the order of magnitude $O(K^2)$, which is polynomial. We will demonstrate this with the ballpoint example of Figure 5.1.

This method proceeds as follows. First we list all the 2-subassemblies, regardless of their geometric feasibility. There are $K$ such subassemblies. For the ballpoint case,

**TABLE 5.5**
**Reverse Method, Applied to Ballpoint Example of Figure 5.1**

| $n = 2$ | | $n = 3$ | | $n = 4$ | | $n = 5$ | |
|---|---|---|---|---|---|---|---|
| AB | + | ABC | + | ABCD | + | ABCDE | + |
| AE | + | ABE | BE not D | ABCF | + | ABCDF | + |
| AF | AF not B | ABF | + | | | | |
| BC | + | BCD | + | | | | |
| CD | + | | | | | | |

$K = N - 1 = 5$. These are listed in the first column of Table 5.5. (Note that only five subassemblies need to be listed here, instead of the combinatorial maximum of 15).

Next, we will check the 2-subassemblies for geometric feasibility. If a subassembly is feasible, a + symbol is placed next to it, otherwise, a selection rule is placed next to it. The set of selection rules guarantees that no deadlocks are encountered. These also reduce the size of the problem. We note that four of the five 2-subassemblies are feasible. The subassembly AF is geometrically infeasible and leads to the selection rule AF not B (see Table 5.5.).

Subsequently, the 2-subassemblies are merged. We would like to stress that the list of topologically feasible subassemblies with $n = 2$ contains all the information that is in the connection diagram. Therefore, checks on topological feasibility are automatically carried out if an $m$-subassembly is merged with a member of a 2-subassembly that has one component in common.

AB can be merged with any 2-subassembly that includes either A or B, which results in ABC, ABE, and ABF (note that AF is subjected to a selection rule and ABF is the only possible merger that can be obtained from this subassembly). Merging of AE gives no additional result. BC can be merged with either AB or CD. However, BCD is the only additional result, and so on. This results in four subassemblies out of a combinatorial maximum of 20. These should be checked against the already obtained selection rules. Geometric feasibility check results in the rejection of ABE, and the addition of an associated selection rule, BE not D (see Table 5.5).

Proceeding along these lines, a complete list of feasible subassemblies is generated (compare this to Figure 5.5(a)). Next, we have to check the cut-sets, which is restricted to the complementary triplets only. These result in a list of six feasible cut-sets as follows:

$$\text{ABCDEF} \rightarrow \text{F,ABCDE}$$
$$\text{ABCDEF} \rightarrow \text{E,ABCDF}$$
$$\text{ABCDE} \rightarrow \text{AE,BCD}$$
$$\text{ABCDF} \rightarrow \text{D,ABCF}$$
$$\text{ABCDF} \rightarrow \text{CD,ABF}$$
$$\text{ABCDF} \rightarrow \text{AF,BCD}$$

The reverse method offers a means for the straightforward generation of all the feasible subassemblies. In most cases, the list remains modest. Carrying out a large

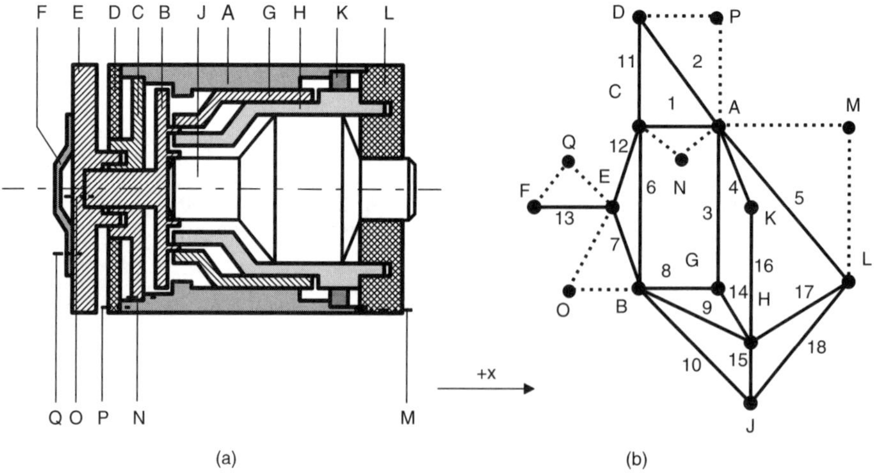


**FIGURE 5.7** Moderately complex axially symmetric example: an automatic transmission (AFI). (a) Assembly drawing and (b) connection diagram. (Source: De Fazio and Whitney, 1987.)

number of automatic checks on potential feasible cut-sets, or the generation of all topologically feasible subassemblies, are avoided.

## 5.5 A MODERATELY COMPLEX EXAMPLE

We apply the two methods (sequential and reverse) to a moderately complex case that was considered by De Fazio and Whitney (1987) (see Figure 5.7). It is a schematized automatic transmission that will be referred to as the assembly from industry (AFI). Initially, we will ignore the fasteners M through Q and only consider the remaining 11 components, A through L. Its assembly drawing is presented in Figure 5.7(a) and the connection diagram in Figure 5.7(b). The dashed lines in Figure 5.7(b) refer to the fasteners. We will consider the fasteners later. It can be seen from Table 4C.1 and Table 4C.2 that the maximum number of potential features in a problem with $N = 11$ are prohibitively large. However, the case is constrained and a full list of subassemblies can be found. With $K = 17$, the index of complexity $\alpha$ is about 3.1 (see Expression 4.23), which is within the range of 2 to 4 as was mentioned in subsection 4.4 for most products in practice.

### 5.5.1 Sequential Method

Applying the sequential method provides us with provisional lists of selection rules, feasible cut-sets, and subassemblies. We will only give an overview of the calculations and present the three provisional lists. In Table 5.6, starting with the complete product, the detachment of every component is considered. If the component is not detachable, a selection rule is returned. If it is detachable, a feasible cut-set is recorded.

From Table 5.6, it is clear that two feasible cut-sets (leading to two subassemblies each with $n = 10$ and two subassemblies each with $n = 1$) and nine selection rules

**TABLE 5.6**
**Sequential Cut-Sets Applied to the AFI**

| | |
|---|---|
| ABCDEFGHJKL → A,BCDEFGHJKL | BG not A |
| ABCDEFGHJKL → B,ACDEFGHJKL | AC not B |
| ABCDEFGHJKL → C,ABDEFGHJKL | AD not C |
| ABCDEFGHJKL → D,ABCEFGHJKL | CE not D |
| ABCDEFGHJKL → E,ABCDFGHJKL | BF not E |
| ABCDEFGHJKL → F,ABCDEGHJKL | F,ABCDEGHJKL |
| ABCDEFGHJKL → G,ABCDEFHJKL | BH not G |
| ABCDEFGHJKL → H,ABCDEFGJKL | GJ not H |
| ABCDEFGHJKL → J,ABCDEFGHKL | HL not J |
| ABCDEFGHJKL → K,ABCDEFGHJL | AL not K |
| ABCDEFGHJKL → L,ABCDEFGHJK | L,ABCDEFGHJK |

are generated. Next, each of the new subassemblies with $n = 10$ is analyzed in the same way. Subsequently, all new feasible subassemblies are analyzed in the order of decreasing $n$. This results in 15 feasible cut-sets and 13 selection rules (see Table 5.7). Additionally, a provisional list of subassemblies is generated (see Table 5.8).

Note that the product is 1-disassemblable and complete disassembly is possible with sequential operations only. However, additional subassemblies do exist but cannot be obtained via sequential disassembly only.

### 5.5.2 Reverse Method

Applying the procedure outlined in subsection 5.4.6 to the AFI (Figure 5.7), a list of potential geometrically feasible subassemblies is generated via merging. With

**TABLE 5.7**
**Feasible Cut-Sets and Selection Rules for the Sequential Approach of the AFI**

| | |
|---|---|
| ABCDEFGHJKL → F,ABCDEGHJKL | BG not A |
| ABCDEFGHJKL → L,ABCDEFGHJK | AC not B |
| ABCDEFGHJK → J,ABCDEFGHK | AD not C |
| ABCDEFGHJK → K,ABCDEFGHJ | CE not D |
| ABCDEGHJKL → E,ABCDGHJKL | BF not E |
| ABCDEFGHK → H,ABCDEFGK | BH not G |
| ABCDGHJKL → D,ABCGHJKL | GJ not H |
| ABCDEFGK → G,ABCDEFK | HL not J |
| ABCGHJKL → C,ABGHJKL | AL not K |
| ABGHJKL → B,AGHJKL | BK not A |
| ABCDEF → A,BCDEF | AH not G |
| AGHJKL → A,GHJKL | BD not C |
| BCDEF → B,CDEF | DF not E |
| GHJKL → G,HJKL | |
| GHJKL → K,GHJL | |

**TABLE 5.8**
**Feasible Subassemblies for the AFI of Figure 5.7**

| $n = 10$ | $n = 9$ | $n = 8$ | $n = 7$ | $n = 6$ | $n = 5$ | $n = 4$ | $n = 3$ | $n = 2$ |
|---|---|---|---|---|---|---|---|---|
| ABCDEFGHJK | ABCDEFGHJ | ABCDEFGH | ABCDEFG | ABCDEF | ABCDE | ABCD | ABC | AG |
| ABCDEGHJKL | ABCDEFGHK | ABCDEFGK | ABCDEFK | ABCDEG | ABCDG | ABCG | ABG | AK |
| | ABCDEGHJK | ABCDEGHJ | ABCDEGH | ABCDEK | ABCDK | ABCK | AGH | BC |
| | ABCDGHJKL | ABCDEGHK | ABCDEGK | ABCDGH | ABCGH | ABGH | AGK | CD |
| | | ABCDGHJK | ABCDGHJ | ABCDGK | ABCGK | ABGK | BCD | GH |
| | | ABCGHJKL | ABCDGHK | ABCGHJ | ABGHJ | AGHJ | CDE | HJ |
| | | | ABCGHJK | ABCGHK | ABGHK | AGHK | GHJ | HK |
| | | | ABGHJKL | ABGHJK | AGHJK | BCDE | GHK | JL |
| | | | | AGHJKL | BCDEF | CDEF | HJK | |
| | | | | | GHJKL | GHJK | HJL | |
| | | | | | | GHJL | | |
| | | | | | | HJKL | | |

**TABLE 5.9**
**Reverse Method, Applied to the AFI**

| $n = 2$ | | $n = 3$ | | $n = 4$ | | $n = 5$ | | $n = 6$ | |
|---|---|---|---|---|---|---|---|---|---|
| AC | AC not B | ABC | + | ABCD | + | ABCDE | + | ABCDEF | + |
| AD | AD not C | ABG | + | ABCG | + | ABCDG | + | ABCDEG | + |
| AG | + | ABH | AH not G | ABCK | + | ABCDK | + | ABCDEK | + |
| AK | + | AGH | + | ABGH | + | ABCGH | + | ABCDGH | + |
| AL | AL not J | AGK | + | ABGK | + | ABCGK | + | ABCDGK | + |
| BC | + | AJL | AL not H | AGHJ | + | ABGHJ | + | ABCGHJ | + |
| BE | BE not C | BCD | + | AGHK | + | ABGHK | + | ABCGHK | + |
| BG | BG not A | CDE | + | BCDE | + | AGHJK | + | ABGHJK | + |
| BH | BH not A | GHJ | + | CDEF | + | AGHJL | AL not K | AGHJKL | + |
| BJ | BJ not H | GHK | + | GHJK | + | BCDEF | + | | |
| CD | + | HJK | + | GHJL | + | GHJKL | + | | |
| CE | CE not D | HJL | + | HJKL | + | | | | |
| EF | + | | | | | | | | |
| GH | + | | | | | | | | |
| HJ | + | | | | | | | | |
| HK | + | | | | | | | | |
| HL | HL not J | | | | | | | | |
| JL | + | | | | | | | | |

only 12 selection rules, we arrive at the same result as in Table 5.8 (see Table 5.9). Note that in Table 5.9, we have only presented a partial list (up to $n = 6$). The rest of the list (from $n = 5$ to $n = 2$) is identical to that in Table 5.8. It should be remarked that no combinatorial explosion took place here! The exercise could even be performed manually within just a few minutes. If automated, however, the CPU time is negligible.

It is noteworthy that an additional subassembly appears here, namely EF, which is the only one that cannot be obtained using sequential disassembly only. In still more complicated products, multiple additional subassemblies could be found when compared to the sequential method.

Using the above results, an AND/OR graph is constructed. The operations are based on the search for complementary triplets. The following additional feasible cut-sets are detected:

$$\text{ABCDEFGHJKL} \rightarrow \text{JL,ABCDEFGHK}$$

$$\text{ABCDEFGHJKL} \rightarrow \text{HJL,ABCDEFGK}$$

$$\text{ABCDEFGHJKL} \rightarrow \text{CDEF,ABGHJKL}$$

$$\text{ABCDEFGHJKL} \rightarrow \text{GHJL,ABCDEFK}$$

$$\text{ABCDEFGHJKL} \rightarrow \text{HJKL,ABCDEFG}$$

$$\text{ABCDEFGHJKL} \rightarrow \text{BCDEF,AGHJKL}$$

$$\text{ABCDEFGHJKL} \rightarrow \text{GHJKL,ABCDEF}$$

The AND/OR graph is presented in four layers in Figures 5.8(a) through 5.8(d). It demonstrates the usefulness of this graphical formalism even in complex cases. Note that in the AND/OR graphs there are slightly less operations and subassemblies than we listed above. This is because we have included the influence of fasteners M through Q in the graphs. Their influence will be elucidated in the next subsection.

### 5.5.3 Fasteners

Fasteners have been discussed previously in chapter 3. Here, we will illustrate their influence on the list of feasible subassemblies and consequently on the list of feasible actions presented in the previous subsection.

We assume that the fasteners M through Q, in Figure 5.7, are screw connections and are quasi-components, i.e., they do influence the precedence constraints but do not significantly add to the materials contents of the subassemblies. We, therefore, do not include these fasteners in the subassemblies, which only consist of the components A through L. This is why we could first solve the problem without fasteners, as was done in previous subsections, and later add additional constraints affected by the fasteners, as will be done here. We proceed as follows.

Let $R_Z$ be the operator that detaches component Z. The fastener connections are subjected to the following precedence relationships that can easily be seen from Figure 5.7(b):

$$\begin{aligned} &R_M \rightarrow R_A \text{ or } R_L \\ &R_N \rightarrow R_A \text{ or } R_C \\ &R_O \rightarrow R_B \text{ or } R_E \\ &R_P \rightarrow R_A \text{ or } R_D \\ &R_Q \rightarrow R_E \text{ or } R_F \end{aligned} \tag{5.7a}$$

Note that the operators $R_M$ and $R_Q$ are unconstrained while the others are subjected to the following precedence relationships:

$$\begin{aligned} &R_D \rightarrow R_N \\ &R_F \rightarrow R_O \\ &R_E \rightarrow R_P \end{aligned} \tag{5.7b}$$

Combination of these two sets of precedence relationships results in:

$$\begin{aligned} &R_D \rightarrow R_A \text{ or } R_C \\ &R_F \rightarrow R_B \text{ or } R_E \\ &R_E \rightarrow R_A \text{ or } R_D \end{aligned} \tag{5.7c}$$

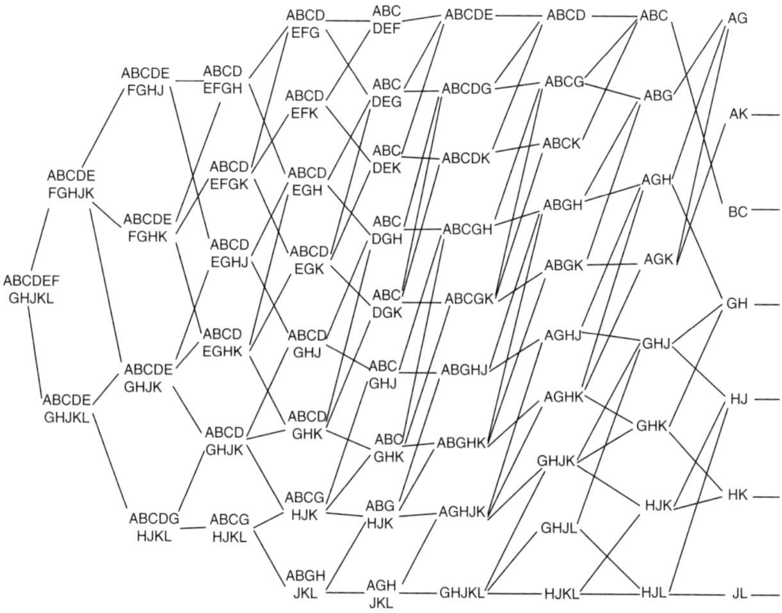


(a)

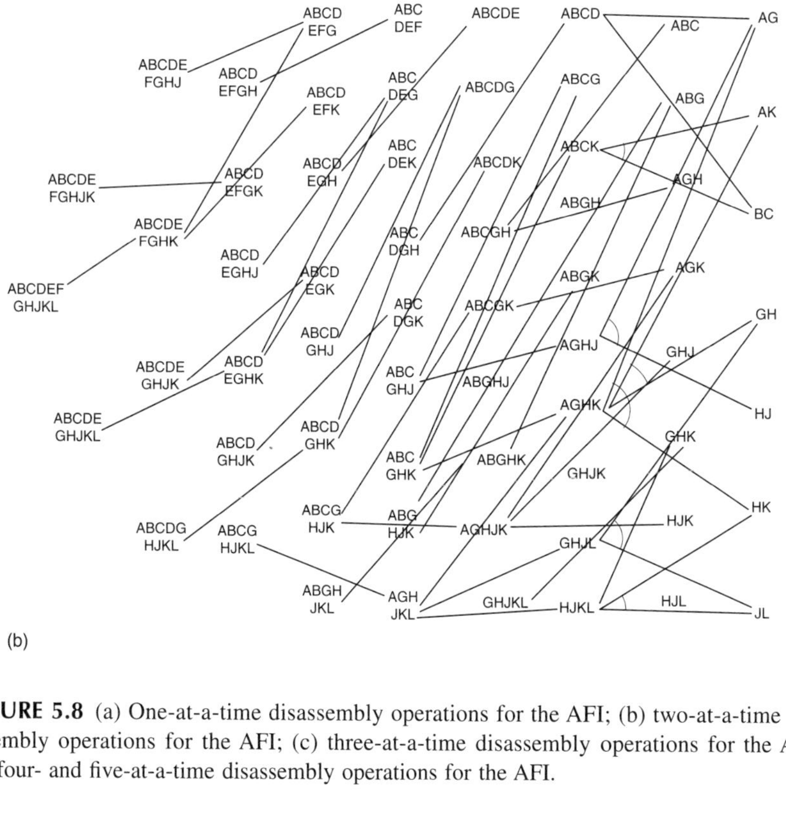


(b)

**FIGURE 5.8** (a) One-at-a-time disassembly operations for the AFI; (b) two-at-a-time disassembly operations for the AFI; (c) three-at-a-time disassembly operations for the AFI; (d) four- and five-at-a-time disassembly operations for the AFI.

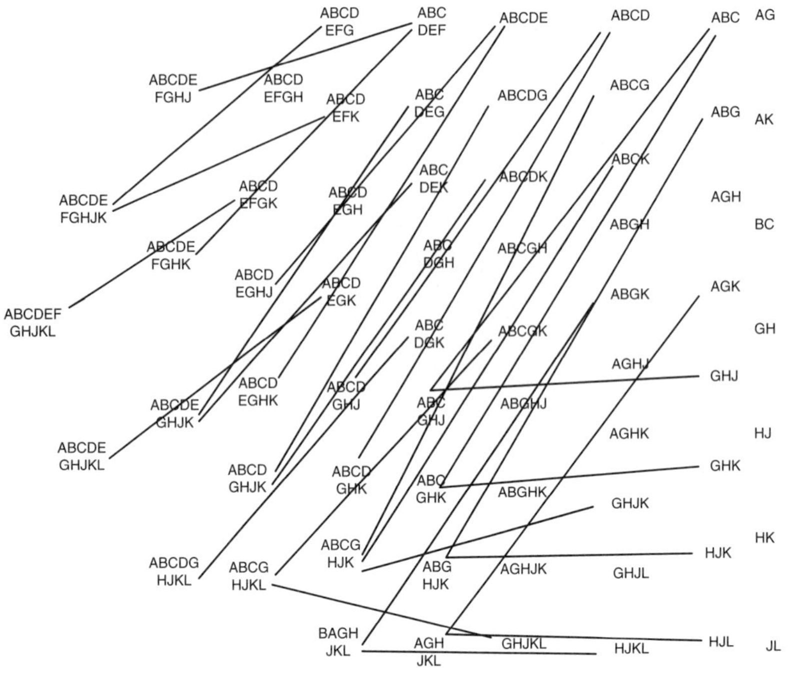

(c)

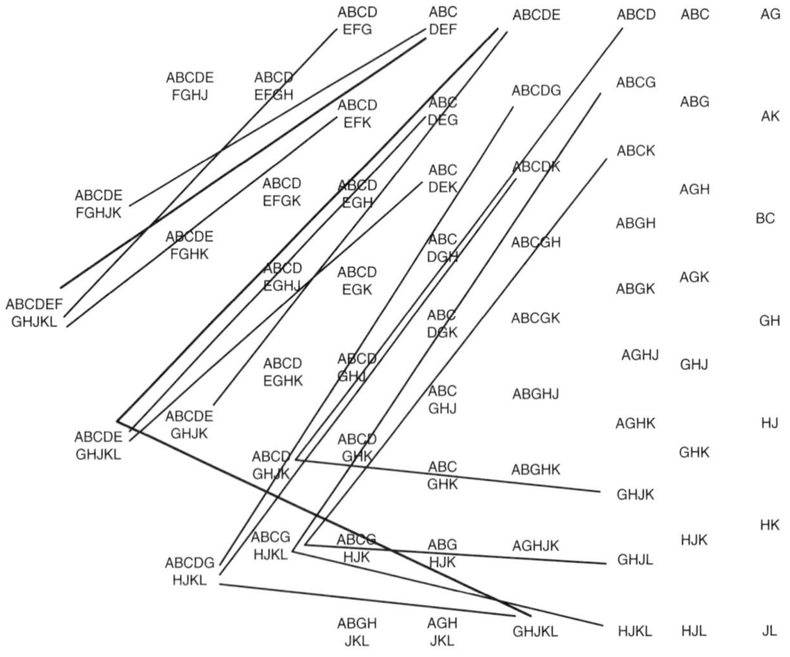

(d)

**FIGURE 5.8** (Continued)

This can be reformulated into a set of six selection rules:

| | | |
|---|---|---|
| AD not C | CD not A | |
| BF not E | EF not B | (5.7d) |
| AE not D | DE not A | |

From these selection rules, AD not C was previously found. CD not A and AE not D are not active, for these do not reject any additional subassembly from Table 5.8. The selection rule CD not A wipes out BCDEF, BCDE, CDEF, CDE, and CD. The selection rule EF not B rejects EF and CDEF (which has already been rejected). The selection rule DE not A rejects BCDEF, BCDE, CDEF, and CDE (all of these subassemblies have already been rejected).

Thus, the inclusion of fasteners results in two extra selection rules and the rejection of six additional subassemblies, which also reduces the list of feasible cutsets. As was mentioned in the last subsection, the AND/OR graphs in Figure 5.8 already include the influence of fasteners.

## 5.6 *m*-DISASSEMBLABLE PRODUCTS

We now present and comment on the results of the previous methods applied to the 2-D product (given in Figure 5.6) that was used, in a different context, by Chen et al. (1997). With $N = 14$ and $K = 33$, this product is slightly more complex than the AFI discussed in the previous section.

It is clear that the sequential method cannot be used here as no single component can be detached from the complete product (because the product is 2-disassemblable). However, the reverse method can be used. It produces a complete list of feasible subassemblies up to $n = 10$ with the exception of AFGHIJKLMN (see Table 5.10). However, AFGHIJKLMN can be found by merging AFGJKLMN and HI, which do not have a component in common. Once AFGHIJKLMN is found, one can subsequently obtain AEFGHIJKLMN and ADEFGHIJKLMN by using the conventional reverse method. Merging the last subassembly with BC leads to the complete product. Feasible subassemblies that cannot be detected via the reverse method only are indicated with an asterisk in Table 5.10. There are four such subassemblies.

We can never be sure if we have the full set of feasible subassemblies or not. Therefore, the study should consider the complete product and investigate which subassemblies of 2 components can be detached from the product. From Table 5.10, it is clear that only 16 of such subassemblies potentially exist. However, only the detachment of BC is feasible, resulting in subassembly ADEFGHIJKLMN. Further detachment of HI leaves us with ADEFGJKLMN, which can be sequentially detached up to AFGK. This can be disassembled only via AF,GK.

A drawback of the reverse method is that we cannot make use of the subset rule. This means that every subassembly that is encountered needs to be manually checked on geometric feasibility. As is observable in the table, in most cases a YES answer

**TABLE 5.10**
**Geometrically Feasible Subassemblies of the Product of Figure 5.6, Reverse Method**

| $n = 2$ | | $n = 3$ | | $n = 4$ | | $n = 5$ | |
|---|---|---|---|---|---|---|---|
| AB | AB not G | AEF | + | ADEF | AD not G | ABFGK | BK not J |
| AE | + | AFK | FK not G | AFGK | + | AEFGK | + |
| AF | + | BCD | + | BCDH | + | AFGKL | + |
| AG | AG not K | BCH | + | BCDI | + | BCDHI | + |
| AH | AH not G | BCI | + | BCDJ | + | BCDHM | + |
| AK | AK not F | BCJ | + | BCHI | + | BCDHN | + |
| BC | + | BHI | + | BCHM | + | BCDIJ | + |
| BD | BD not C | BHM | + | BCHN | + | BCDJN | + |
| BH | + | BHN | + | BCIJ | + | BCHIJ | + |
| BI | + | BIJ | + | BCJN | + | BCHIM | + |
| BJ | + | BJN | + | BHIJ | + | BCHIN | + |
| CD | + | CDJ | + | BHIM | + | BCHMN | + |
| DE | DE not F | DEF | + | BHIN | + | BCIJN | + |
| DF | DF not E | DIJ | + | BHMN | + | BCJMN | + |
| DJ | + | DJN | + | BIJN | + | BHIJM | + |
| DK | DK not J | GKL | + | BJMN | + | BHIJN | + |
| EF | EF not A | GLM | + | CDIJ | + | BHIMN | + |
| FK | FK not A | HIJ | + | CDJN | + | BIJMN | + |
| GH | GH not M | HIM | + | DHIJ | + | CDHIJ | + |
| GK | + | HIN | + | DIJN | + | CDIJN | + |
| GL | + | HMN | + | DJKN | JK not G | CDJMN | + |
| GM | GM not L | IJN | + | DJMN | + | DHIJM | + |
| HI | + | JMN | + | GJKL | GJ not M | DHIJN | + |
| HJ | HJ not I | | | GHLM | HL not I | DIJMN | + |
| HM | + | | | GKLM | + | GHILM | HL not J |
| HN | + | | | GLMN | + | GJKLM | JM not N |
| IJ | + | | | HIJM | + | GKLMN | + |
| JK | JK not L | | | HIJN | + | HIJMN | + |
| JL | JL not K | | | HIMN | + | | |
| JN | + | | | IJMN | + | | |
| KL | KL not G | | | | | | |
| LM | LM not G | | | | | | |
| MN | + | | | | | | |

| $n = 6$ | | $n = 7$ | | $n = 8$ | | $n = 9$ | |
|---|---|---|---|---|---|---|---|
| AEFGKL | + | AEFGKLM | + | AEFGKLMN | + | AEFGJKLMN | + |
| AFGJKL | + | AFGKLMN | + | AFGJKLMN | + | BGHIJKLMN | + |
| AFGKLM | + | BCDHIJM | + | BCDHIJMN | + | DGHIJKLMN | + |
| BCDHIJ | + | BCDHIJN | + | CDGJKLMN | + | | |
| BCDHIM | + | BCDHIMN | + | GHIJKLMN | + | | |
| BCDHIN | + | BCDIJMN | + | | | | |
| BCDHMN | + | BCHIJMN | + | | | | |
| BCDIJN | + | BGJKLMN | BG not H | | | | |
| BCDJMN | + | CDHIJMN | + | | | | |
| BCHIJM | + | DGJKLMN | + | | | | |
| BCHIJN | + | GIJKLMN | GI not H | | | | |
| BCHIMN | + | | | | | | |
| BCIJMN | + | | | | | | |
| BHIJMN | + | | | | | | |
| CDHIJM | + | | | | | | |
| CDHIJN | + | | | | | | |
| CDIJMN | + | | | | | | |
| DHIJMN | + | | | | | | |
| GJKLMN | + | | | | | | |

*(Continued)*

**TABLE 5.10 (Continued)**

| *n* = 10 | | *n* = 11 | | *n* = 12 |
|---|---|---|---|---|
| ADEFGJKLMN | + | ABFGHIJKLMN | AB not C * | ADEFGHIJKLMN* |
| AFGHIJKLMN | * | ACDEFGJKLMN | AC not B | |
| BCGHIJKLMN | + | AEFGHIJKLMN | * | |
| CDGHIJKLMN | + | BCDGHIJKLMN | + | |
| *n* = 13 | | *n* = 14 | | |
| — | — | ABCDEFGHIJKLMN * | | |

*Cannot be detected by reverse method only.

is returned, which means that not much more queries are put to the user than there are feasible subassemblies. In case of geometric infeasibility, the "NO" answer should be accompanied by an additional selection rule (see Table 5.10).

Starting with the complete list of subassemblies, the AND/OR graph can be constructed. The search for complementary triplets is essential here. We also find all the parallel operations. Using the subset rule, we have to complete a list with feasible cut-sets, as in the preceding section 5.5. For the 2-D example, following is the list of feasible cut-sets:

$$\text{ADEFGHIJKLMN} \rightarrow \text{D,AEFGHIJKLMN}$$

$$\text{BCDGHIJKLMN} \rightarrow \text{D,BCGHIJKLMN}$$

$$\text{AEFGHIJKLMN} \rightarrow \text{E,AFGHIJKLMN}$$

$$\text{AEFGJKLMN} \rightarrow \text{J,AEFGKLMN}$$

$$\text{BCDHIJMN} \rightarrow \text{J,BCDHIMN}$$

$$\text{AEFGKLMN} \rightarrow \text{N,AEFGKLM}$$

$$\text{BCDHIJMN} \rightarrow \text{N,BCDHIJM}$$

$$\text{AEFGKLM} \rightarrow \text{M,AEFGKL}$$

$$\text{BCDHIJMN} \rightarrow \text{M,BCDHIJN}$$

$$\text{AEFGKL} \rightarrow \text{L,AEFGK}$$

$$\text{GKLMN} \rightarrow \text{K,GLMN}$$

$$\text{BCDGHIJKLMN} \rightarrow \text{B,CDGHIJKLMN}$$

$$\text{BCGHIJKLMN} \rightarrow \text{C,BGHIJKLMN}$$

$$\text{CDHIJMN} \rightarrow \text{C,DHIJMN}$$

$$BCDHIJMN \rightarrow H,BCDIJMN$$

$$BCDIJMN \rightarrow I,BCDJMN$$

$$BCDHIMN \rightarrow I,BCDHMN$$

$$ABCDEFGHIJKLMN \rightarrow BC,ADEFGHIJKLMN$$

$$AFGHIJKLMN \rightarrow AF, GHIJKLMN$$

$$ADEFGHIJKLMN \rightarrow HI, ADEFGJKLMN$$

$$CDGHIJKLMN \rightarrow HI,CDGJKLMN$$

$$AEFGJKLMN \rightarrow JN,AEFGKLM$$

$$GJKLMN \rightarrow JN,GKLM$$

$$AEFGKLMN \rightarrow MN,AEFGKL$$

$$AEFGK \rightarrow GK,AEF$$

$$DGJKLMN \rightarrow DJ,GKLMN$$

$$ABCDEFGHIJKLMN \rightarrow AEF,BCDGHIJKLMN$$

$$ABCDEFGHIJKLMN \rightarrow BCD,AEFGHIJKLMN$$

$$BCDGHIJKLMN \rightarrow GKL,BCDHIJMN$$

$$AEFGHIJKLMN \rightarrow HIJ,AEFGKLMN$$

All other operations can be checked from this list, together with the subset rule.

## 5.7 COMPLEX AND/OR RELATIONSHIPS

So far we have expressed precedence relationships (selection rules) as defined in Expression 5.4, i.e., they are expressed as follows:

AB not I

where A, B, and I are components. This means that there cannot exist any subassembly that includes both A and B but not I. Note that A and B are not necessarily connected. For example, in the selection rule BG not H listed in Table 5.10, components B and H are not connected.

Often, a *combination* selection rule is convenient to describe geometric feasibility in disassembly operations. A combination is obtained when two or more of the elementary selection rules are combined. This frequently occurs in disassembly theory. The combination is called a *complex AND/OR relationship* (Gungor and

Gupta, 2001a; Moore et al., 2001). For example, in Table 5.10, we encountered the following precedence relationships:

AK not F

DE not F

These can be combined into

(AK or DE) not F

Similarly, from the same table, we can derive the following combination:

(AB or AH or KL or LM or FK or AD or JK) not G

which can be rewritten as

((A and (B or D or H)) or (K and (F or J or L) or LM)) not G

Figure 5.9 presents several example cases of precedence relationships that are described below:

*Case (a)* depicts the basic features of the conventional geometric constraints. It is evident here that A or B has to be detached prior to the detachment of I. This is expressed by the elementary selection rule:

AB not I

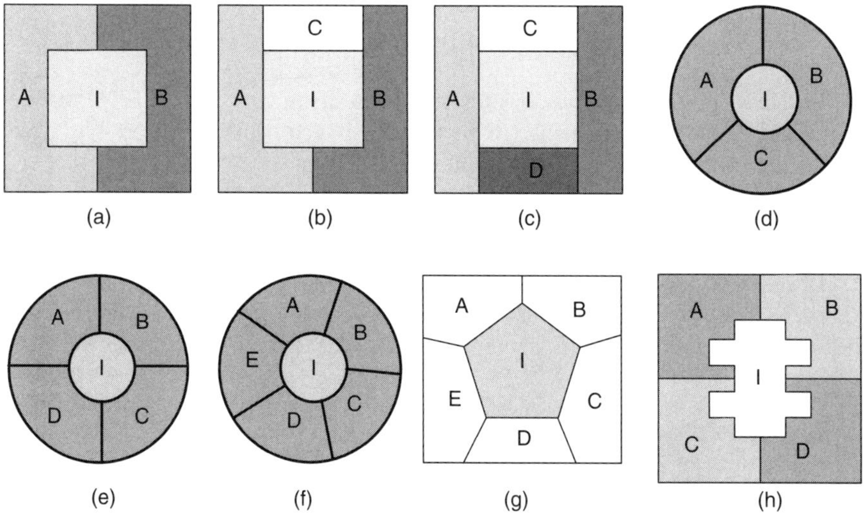


**FIGURE 5.9** Example cases of precedence relationships.

*Case (b)* requires the consideration of all individual components that enclose I. We observe that an arbitrary peripheral component has to be detached prior to the detachment of I, which is expressed via the selection rule:

ABC not I

*Case (c)* is treated analogously. We therefore have the following selection rule:

ABCD not I

*Case (d)* is the same as the example in Figure 5.4. Here, the detachment of any single component does not result in the detachability of component I. Therefore, one has to consider every possible combination of components that directly obstructs the motion of I. These are AB, AC, and BC. Only the detachment of a combination enables the detachment of I. The corresponding selection rule is as follows:

(AB or AC or BC) not I

*Case (e)* is an extension of case (d), with four components surrounding component I. Component I cannot be removed as long as at least one pair of opposite peripheral components is still attached. This is expressed by the following complex AND/OR selection rule:

(AC or BD) not I

*Case (f)* is similar to case (d), but with five components surrounding component I. Detachment of component I is obstructed if any of the pairs AC or AD or BD or BE or CE is still present. This is expressed by the following complex AND/OR selection rule:

(AC or AD or BD or BE or CE) not I

In *Case (g)*, detachment of component I is obstructed only if at least three strategically placed components are present. Five such triplets are possible, which lead to the following complex AND/OR selection rule:

(ABD or ACD or ACE or BCE or BDE) not I

*Case (h)* is due to Gungor and Gupta (2001a). The presence of at least one of four pairs, viz., AB, AD, BC, or CD results in obstruction of the detachment of I. This can also be expressed in a complex AND/OR selection rule as follows:

(AB or AD or BC or CD) not I

## 5.8 THREE-DIMENSIONAL APPLICATIONS

As an example of a simple 3-dimensional product, consider a single-piston engine like the one used in model airplanes (Figure 5.10). It has 15 components and 21

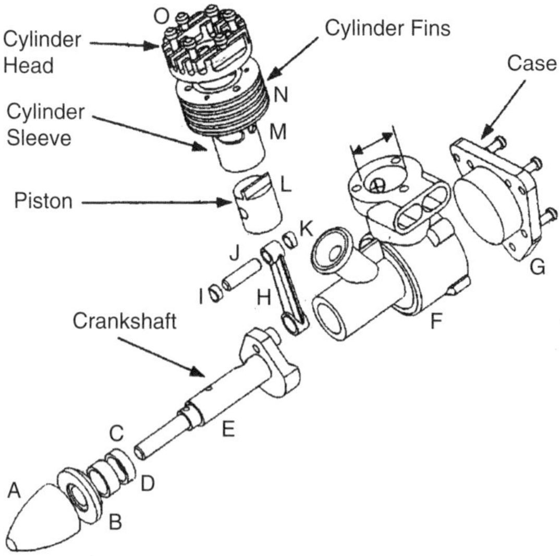


**FIGURE 5.10** Single-piston engine. (Source: Silva and Chang, 2002.)

connections (i.e., $N = 15$ and $K = 21$) as shown in Figure 5.11(a). The piston system, which consists of the piston L, the piston rod with bearings H, a rod J, and two fasteners I and K, can only be attached or detached as a whole unit. This is a kind of technical constraint and therefore the piston system HIJKL can be consolidated into a module or a super-component Q. With the module Q in place, we are now left with 11 components and 16 connections. A more comprehensive treatment on modularization is given in chapter 6.

For the single piston engine of Figure 5.10 together with the consolidation of HIJKL into module Q, we have listed the feasible subassemblies and the selection rules for infeasible subassemblies using the reverse method in Table 5.11.

The disassembly of the piston system is subjected to some technical constraint of its own (Figure 5.11(b)). It is clear that a possible disassembly sequence of the piston system could follow the detachment of I, detachment of J, and detachment of K. We notice that after that it is obligatory to perform at least one ternary operation (see subsection 4.3.4.4), because when J is detached then L and H are also detached, as they are not connected to each other. In connection diagram, this can be circumvented with a quasi-connection between H and L, as shown in Figure 5.11(a) with a dashed line.

The quasi-connection between H and L enables the existence of subassembly HL. With this assumption, the reverse method generates the subassemblies shown in Table 5.12. Note that the original product, HIJKL (= Q), can lead to the following possibilities: the detachment of I, the detachment of K, the detachment of JK, the detachment of IJ. Similarly, the subassembly HIJL can lead to the following possibilities: IJ and KL; and the subassembly HJKL can lead to the following possibilities: HL and JK. Some operations may be restricted in practice by additional technical constraints, but they are not considered here.

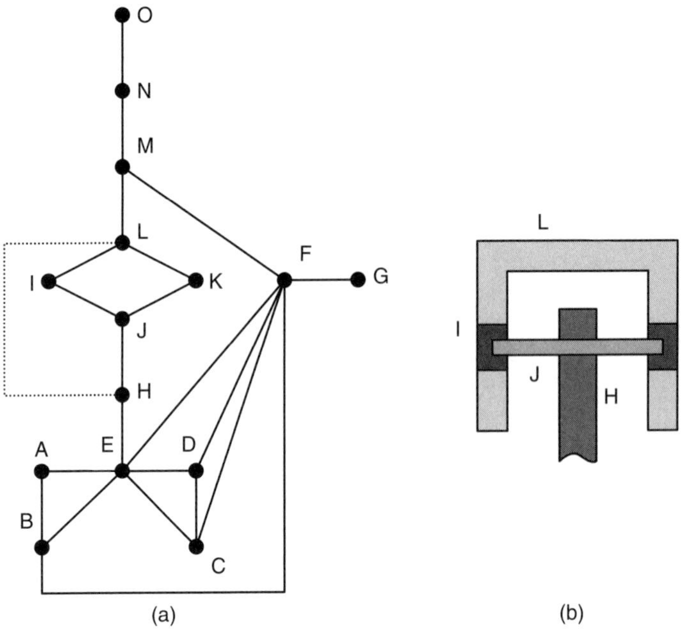


**FIGURE 5.11** (a) Connection diagram of the single-piston engine; (b) piston system configuration.

If we start with the cut-set method, applying the sequential method to the complete product leads to the list of feasible cut-sets and selection rules shown in Table 5.13.

It is interesting to note that the subassembly MNOQ cannot be found by the sequential cut-set method, because this subassembly, although feasible, is not the result of any sequential disassembly process. It results from the separation of CDEFMNOQ into CDEF and MNOQ. It is revealed via the reverse method because sequential assembly can reach it. However, it stops after that because merging MNOQ with any of the 2-subassemblies does not result in a feasible 5-subassembly.

The sequential cut-set method results in 39 queries, which refer to 17 YES answers, resulting in 17 feasible cut-sets and 22 NO answers, accompanied by 22 selection rules. Apart from this, the following cut-sets can be found via a search for complementary triplets that refer to parallel disassembly (see also Table 5.11):

ABCDEFGMNOQ → AB,CDEFGMNOQ

ABCDEFGMNOQ → NO,ABCDEFGMQ

ABCDEFGMNOQ → MNO,ABCDEFGQ

CDEFMNOQ → CDEF,MNOQ

**TABLE 5.11**
**Geometrically Feasible Subassemblies of the Single-Piston Engine of Figure 5.10, Reverse Method**

| *n* = 2 | | *n* = 3 | | *n* = 4 | | *n* = 5 | |
|---|---|---|---|---|---|---|---|
| AB | + | ABF | + | ABCF | + | ABCDF | + |
| AE | AE not B | BCF | + | ABFM | + | ABCFM | + |
| BE | BE not F | BDF | DF not C | BCDF | + | ABFMN | + |
| BF | + | BFM | + | BCFM | + | BCDEF | + |
| CD | + | CDE | + | BFMN | + | BCDFM | + |
| CE | CE not D | CDF | + | CDEF | + | BCFMN | + |
| CF | + | CEF | EF not D | CFMN | + | CDEFG | FG not Q |
| DE | + | CFM | + | FMNO | FO not Q | CDEFQ | + |
| DF | DF not C | FMN | + | MNOQ | + | CDEFM | + |
| EF | EF not C | FMQ | FQ not E | | | CDFMN | + |
| EQ | EQ not F | MNO | + | | | | |
| FG | FG not E | MNQ | + | | | | |
| FM | + | | | | | | |
| MN | + | | | | | | |
| MQ | + | | | | | | |
| NO | + | | | | | | |

| *n* = 6 | | *n* = 7 | | *n* = 8 | | *n* = 9 | |
|---|---|---|---|---|---|---|---|
| ABCDEF | + | ABCDEFM | + | ABCDEFGQ | + | ABCDEFGMQ | + |
| ABCDFM | + | ABCDEFQ | + | ABCDEFMN | + | ABCDEFMNQ | + |
| ABCFMN | + | ABCDFMN | + | ABCDEFMQ | + | BCDEFGMNQ | + |
| BCDEFM | + | BCDEFGQ | + | BCDEFGMQ | + | BCDEFMNOQ | + |
| BCDEFQ | + | BCDEFMN | + | BCDEFMNQ | + | CDEFGMNOQ | + |
| BCDFMN | + | BCDEFMQ | + | CDEFGMNQ | + | | |
| CDEFGQ | + | CDEFGMQ | + | CDEFMNOQ | + | | |
| CDEFMN | + | CDEFMNQ | + | | | | |
| CDEFMQ | + | | | | | | |

| *n* = 10 | |
|---|---|
| ABCDEFGMNQ | + |
| ABCDEFMNOQ | + |
| BCDEFGMNOQ | + |

**TABLE 5.12**
**Geometrically Feasible Subassemblies of the Piston System Q, Reverse Method**

| *n* = 2 | | *n* = 3 | | *n* = 4 | | *n* = 5 | |
|---|---|---|---|---|---|---|---|
| HJ | HJ not L | HIL | + | HIJL | + | HIJKL | + (= Q) |
| HL | + | HJL | + | HIKL | IK not J | | |
| IJ | + | HKL | + | HJKL | + | | |
| IL | + | IJK | IK not H | | | | |
| JK | + | IJL | JL not H | | | | |
| KL | + | | | | | | |

**TABLE 5.13**
**Feasible Cut-Sets and Selection Rules for the Single-Piston Engine of Figure 5.10**

| Feasible Cut-Set | Selection Rule |
|---|---|
| ABCDEFGMNOQ → A,BCDEFGMNOQ | AE not B |
| ABCDEFGMNOQ → G,ABCDEFMNOQ | EF not C |
| ABCDEFGMNOQ → O,ABCDEFGMNQ | CE not D |
| ABCDEFGMNQ → N,ABCDEFGMQ | FG not E |
| BCDEFGMNOQ → B,CDEFGMNOQ | BE not F |
| ABCDEFGMQ → M,ABCDEFGQ | FO not M |
| ABCDEFMNQ → Q,ABCDEFMN | FO not N |
| ABCDEFMN → E,ABCDFMN | FO not Q |
| ABCDFMN → D,ABCFMN | FN not M |
| ABCFMN → C,ABFMN | EG not Q |
| MNOQ → Q,MNO | FQ not E |
| CDEF → F,CDE | EQ not F |
| ABF → F,AB | AF not B |
| FMQ → F,MQ | DF not C |
| MNO → M,NO | CM not F |
| CDE → C,DE | EG not F |
| FMN → F,MN | FN not M |
| | BC not F |
| | BM not F |
| | NQ not M |
| | MO not N |
| | BD not F |

## 5.9 DISASSEMBLY PRECEDENCE GRAPHS

### 5.9.1 General

The concept of precedence graph was introduced in subsection 4.2.1. In this section, we will have a closer look at this method of representation.

Disassembly precedence graphs are essentially derived from task planning and might contain tasks other than disassembly operations such as testing, cleaning, etc. However, in the majority of published papers that deal with disassembly precedence graphs, the tasks considered are confined to the (dis-)assembly of a single component (see, for example, Fox and Kempf, 1985; Woo and Dutta, 1991; Delchambre and Gaspart, 1992; Lin and Chang, 1993; Kroll, 1993; and Dutta and Woo, 1995). Chen et al. (1997) discuss the use of precedence graphs in parallel disassembly. Swaminathan and Barber (1996), and Swaminathan et al. (1998), apply this concept in discussing assembly sequencing via case-based reasoning. Tseng and Liou (2000) discuss the use of precedence graphs to integrate assembly and machining planning. A comprehensive discussion of precedence graphs aimed at assembly line balancing is due to Bratcu

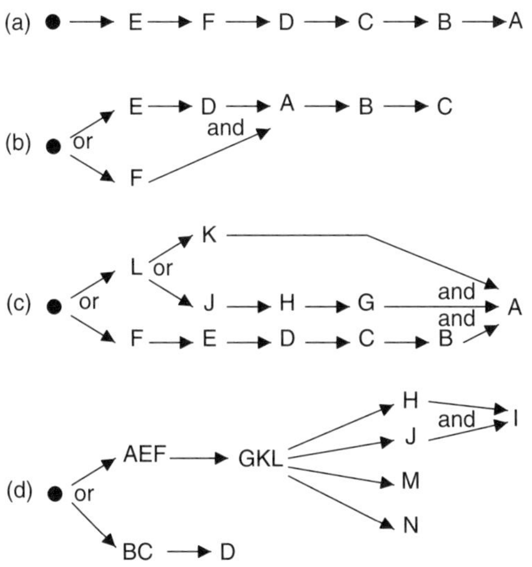


**FIGURE 5.12** Disassembly precedence graphs. (a) Ballpoint example of Figure 5.1 (total ordering); (b) ballpoint example of Figure 5.1 (partial ordering); (c) AFI example of Figure 5.7 (partial ordering); (d) two-dimensional example of Figure 5.6 (partial ordering).

(2001). Gungor and Gupta (2001b, 2002) use precedence graphs to solve the disassembly line balancing problem.

There are two types of disassembly precedence graphs, viz., totally ordered and partially ordered. A *totally ordered* disassembly precedence graph exhibits an exclusively sequential structure, thus representing a single disassembly sequence (see Figure 5.12(a)). A *partially ordered* disassembly precedence graph exhibits one or more parallel structures, thus representing multiple sequences (see Figure 5.12(b)).

We will illustrate the use of precedence graphs by applying them to the ballpoint example of Figure 5.1, the AFI example of Figure 5.7, and the 2D example of Figure 5.6 (see Figure 5.12). It should be remarked here that one single disassembly precedence graph typically does not represent all the possible disassembly sequences of the given product.

A letter in Figure 5.12 refers to the operation of removing a component represented by the letter. The OR relationship refers to a choice that can be made while the AND relationship refers to the condition that every preceding operation is performed prior to the execution of the operation to which the arcs are pointing.

The sequences that are represented by Figure 5.12(a) and Figure 5.12(b) correspond to part of the sequences that are in Figure 5.5(a), thus without technical constraints. Figure 5.12(a) represents one single sequence while Figure 5.12(b) represents three different sequences. One of them is as follows:

$$ABCDEF \rightarrow F,ABCDE \rightarrow E,F,ABCD \rightarrow D,E,F,ABC \rightarrow A,D,E,F,BC \rightarrow A,B,C,D,E,F$$

*Note that in the sequel of section 5.9 we will use an alternate notation. The letters are used here for representing the detachment of the corresponding component. Consequently, the symbol A is used for the operator $R_A$, and so on.*

Using this short notation, the previously mentioned sequence of Figure 5.12(b) is

FEDABC

The other two sequences of Figure 5.12(b) are

EFDABC and EDFABC

Figure 5.12(c), which refers to the AFI example, represents a multitude of sequences. However, some of those sequences might result in disjoint subassemblies. For instance, the sequence FEDCLJHG results in a ternary disassembly operation, resulting in component B becoming disjoint from the remaining components (see Figure 5.7). This is one of the drawbacks of disassembly graph representation. Other drawbacks include inflexibility and the fact that often multiple graphs are required for representing all the possible disassembly sequences of a product. The inflexibility manifests itself if a modification has to be applied, e.g., an additional technical constraint. The need for multiple graphs and the determination of the number of sequences that is represented by a single disassembly precedence graph will be discussed in the next subsection. Even so, the disassembly precedence graph is an extremely useful and compact representation, and comes in very handy when dealing with scheduling issues.

Figure 5.12(d), which refers to the 2-D example, represents many sequences. It also shows a major advantage of this method. Since this subassembly is 3-disassemblable, some parallel disassembly has been included in this graph. It deals with the permutations in the ordering of equivalent disassembly actions in an elegant way, e.g., in the disassembly of HIJMN. It can be seen from Figure 5.6 that the subassembly HIJMN can be disassembled in many ways, because multiple components, namely H, J, M, and N can be detached. This means that these components can be detached in an arbitrary order, e.g., HJMN or HJNM, etc. All these permutations are combined in a single graph.

### 5.9.2 Number of Sequences

As mentioned before, a disassembly precedence graph can represent multiple disassembly sequences. So far, there is no method available for determining the exact number of sequences in a generalized graph. However, methods for deriving the exact number of sequences in some special types of graphs are available. Two such methods for calculating the number of complete disassembly sequences will be discussed. The first one is an iterative method that can be used for precedence graphs that combine both divergent and convergent characteristics. A second one is a direct method that is restricted to strictly divergent disassembly precedence graphs.

#### 5.9.2.1 Iterative Method

An algorithm for calculating the number of sequences in a disassembly precedence graph was discussed by Miller and Stockman (1990). The disassembly precedence graphs are derived from the set of precedence relationships. However, problems are encountered if this set is complicated, which particularly occurs if a connection-oriented approach is used, and generally occurs if a product is characterized by many constraints, each consisting of numerous AND and OR relationships. The calculation for the number of possible sequences that is represented by a disassembly precedence graph proceeds as follows.

*Totally ordered graph:* First a sequential, totally ordered, graph is considered (see Figure 5.12(a)). The filled dot, which represents the product, is the starting position. The graph is characterized by two parameters, viz., the number of precedence constraints $\eta$, and the number of orderings $\sigma$. Precedence constraints are expressions such as: detachment of B must precede detachment of A, etc. The directed arc from the starting position to the first action (removal of E) is *not* considered a precedence constraint, thus $\eta = 5$. Only one ordering (EFDCBA) is possible, thus $\sigma = 1$.

*Partially ordered graph:* Consider Figure 5.12(b). We observe that the two ordered sequences E→D→A and F→A are in parallel. They share the same starting position and the same final position (detachment of A). If we represent the two branches with indices 1 and 2, respectively, and the combination with the index *c*, we have $\eta_1 = 2$, $\eta_2 = 1$, $\sigma_1 = 1$, $\sigma_2 = 1$. For the *parallel portion* of the graph, the following formulae are applicable:

$$\eta_c = \eta_1 + \eta_2 \tag{5.8a}$$

$$\sigma_c = \frac{(\eta_1 + \eta_2)!}{\eta_1!\eta_2!}\sigma_1\sigma_2 \tag{5.8b}$$

Consequently $\eta_c = 3$ and $\sigma_c = 3$ (viz. EDF, EFD, and FED). This parallel graph is replaced with a single node, with $\eta = 3$ and $\sigma = 3$.

Next, the sequential portion of the graph ABC is attached, with $\eta = 2$ and $\sigma = 1$. If the two graphs (i.e., the parallel and the sequential) are again represented by the indices 1 and 2, we make use of the following property:

If two graphs, one with $\eta_1$ and $\sigma_1$, the other with $\eta_2$ and $\sigma_2$, are *sequentially attached*, the resulting degree of freedom and number of orderings are given as follows:

$$\eta_c = \eta_1 + \eta_2 \tag{5.9a}$$

$$\sigma_c = \sigma_1\sigma_2 \tag{5.9b}$$

For the graph of Figure 5.12(b) we end up with $\eta_c = 5$ and $\sigma_c = 3$. Thus there are five constraints and still three orderings, which is evident.

*Combined graph:* Consider the combined graph of Figure 5.13. We will calculate the number of sequences represented by this graph.

Consider the branches D→E→F→C and D→G→C, which are represented by the indices 1 and 2, respectively. For now, assume node D to be a local starting

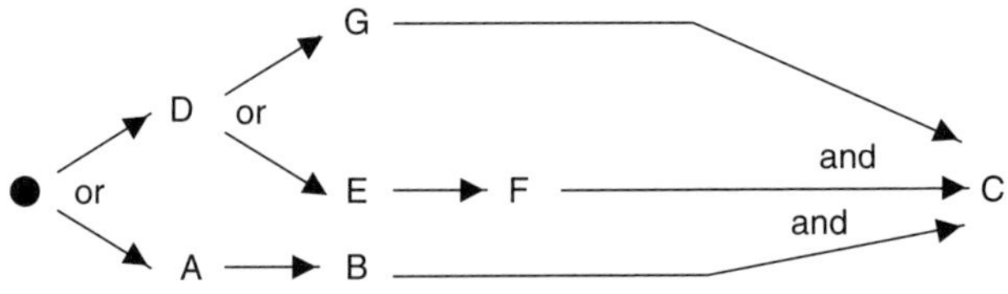


**FIGURE 5.13** Combined disassembly precedence graph.

point. Therefore, the constraints D→G and D→E are not counted at this point. Thus, $\eta_1 = 2$, $\sigma_1 = 1$, $\eta_2 = 1$, $\sigma_2 = 1$ and using Expression 5.8a and Expression 5.8b we get $\eta_c = 3$, $\sigma_c = 3$.

Next the global starting point (which is the complete product) is accounted for. The graph that was previously represented by index *c* is now represented by the index 2. The sequence from the product via D to the combined node (with index 2) is represented by the index 1. The parameters for the new combined graph *c* can now be calculated using Expression 5.9a and Expression 5.9b. With $\eta_1 = 1$, $\sigma_1 = 1$, $\eta_2 = 3$, $\sigma_3 = 3$, one arrives at $\eta_c = 4$, $\sigma_c = 3$.

Finally, the remaining branch that connects the product to node C via the nodes A and B is accounted for. If this totally ordered subgraph is represented by index 1 and the graph that was previously represented by index *c* is now represented by the index 2, we have $\eta_1 = 2$, $\sigma_1 = 1$, $\eta_2 = 4$, $\sigma_2 = 3$. Consequently, using Expression 5.8a and Expression 5.8b we get $\eta_c = 6$, $\sigma_c = 45$.

The 45 possible sequences that are represented by the disassembly precedence graph of Figure 5.13 are listed as follows:

| | | | | |
|---|---|---|---|---|
| ABDEFG | ADEFGB | DAEBFG | DEABFG | DEGABF |
| ABDEGF | ADEGBF | DAEBGF | DEABGF | DEGAFB |
| ABDGEF | ADEGFB | DAEFBG | DEAFBG | DEGFAB |
| ADBEFG | ADGBEF | DAEFGB | DEAFGB | DGABEF |
| ADEBGF | ADGEBF | DAEGBF | DEAGBF | DGAEBF |
| ADBGEF | ADGEFB | DAEGFB | DEAGFB | DGAEFB |
| ADEBFG | DABEFG | DAGBEF | DEFABG | DGEABF |
| ADEBGF | DABEGF | DAGEBF | DEFAGB | DGEAFB |
| ADEFBG | DABGEF | DAGEFB | DEFGAB | DGEFAB |

Note that in all the above sequences, the last component that is retrieved is component C.

### Example 6

Consider the disassembly precedence graph of a gearbox shown in Figure 5.14, which has been taken from Miller and Stockman (1990). We will calculate the number of possible sequences for the gearbox.

Consider first the parallel configuration with branches B→F→J and B→C→D→E→J and B considered as the local starting point, with $\eta_1 = 1$, $\sigma_1 = 1$, $\eta_2 = 3$, $\sigma_2 = 1$ respectively. Thus, using Expression (5.8a) and Expression (5.8b), we have $\eta_c = 4$, $\sigma_c = 4$.

Consider next the series configuration with AB as configuration 1, and FCDEJ as configuration 2, with $\eta_1 = 2$; $\sigma_1 = 1$; $\eta_2 = 4$; $\sigma_2 = 4$. Thus, using Expression (5.9a) and Expression (5.9b), we have $\eta_c = 6$, $\sigma_c = 4$.

Next, consider the parallel configuration up to J with ABCDEFJ as configuration 1, and GHIJ as configuration 2, with $\eta_1 = 6$, $\sigma_1 = 4$, $\eta_2 = 3$, $\sigma_2 = 1$. Thus, using Expression (5.8a) and Expression (5.8b), we have $\eta_c = 9$, $\sigma_c = 336$.

Adding K can be realized by considering a series configuration again. We finally end up with $\eta_c = 10$, $\sigma_c = 336$ for the product.

At this point some remarks on this topic are in order:

1. The number of sequences that is calculated here does not account for topological constraints. Therefore, the number of feasible sequences that is realized with a specific disassembly precedence graph can actually be less than the theoretical maximum number calculated here.
2. It should be emphasized that the number thus calculated generally does not include the complete set of sequences, for precedence relationships can be far more complicated than in the above-mentioned cases and as of now discovery of all possible sequences cannot be guaranteed.
3. Note that when calculating the number of possible sequences for the disassembly precedence graphs of the type represented in Figure 5.12(d), a virtual node should be added to facilitate the analysis. For example, in Figure 5.12(d), the nodes D, N, M, and I are connected to a virtual node. These newly added arcs should not be counted as extra constraints.

Miller and Stockman (1990) investigated still more complex configurations of disassembly precedence graphs, including *N patterns* and *X patterns*, which might be entrenched in disassembly precedence graphs. For example, in Figure 5.14, if the precedence relationship A→H is added to the graph, an N-pattern results. If both A→H and B→G are added, an X pattern is encountered. Elaboration of these patterns results in rather complicated formulae that still do not cover all possible structures of precedence graphs. As we have already pointed out, it is not always possible to analyze an arbitrary product with a disassembly precedence graph. Representation with state diagrams or AND/OR graphs might therefore be preferable, except for those products that are characterized by a simple structured set of precedence relationships. This is particularly true in hierarchical tree structures.

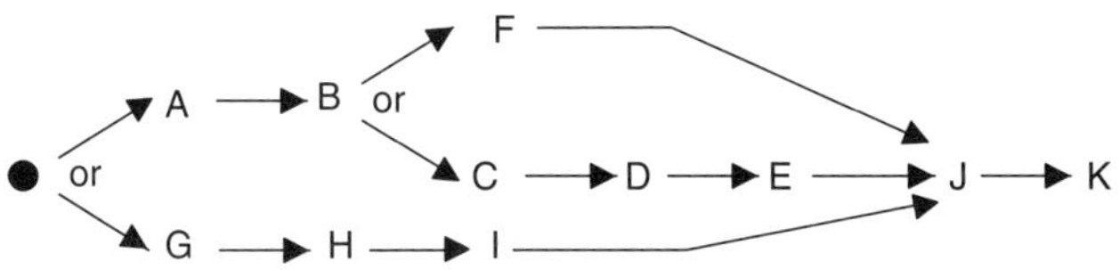


**FIGURE 5.14** Disassembly precedence graph of a gearbox (see Miller and Stockman, 1990).

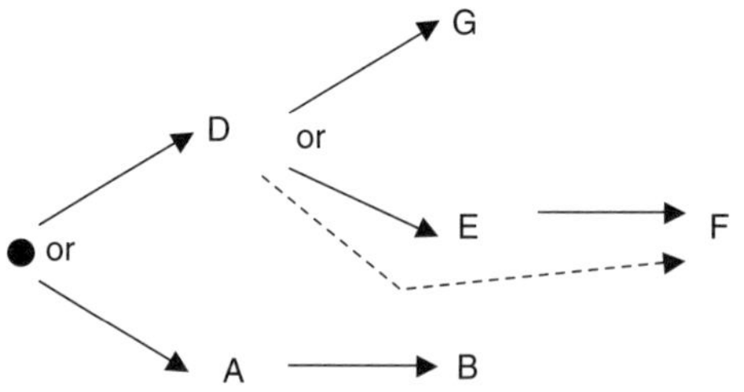


**FIGURE 5.15** Disassembly precedence graph of the divergent type.

### 5.9.2.2 Direct Method

The direct method, which is due to Uchiyama et al. (1994), is simpler to apply, but its domain of applications is restricted to disassembly precedence graphs of the divergent type, i.e., the graphs that have OR relationships only. Let us illustrate this for the case of Figure 5.13. It is reduced to a disassemly precedence graph of drivergent type, as shown in Figure 5.15.

The graph is made *transitive* by connecting every node $C_i$ to all the nodes that can only be detached if component $C_i$ is detached first (see the dashed line in Figure 5.15). For instance: F cannot be detached prior to the detachment of E, and E cannot be detached prior to the detachment of D, thus also F cannot be detached prior to the detachment of D. This is called the transitive property.

If $N$ is the number of components that appears in a disassembly precedence graph, the number of sequences, with no precedence constraints present, equals $N!$, provided this number is sensitive to the ordering of the components that are separated (see subsection 4.3.7). However, if constraints are present, the following is valid:

$$\text{The number of sequences} = \frac{N!}{\prod_{i=1}^{R}(r_i+1)} \tag{5.10}$$

where $R$ is the number of nodes in the graph, and $r_i$ is the number of arcs representing the precedence relationships that are leaving the $i$th node after the graph is made transient. Thus, for Figure 5.15, $N = 6$, $R = 3$, $r_A = 1$, $r_D = 3$, $r_E = 1$. Inserting these values in Expression 5.10 results in 45 sequences, which agrees with the result obtained using the iterative method.

**Example 7**

The disassembly precedence graph of the gearbox (Figure 5.14) can be converted into a parallel graph by leaving out the components J and K and the related arcs. One is left with the components A through I and thus $N = 9$. If one makes this graph transitive, the following parameters are revealed: $r_A = 5$, $r_B = 4$, $r_C = 2$, $r_D = 1$, $r_G = 2$, $r_H = 1$. Inserting these values in Expression (5.10) results in 336 sequences, which coincides with the result that was found in the previous subsection.

Note that a divergent precedence graph cannot as such be investigated by the direct method. However, it may be possible to convert it into a convergent type of disassembly precedence graph, which can be treated with the direct method. Mixed convergent/divergent graphs, on the other hand, cannot be dealt by this method.

### 5.9.3 Number of Selection Rules

In subsection 5.4.4, we encountered the problem of redundancy in determining the number of precedence relationships or the equivalent selection rules that are connected with geometrical constraints. Determining this number exactly for an arbitrary product is not a trivial task. Even in a relatively simple product as the ballpoint (Figure 5.1), we found 10 possible selection rules. However, because of redundancy only two of these were required for the derivation of the feasible disassembly sequences. As of now, no methods are available to determine the number of selection rules for an arbitrary product.

If the disassembly precedence graph of Figure 5.12(b), which refers to ballpoint example of Figure 5.1, is considered, the following selection rules can be derived after making the graph transitive:

E not A, E not B, E not C, E not D, F not A, F not B, F not C,
D not A, D not B, D not C, A not B, A not C, B not C.

The selection rule E not A, for example, implies that those subassemblies are infeasible that contain component E, but not component A. This is a stricter constraint than the usual selection rules that have two components on the left. Evidently, the selection rule E not A is a combination of the following selection rules:

BE not A, CE not A, DE not A, EF not A

The same can be done for the other selection rules. The complete list of selection rules is given in Table 5.14.

**TABLE 5.14**
**Selection Rules Derived from Disassembly Precedence Graph of Figure 5.12(b)**

| | | | |
|---|---|---|---|
| BD not A | AC not B | AB not C | AE not D |
| BE not A | AD not B | AD not C | BE not D |
| BF not A | AE not B | AE not C | CE not D |
| CD not A | AF not B | AF not C | EF not D |
| CE not A | CD not B | BD not C | |
| CF not A | CE not B | BE not C | |
| DE not A | CF not B | BF not C | |
| DF not A | DE not B | DE not C | |
| EF not A | DF not B | DF not C | |
| | EF not B | EF not C | |

**TABLE 5.15**
**Selection Rules from Each of the Precedence Graphs of Figure 5.20**

| | | | |
|---|---|---|---|
| *B not D* | *A not D* | A not C | *A not D* |
| B not A | *C not D* | B not C | *B not D* |
| *C not D* | A not B | D not C | *C not D* |
| C not A | C not B | *A not D* | |
| D not A | D not B | *B not D* | |

It is clear that a lot of selection rules are actually associated with a precedence graph. In practice, the disassembly sequences of a product may be represented by multiple precedence graphs (see next subsection). A set of selection rules can then be generated for every precedence graph. The intersection of these sets uncovers the selection rules that refer to the product. In general, there still remain many more selection rules.

As an example, consider the multiple graphs shown in Figure 5.20. Every element in each of the four precedence graphs is associated with a set of selection rules (see Table 5.15). Only the selection rules with component D on the right will appear in the intersection of these four sets. These are indicated in italics in Table 5.15. If expanded to selection rules with two components on the left, the intersection of the four sets of selection rules reveals:

BC not D, AC not D, AB not D

This agrees with the results of case (d) in section 5.7, see Figure 5.9(d). Consequently, the complete set of disassembly sequences for this product is represented by Figure 5.20. We will discuss this to some more extent in the next subsection.

A further discussion of selection rules will take place in chapter 6 in connection with a restricted set of detachment directions. Obstruction diagrams and disassembly precedence matrices are appropriate tools in such cases for the generation of those rules.

### 5.9.4 Construction of Disassembly Precedence Graphs

Disassembly precedence graphs can be constructed provided a set of disassembly sequences has been previously obtained, perhaps in an intuitive way or by a set of heuristic rules. We will discuss a method due to Minzu et al. (1999), which returns one or more disassembly precedence graphs.

For a better understanding of this theory, let us consider the set of sequential disassembly sequences, which can be found via the following procedure:

1. Give a uniquely ordered value (e.g., alphabetical) to (the detachment of) every component.
2. Start with the sequence that has been arranged in ascending order. Put this sequence on the top line.
3. Identify all adjacent pairs of operators that are in ascending order, if the higher of the two values is to the right of the pair.

4. Exchange every adjacent pair of components. Put all of these on the next line. Delete any duplicates.
5. Repeat the process in step 3 with all the sequences written on the newest line and place the result on the line beneath.
6. Continue this procedure with the new line till the original sequence is arranged in descending order. This will go at the bottom line.
7. Connect those sequences that are different in only one adjacent pair.

The above procedure is illustrated for a product with four components (i.e., $N = 4$) (see Figure 5.16).

In step (1) we name the four components, A, B, C, and D.
In step (2) we have the sequence ABCD.
In step (3) we identify the pairs CD, BC, and AB.
In step (4) we have the sequences ABDC, ACBD, and BACD.
In step (5), this procedure is repeated for each of the three sequences.

Viswanathan and Allada (2001) define the *distance* $\delta$ between sequential sequences as the number of adjacent pairs exchanges required to reach one sequence from the other. For example, the distance between ABCDE and ACBDE is 1 because only one adjacent pair (i.e., BC) needs to be exchanged in ABCDE to reach ACBDE. Similarly the distance between the sequences ABCDE and CBDEA is 5, because starting from ABCDE, one has to go through the following exchanges: BACDE, BCADE, BCDAE, BCDEA, and CBDEA.

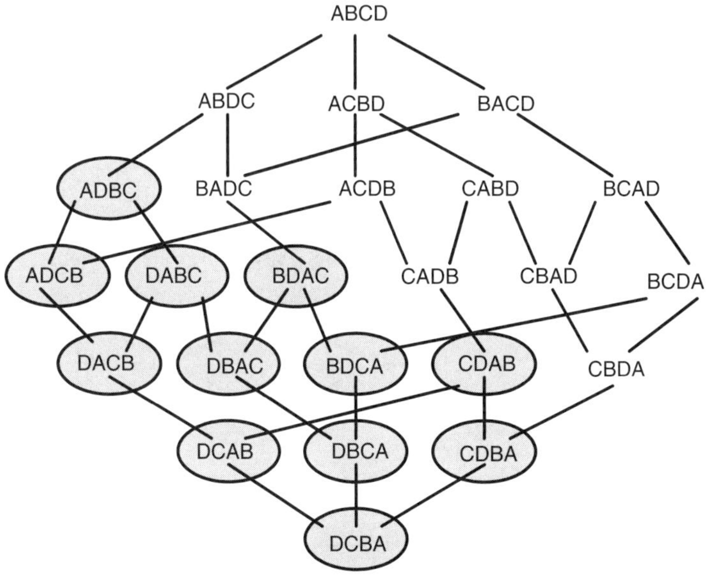


**FIGURE 5.16** Set of sequences for a product with $N = 4$. For the special case of the product in Figure 5.4, all infeasible sequences are indicated with shaded ellipses.

The set of sequences is depicted in the sequence space, which can be represented with an undirected graph. Nodes are sequences; arcs are between sequences that have a distance of one. If all sequences are possible, there will be $N!$ sequences. Every sequence in the graph has $N - 1$ neighbors. By induction it follows that the maximum distance between two sequences is equal to

$$1+2+\cdots+(N-1)=\tfrac{1}{2}(N^2-N) \tag{5.11}$$

This can be easily shown by induction. Thus if $N=4$, the maximum distance between two sequences is 6 (e.g., between ABCD and DCBA). The graphical representation of the set of sequences for a product with $N = 4$ is given in Figure 5.16.

Minzu et al. (1999) describe a method for finding the disassembly precedence graph from a list of disassembly sequences. Such sequences may be derived heuristically and thus the list may not be complete. Nevertheless, it is possible to discern between pairs of components from which the ordering can be reverted, and those from which the ordering cannot be reverted. For example, if one has only one sequence ABCD the pairs AB, AC, AD, BC, BD, and CD are ordered. There is no sequence available that has a reverted ordering. Consequently, these pairs of components are subjected to precedence relationships.

If one considers the sequences ABCD and ACBD, it is evident that one pair of adjacent components, namely BC is *not* subjected to a precedence relationship because it has both an ordered and a reverted ordering. With this knowledge, a disassembly precedence graph can now be constructed (see Figure 5.17(a)).

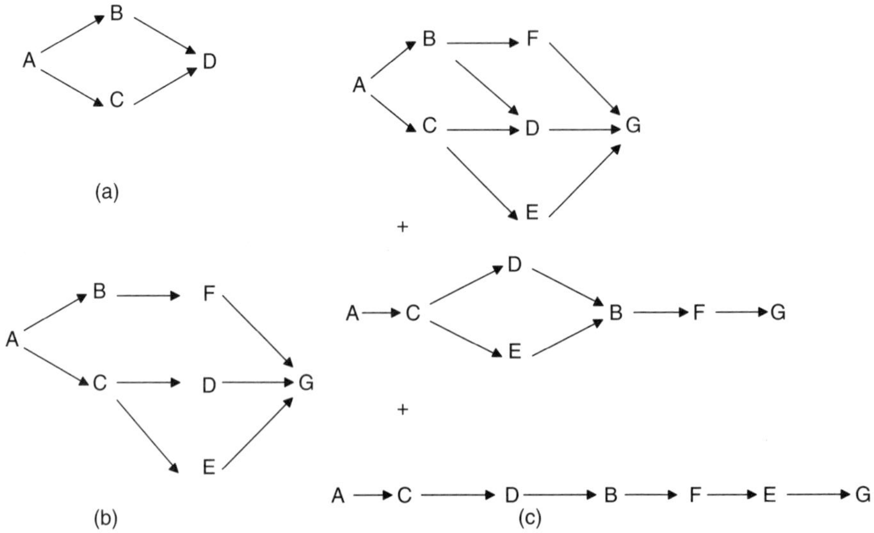


**FIGURE 5.17** Disassembly precedence graphs. (a) Simple case; (b) the listing from the table; (c) the same set, with one missing sequence.

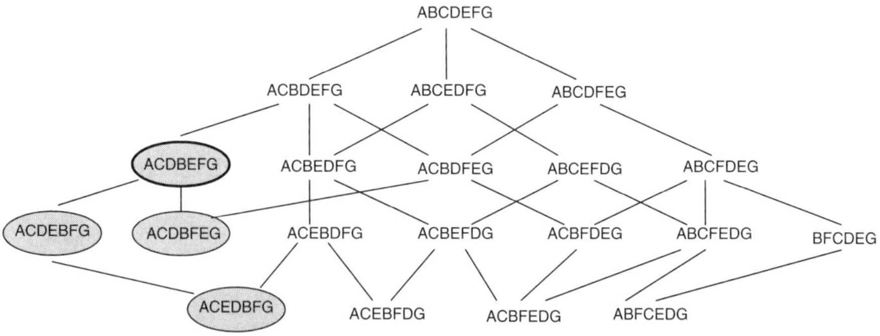


**FIGURE 5.18** Set of sequences, see text. Those in ellipses have B and D reverted.

If one considers the sequences ABCD and CBAD, it follows from Figure 5.16 that their distance equals 3. This implies that three pairs of components, namely AB, AC, and BC are not subjected to precedence relationships, because these appear in both orderings. Reverting, however, of an arbitrary pair of those three might result in a sequence that was originally not given (e.g., BACD). Therefore, representing this set of two sequences in a single disassembly precedence graph is not possible.

Let us proceed with the following set of 20 sequences:

| | | | |
|---|---|---|---|
| ABCDEFG | ABCFEDG | ACBDFEG | ACDEBFG |
| ABCDFEG | ABFCDEG | ACBEFDG | ACDBFEG |
| ABCEDFG | ABFCEDG | ACBFDEG | ACEBDFG |
| ABCEFDG | ACBDEFG | ACBFEDG | ACEBFDG |
| ABCFDEG | ACBEDFG | ACDBEFG | ACEDBFG |

It can be noticed that the pairs BC, BD, BE, CF, DE, DF, and EF are not subjected to precedence relationships. If these are adjacent, they may be reverted (see, e.g., the pairs ABCDEFG and ACBDEFG).

This can be put in the form of a scheme shown in Figure 5.18. All the permutations of adjacent pairs result in sequences that are in the set of 20 sequences. Consequently, they can be put in one disassembly precedence graph (see Figure 5.17(b)).

Now, consider the case where the set of sequences listed above has the sequence ACDBEFG missing. From Figure 5.18 it is evident that the missing sequence results from ACBDEFG by reverting the adjacent pair BD. This means that reverting this pair might lead to an invalid sequence. Therefore, the pair BD is rejected from the above-mentioned set of pairs and a precedence relationship from B to D has to be assumed, which is added to the original disassembly precedence graph of Figure 5.17(b), (see the first part of Figure 5.17(c)).

However, by doing this, not only the sequence ACDBEFG is excluded, but other sequences are also excluded that were in the original list, namely those with B and D reverted, i.e., ACDEBFG; ACEDBFG; ACDBFEG. It can be seen from Figure 5.18, that only ACDEBFG and ACEDBFG are related and can thus be represented with a single disassembly precedence graph, with the pair DE the only one that is not subjected to a precedence relationship (see the second part of Figure 5.17(c)). The remaining sequence ACDBFEG, is represented with an additional disassembly precedence graph (see the third part of Figure 5.17(c)).

A thorough mathematical formulation of this theory can be found in Minzu et al. (1999).

It is now clear that the disassembly precedence graphs represent ordered sets. By performing successive permutations, it is possible to derive systematically the full list of disassembly sequences that are represented by a given disassembly precedence graph. On the other hand, it appears possible to express a set of given disassembly sequences by the corresponding set of disassembly precedence graphs. In the most pessimistic case, every given disassembly sequence corresponds to a totally ordered disassembly precedence graph, which results in as much totally ordered sets as there are sequences. However, in most cases multiple sequences can be combined to a single partially ordered graph, particularly if many ordered pairs are present. This corresponds to cases that have relatively few precedence constraints. In this way, one arrives at a significantly reduced set of disassembly precedence graphs.

If there are no constraints, one ends up with the full set of $N!$ sequences represented by one disassembly precedence graph, with all the components in parallel. It should be stressed, that the disassembly precedence graph does not account for disjoint subassemblies. These can, however, automatically be rejected *a posteriori*.

### Example 8

Consider the product of Figure 5.4. By applying sequential disassembly operations, the AND/OR graph can be readily derived. This AND/OR graph can be converted into a set of sequences (see Figure 5.19), which represents a subset of the sequences that are depicted in Figure 5.16. From this, a set of disassembly graphs is derived as follows:

There are 12 permitted sequences, which means that this set can be represented by 12 separate disassembly precedence graphs, each representing one totally ordered set. There is no ordered pair in the set of Figure 5.19. However, there are three pairs, viz. AD, CD, and BD, each of which appear only twice in reverse ordering. Sequences that contain such reverted pairs are shaded in Figure 5.19. If these sequences are taken apart, quasi-ordering is added to the remaining sequences. For these, there is still the

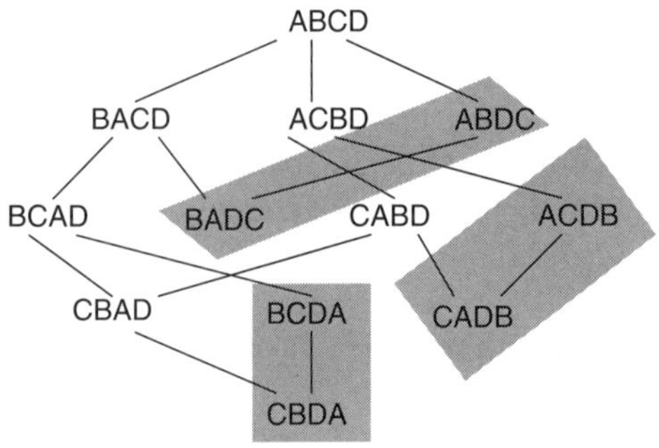


**FIGURE 5.19** Set of sequences represented by Figure 5.4.

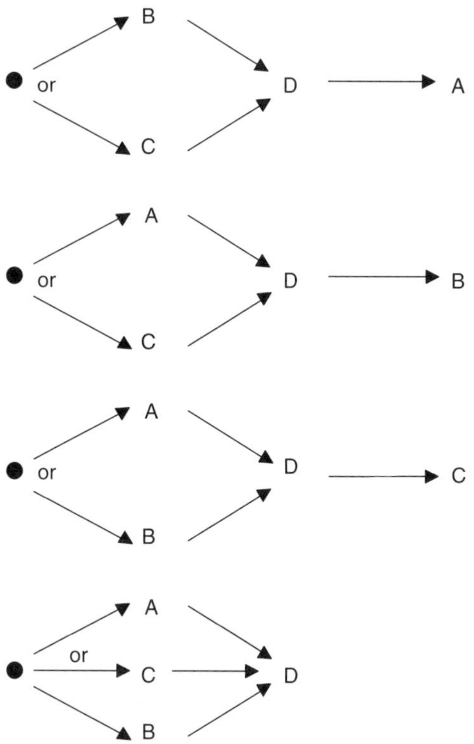


**FIGURE 5.20** Disassembly precedence graphs representing the set of sequences of Figure 5.19.

freedom in reverting the pairs BC, AC, and AB. Precedence relationships exist for the pairs AD, CD, and BD. The set of sequences thus can be represented by four separate disassembly precedence graphs (see Figure 5.20).

Although much theoretical work is still to be done on disassembly precedence graphs, the discussion in this subsection can be considered a synopsis of the basic features of this method.

## 5.10 CONSTRAINED CONNECTION DIAGRAMS

By now we have discussed all the basic methods for representing the precedence relationships and the disassembly sequences. However, we are still left with an alternative method that has also been used by many authors. This is based on the connection diagram with precedence constraints added to it, thus making use of the directed arcs and is known as a *constrained connection diagram.* Apart from information on operations precedence, many authors append additional information to the connections, such as the permitted direction of motion, the fastener type, complexity and cost metrics, etc. Apart from this, distinction may be made between the static and the dynamic components. We will only discuss precedence information here.

Constrained connection diagrams usually show a hybrid structure, because they contain both undirected arcs (connections) and directed arcs (precedence relationships). Although different authors follow different conventions, a precedence relationship is usually represented with an arrow pointing from a particular component (node) to those components that obstruct its removal. This means that a component that has only incoming arcs has to be detached first. Once a component is removed, the arcs that are connected to it are also removed, thus making new components available for detachment. This procedure continues till all the components are detached.

Among the authors that use such diagrams include: Frommherz and Hornberger (1988), De Floriani and Nagy (1989), Lee (1989), Lee and Shin (1990ab), Wolter (1989), Delchambre (1990), Subramani and Dewhorst (1991), Dini and Santochi (1992), Yokota and Brough (1992), Wilson and Schweikard (1992), Lin and Chang (1993), Uchiyama et al. (1994), Murayama et al. (1994, 2001), Vujosevic (1995), Vujosevic et al. (1995), Bhatia et al. (1996), Swaminathan and Barber (1996), Armillotta and Semeraro (1997), Choi et al. (1998), Srinivasan et al. (1998b, 1999, 2000, 2002), Lee and Moradi (1999), Ong and Wong (1999), Garcia et al. (2000), Gungor and Gupta (2001b), and Viswanathan and Allada (2001).

These authors have utilized a multitude of notations and it obviously does not make sense to discuss them all. We will discuss a notation that is close to the selection rules method. It combines many of the advantages that are inherent to the solutions given by the above-mentioned authors, without getting stranded in too much complexity.

Let us consider the ballpoint example of Figure 5.1 again. We have seen that not *all* the possible selection rules have to be considered, but only those that are revealed by the cut-set method (see Table 5.3). If we consider a selection rule, such as BE not D, it means that either B or E have to be removed prior to the removal of D. This can be represented by a hyperarc, starting from D with directed arcs pointed to B and E. In contrast with the AND/OR graph that has already been discussed as a disassembly graph representation, both branches of the hyperarc reflect an OR relationship here. Figure 5.21 presents the constrained connection diagram for the ballpoint, with both the geometrically determined selection rules included, namely BE not D, and AF not B. Apart from these, the technical constraint for disassembly, BC not D, has also been added. This means that two hyperarcs are departing from D. This corresponds to an AND relationship. From Figure 5.21 it follows that only E and F can be detached as first components. Note that component

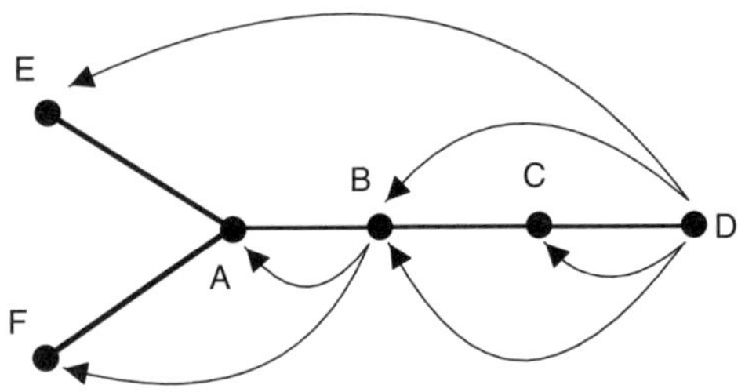


**FIGURE 5.21** Constrained connection diagram for the ballpoint example of Figure 5.1.

A cannot be detached because of topological reasons. Let us assume that we start by detaching E first. With E detached, both branches of the associated hyperarc are simultaneously removed.

It is evident that the next component to be detached will be F, followed by A, B, and finally C or D. If the technical constraint BC not D were not there, one could also start with the detachment of D after that of E, because there would be no hyperarc left departing from D.

The same graph also helps visualize some aspects of parallel disassembly operations. For instance, the two components that appear at one side of a selection rule, e.g., A and F in AF not B, cannot be simultaneously detached. Thus, once E is detached, the detachment of AF is not possible because AF is outruled via the selection rule. If both E and F are removed, the detachment of AB is possible, because this subassembly is connected and it has no outgoing arcs left. It can also be seen that, once component E is detached, the subsequent detachment of subassembly ABF is also possible. The hyperarc that corresponds to AF not B is completely embedded in this subassembly and thus causes no constraint here. The hyperarc that belongs to BC not D has an arc that points toward ABF, which does not represent an obstruction.

It should be stressed that this type of constrained connection diagram is a streamlined version. We will apply this to the subassembly AFGK (Figure 5.22(a)), which is part of the 2D-product of Figure 5.6(a). From subsection 5.6 it became clear that this subassembly was not 1-disassemblable. Either GK or AF has to be removed. This is visualized with a constrained connection diagram in Figure 5.22(b).

The selection rules: AK not F, AK not G, FK not A, and AG not K are depicted. There is no component without departing hyperarcs, thus the subassembly is not 1-disassemblable. Consequently, this subassembly must be 2-disassemblable. If it were not, then it would not be disassemblable at all, because it consists of four components only, and this would imply that AFGK should be treated as a single component, which is not true. A disassembly operation that results in FK, AG, or AK is not feasible, because these subassemblies appear on one side of a selection rule. The only potential separation left is in AF and GK, which are feasible subassemblies. From the drawing of the subassembly it is clear that this separation is

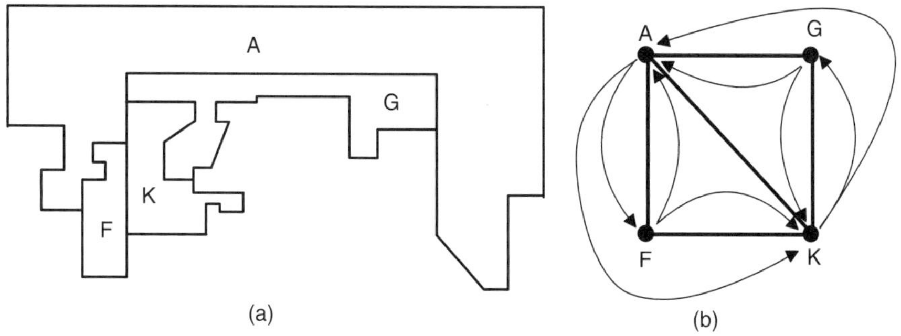


**FIGURE 5.22** Subassembly of the two-dimensional case. (a) Subassembly drawing; (b) constrained connection diagram.

feasible indeed. Obviously, the hyperarcs from which one branch is pointing from the one to the other of the two components of a subassembly that is considered as a unit, disappear as a constraint. For instance, if the detachment of component A is obstructed if K and F are still present, this means that the detachment of AF is no longer obstructed, unless K also obstructs the detachment of F. In this case, although the selection rule AK not F exists, there is no obstruction, because A moves with F.

This type of notation therefore combines the selection rule formalism with the graphical method of the constrained connection diagrams. An extension of this method to include information on directions of detachment will be discussed in chapter 6.

## 5.11 CONCLUSION

The intention of chapter 5 has been the presentation of a formalism that is applied to the generation of the AND/OR graph of mechanical products, which accounts for topological and geometrical constraints, and which is applicable to both disassembly and assembly studies. Technical constraints can be added if desired. The selection rule formalism in the subassembly-oriented approach appears to be the most efficient and universally applicable.

The theory has been explained with the help of a number of examples. It was shown that the consideration of topological and geometric constraints could considerably mitigate the combinatorial explosion in disassembly problems. The geometric constraints can be condensed in a modest set of selection rules. This is used for a smooth derivation of the feasible subassemblies. From these, the disassembly graph is derived. If an assembly problem is considered, this graph is reverted. In this case, technical constraints, which are in general asymmetric with respect to assembly and disassembly, are added once the reversion has taken place. If disassembly problems are considered, the technical constraints might be considered together with the geometric constraints.

Various methods have been discussed for addressing geometric constraints, viz., the complete cut-set method, the sequential cut-set method, and the reverse method. The basic idea behind these methods is that the subassembly-oriented approach should be carried through. This results in a set of selection rules, which can be dealt with straightly. The selection rules, which are powerful due to the superset rule, are complemented with a set of feasible cut-sets that account for the feasibility of disassembly operations, making use of the subset rule. From this, a list of feasible subassemblies can be derived, from which a list of feasible disassembly operations follows. This list is obtained using complementary triplets.

The methods were applied on some examples for demonstrating their universality and ease. These methods form the basis for developing software that is able to determine AND/OR graphs that represent all possible disassembly sequences.

Apart from the AND/OR graph formalism, the disassembly precedence graph formalism was discussed, which is used alternatively by many authors. Some theory on this operation-oriented approach has been explained. The relationship with the selection rules was discussed. Alternative formalisms, such as the constrained connection diagram that is commonly used in heuristic approaches, have also been elaborated. Here also, the selection rule formalism turns out to be applicable.

Because of the simplicity of the method, manual calculations became possible for some standard examples that usually required software aids. The method has been formalized so that semi-automatic operation is possible, which is beneficial for larger products. A modest number of queries still remains necessary. YES answers automatically generate a feasible cut-set, and NO answers need selection rules accompanying them.

Visual inspections can even be circumvented completely if the number of directions of detachment is restricted. In this case, a fully automated calculation is possible. Automatic generation of selection rules is also viable if more detailed information on component is available. This will be discussed in chapter 6.

## REFERENCES


Armillotta, A. and Semeraro, Q., 1997, Assessment of destructive operations in disassembly process planning. *International Journal of Flexible Automation and Integrated Manufacturing*, **5**(1&2), 57–78.

Baldwin, D.F., Abell, T.E., Lui, M.M., De Fazio, T.L., and Whitney, D.E., 1991, An integrated computer aid for generating and evaluating assembly sequences for mechanical products. *IEEE Transactions on Robotics and Automation*, **7**, 78–94.

Bhatia, P., Karnick, H., and Ghosh, A., 1996, A configuration space based approach for robot assembly sequencing. *Robotica*, **14**, 633–645.

Bonneville, F., Henrioud, J.M., and Bourjault, A., 1995, Generation of assembly sequences with ternary operations. *Proceedings of 1995 IEEE International Symposium on Assembly and Task Planning*, 245–249.

Bourjault, A., 1984, *Contribution à une approche méthodologique de l'assemblage automatisé: elaboration automatique des séquences opératoires* (Contribution to a systematic approach of automatic assembly: automatic generation of task sequences). Ph.D. Thesis, Besançon, France: Université de Franche-Comté, (in French).

Bourjault, A., 1987, Methodology of assembly automation: a new approach. *Abstracts of 2nd International Conference on Robotics and Factories of the Future*, San Diego, CA, 37–45.

Bratcu, A., 2001, *Détermination systématique des graphes de précédence et équilibrage des lignes d'assemblage* (Systematic determination of precedence graphs and assembly line balancing). Ph.D. Thesis, Besançon, France: Université de Franche-Comté, (in French).

Chen, S.-F., Oliver, J.H., Chou, S.-Y. and Chen, L.-L., 1997, Parallel disassembly by onion peeling. *Journal of Mechanical Design*, **119**, 267–274.

Choi, C.K., Zha, X.F., NG, T.L., and Lau, W.S., 1998, On the automatic generation of product assembly sequences. *International Journal of Production Research*, **36**(3), 617–633.

De Fazio, T.L. and Whitney, D.E., 1987, Simplified generation of all mechanical assembly sequences. *IEEE Journal of Robotics and Automation*, **RA-3**(6), 640–658.

De Floriani, L. and Nagy, G., 1989, *A* graph model for face-to-face assembly. *Proceedings of 1989 IEEE International Conference on Robotics and Automation*, 75–78.

Delchambre, A., 1990, A pragmatic approach to computer-aided assembly planning. *Proceedings of IEEE International Conference on Robotics and Automation*, 1600–1605.

Delchambre, A. and Gaspart, P., 1992, KBAP: an industrial prototype of knowledge-based assembly planner. *Proceedings of IEEE International Conference on Robotics and Automation*, 2404–2411.

Dini, G. and Santochi, M., 1992, Automated sequencing and subassembly detection in assembly planning. *Annals of the CIRP*, **41**(1), 1–4.

Dutta, D. and Woo, T.C., 1995, Algorithm for multiple disassembly and parallel assemblies. *Journal of Engineering for Industry*, **117**, 102–109.

Erdos, G., Kis, T., and Xirouchakis, P., 2001, Modelling and evaluating product end-of-life options. *International Journal of Production Research*, **39**(6), 1203–1220.

Fox, B.R. and Kempf, K.G., 1985, Opportunistic scheduling for robotic assembly. *Proceedings of 1985 IEEE International Conference on Robotics and Automation*, 880–889.

Frommherz, B. and Hornberger, J., 1988, Automatic generation of precedence graphs. *Proceedings of 13th International Symposium on Industrial Robots*, Lausanne, CH, 453–466.

Garcia, M.A., Larré, A., López, B., and Oller, A., 2000, Reducing the complexity of geometric selective disassembly. *Proceedings of IEEE/RSJ International Conference on Intelligent Robots and Systems*, 1474–1479.

Gungor, A. and Gupta, S.M., 2001a, Disassembly sequence plan generation using a branch-and-bound algorithm. *International Journal of Production Research*, **39**(3), 481–509.

Gungor, A. and Gupta, S.M., 2001b, A solution approach to the disassembly line balancing problem in the presence of task failures. *International Journal of Production Research*, **39**(7), 1427–1467.

Gungor, A. and Gupta, S.M., 2002, Disassembly line in product recovery. *International Journal of Production Research*, **40**(11), 2569–2589.

Homem De Mello, L.S. and Sanderson, A.C., 1990, AND/OR graph representation of assembly plans. *IEEE Transactions on Robotics and Automation*, **6**(2), 188–189.

Homem De Mello, L.S., and Sanderson, A.C., 1991, A correct and complete algorithm for the generation of mechanical assembly sequences. *IEEE Transactions on Robotics and Automation*, **7**(2), 228–240.

Huang, Y.F. and Lee, C.S.G., 1989, Precedence knowledge in feature mating operation assembly planning. *Proceedings of 1989 IEEE International Conference on Robotics and Automation,* 216–221.

Huang, Y.F. and Lee, C.S.G., 1990, An automatic assembly planning system. *Proceedings of 1990 IEEE International Conference on Robotics and Automation*, 1594–1599.

Huang, Y.F. and Lee, C.S.G., 1991, A framework of knowledge-based assembly planning. *Proceedings of 1989 IEEE International Conference on Robotics and Automation,* 599–604.

Kroll, E., 1993, Modelling and reasoning for computer-based assembly planning. In: A. Kusiak (ed), *Concurrent engineering: Automation, Tools, and Techniques.* New York: Wiley-Interscience, chapter 8, pp. 177–205.

Lambert, A.J.D., 2000, Optimum disassembly sequence generation. *Proceedings of 2000 SPIE Conference of Environmentally Conscious Manufacturing*, 56–67.

Lambert, A.J.D., 2001, Automatic determination of transition matrices in optimal disassembly sequence generation. *Proceedings of 4th IEEE International Symposium on Assembly and Task Planning*, 220–225.

Lambert, A.J.D., 2002a, Generation of assembly graphs by systematic analysis of assembly structures. *Proceedings of IFAC 15th World Congress on Automatic Control*, Barcelona, Spain (CD-ROM).

Lambert, A.J.D., 2002b, Determining optimum disassembly sequences in electronic equipment. *Computers and Industrial Engineering*, **43**, 553–575.

Lee, S., 1989, Disassembly planning based on subassembly extraction. *Proceedings of* 3rd *ORSA/TIMS Conference on Flexible Manufacturing Systems: Operations Research Models and Applications*, 383–388.

Lee, S. and Shin, Y.G., 1990a, A cooperative planning system for flexible assembly. *Proceedings of Rensselaer's 2nd International Conference on Computers and Intelligent Manufacturing*, 306–313.

Lee S. and Shin, Y.G., 1990b, Assembly planning based on subassembly extraction. *Proceedings of 1990 IEEE Conference on Robotics and Automation*, 1606–1611.

Lee S. and Moradi, H., 1999, Disassembly sequencing and assembly sequence verification using force flow networks. *Proceedings of 1999 IEEE International Conference on Robotics and Automation*, 2762–2767.

Lin, A.C. and Chang, T.C., 1993, An integrated approach to automated assembly planning for three-dimensional mechanical products. *International Journal of Production Research*, **31**(5), 1201–1227.

Miller, J.M. and Stockman, G.C., 1990, Precedence constraints and tasks: how many task orderings? *Proceedings of IEEE International Conference on Systems Engineering*, 408–411.

Minzu, V., Bratcu, A., and Henrioud, J.M., 1999, Construction of the precedence graphs equivalent to a given set of assembly sequences. *Proceedings of 1999 IEEE International Conference on Assembly and Task Planning*, 14–19.

Moore K.E., Gungor, A., and Gupta, S.M., 2001, Petri net approach to disassembly process planning for products with complex AND/OR precedence relationships. *European Journal of Operations Research*, **135**, 428–449.

Murayama, T., Oba, F., and Abe, S., 1994, Assembly partitioning by genetic algorithm for generating assembly sequences efficiently. *Advancement of Intelligent Production*, 695–700.

Murayama, T., Oba, F., Abe, S., and Yamamichi, Y., 2001, Disassembly sequence generation using information entropy and heuristics. *Proceedings of 4th IEEE International Symposium on Assembly and Task Planning*, 208–213.

Ong, N.S. and Wong, Y.C., 1999, Automatic subassembly detection from a product model for disassembly sequence generation. *International Journal of Manufacturing Technology*, **15**, 425–431.

Rajan, V.N. and Nof, S.Y., 1996, Minimal precedence constraints for integrated assembly and execution planning. *IEEE Transactions on Robotics and Automation*, **12**(2), 175–186.

Seow K.T. and Devanathan R., 1994, A temporal framework for assembly sequence representation and analysis. *IEEE Transactions on Robotics and Automation*, **10**(2), 220–229.

Silva, J. and Chang, K.H., 2002, Design parameterization for concurrent design and manufacturing of mechanical systems. *Concurrent Engineering*, **10**(1), 3–14.

Srinivasan, H. and Gadh, R., 1998a, Complexity reduction in geometric selective disassembly using the wave propagation abstraction. *Proceedings of IEEE International Conference on Robotics and Automation*, vol. **2**, 1478–1483.

Srinivasan, H. and Gadh, R., 1998b, A geometric algorithm for single selective disassembly using the wave propagation abstraction. *Computer-Aided Design*, **30**(8), 603-613.

Srinivasan, H., Figueroa, R., and Gadh, R., 1999, Selective disassembly for virtual prototyping as applied to de-manufacturing. *Robotics and Computer-Integrated Manufacturing*, **15**, 231–245.

Srinivasan, H. and Gadh, R., 2000, Efficient geometric disassembly of multiple components from an assembly using wave propagation. *Journal of Mechanical Design*, **122**, 179–184.

Srinivasan, H. and Gadh, R., 2002, A non-interfering selective disassembly sequence for components with geometric constraints. *IIE Transactions*, **34**, 349–361.

Subramani, A.K. and Dewhurst, P., 1991, Automatic generation of product disassembly sequences. *Annals of the CIRP*, **40**(1), 115–118.

Swaminathan, A. and Barber, K.S., 1996, An experience-based assembly sequence planner for mechanical assemblies. *IEEE Transactions on Robotics and Automation*, **12**(2), 252–267.

Swaminathan, A., Shaikh, S.A., and Barber, K.S., 1998, Design of an experience-based assembly sequence planner for mechanical assemblies. *Robotica*, **16**, 265–283.

Tseng, Y.J. and Liou, L.C., 2000, Integrating assembly and machining planning using graph-based representation models. *International Journal of Production Research*, **38**(12), 2619–2641.

Uchiyama, N., Arai E., and Igoshi, M., 1994, Generation of mechanical assembly sequences considering different evaluation viewpoints. In: E. Usui (ed.), *Advancement of Intelligent Production*, Elsevier Science, Amsterdam 701–706.

Viswanathan, S. and Allada, V., 2001, Configuration analysis to support product redesign for end-of-life disassembly. *International Journal of Production Research*, **39**(8), 1733–1753.

Vujosevic, R., 1995, Maintainability analysis in concurrent engineering of mechanical systems. *Concurrent Engineering*, **3**(1), 61–73.

Vujosevic, R., Raskar, R., Yetekuri, N.V., Jothishankar, M.C., and Juang, S.H., 1995, Simulation, animation, and analysis of design disassembly for maintainability analysis. *International Journal of Production Research*, **33**(11), 2999–3022.

Waarts, J.J., Boneschanscher, N., and Bronsvoort W.F., 1992, A semi-automatic assembly sequence planner. *Proceedings of 1992 IEEE International Conference on Robotics and Automation*, 2431–2438.

Wilson, R.H., 1995, Minimising user queries in interactive assembly planning. *IEEE Transactions on Robotics and Automation*, **11**(2), 308–312.

Wilson, R.H. and Rit, J.F., 1990, Maintaining geometric dependencies in an assembly planner. *Proceedings of 1990 IEEE International Conference on Robotics and Automation*, 890–895.

Wilson, R.H. and Schweikard, A., 1992, Assembling polyhedra with single translations. *Proceedings of 1992 IEEE International Conference on Robotics and Automation*, 2392–2397.

Wolter, J.D., 1989, On the automatic generation of assembly plans. *Proceedings of 1989 IEEE International Conference on Robotics and Automation*, 62–68.

Woo, T.C. and Dutta, D., 1991, Automatic disassembly and total ordering in three dimensions. *Journal of Engineering for Industry*, **113**, 207–213.

Yokota, K. and Brough, D.R., 1992, Assembly/disassembly sequence planning. *Assembly Automation*, **12**(3), 31–38.

# 6 Surface- and Direction-Oriented Analysis, and Modularity

## 6.1 INTRODUCTION

The preceding chapter focused on topological and geometric properties of a product. While the process of deriving all the feasible disassembly sequences from an assembly drawing has more or less been automated and an efficient use of subset and superset rules results in minimum human interaction, some detachability analysis based on *visual inspection* remains indispensable. However, the extent of visual inspection can be reduced by the use of direction-oriented representations. Two different approaches can be distinguished here, surface-oriented analysis and direction-oriented analysis. The *surface-oriented analysis* starts with detailed information on product geometry, including the distinct surfaces of different components, thus analyzing the constraints in degrees of freedom of motion, which refer to infinitesimal relocations of components with respect to each other. This is called *movability analysis*. Translational or rotational movability is often required for the functionality of a product. Usually, this motion is constrained. The process of searching for the existence of a path along with a component or subassembly that can be detached from the rest of the product (implying motion to infinity) is called *detachability analysis*. Obstruction of the component or subassembly by components that are still present in the product might hinder the detachment. Infinity is usually approximated by the minimum distance of separation where the convex hulls of the two child subassemblies do not intersect (Oliver and Huang, 1994; Srinivasan et al., 1997). Prior to calculating this distance, the direction along which the dynamic subassembly is moved away from the static one has to be determined. This direction is called the *separation direction vector*.

In many products there is a restricted number of preferred directions (which is due to the product design) that accounts for assembly. In mechanized or automated assembly processes, this reduces the number of tool and product reorientations. Graphical tools such as obstruction diagrams, and formal tools such as disassembly precedence matrices, can reduce the amount of human interference in obtaining the set of feasible subassemblies and disassembly operations via automatic generation of the precedence relationships and the associated selection rules as well as the automatic generation of feasible cut-sets. This type of analysis is called *direction-oriented* analysis.

We will make an explicit distinction between *static* and *dynamic subassemblies*. This is referred to as the asymmetric approach (see subsection 4.3.7). In chapter 5 this was implicitly done by assuming fixed coordinates for parent subassemblies, including the original product. There, whether one component or subassembly could be moved away, leaving us with a static subassembly and a dynamic child subassembly, was examined. In practice, this distinction is effectuated by the fact that one (or more) component(s) of the product is immobilized by a *fixture*. The subassemblies that have to be detached are, according to a particular ordering, grasped by a gripper or specialized tool and moved away. Motion of the fixture, e.g., by rotating over a specific angle, might be required in some phase of the disassembly process, which is called *reorientation*.

Within the above framework we are now ready to discuss the accessibility, stability, and modularity analyses.

*Accessibility analysis* explores the possibility to reach a component with a tool. Usually, forces have to be applied to the component or subassembly that is to be detached. A minimum presupposition is the existence of free surfaces that can be reached from the outer space. Accessibility depends on the set of tools that is available (such as a specific type of gripper or a sucking disk). Initially, only the requirement of a free surface is accounted for. A restricted set of tools might put additional constraints on accessibility. Because internal forces might play a crucial role in disassembly operations, these will have to be incorporated in the analysis, e.g., via *force-flow analysis*.

*Stability analysis* refers to the stability of the subassemblies that are involved in the disassembly (or assembly) operations. This deals with investigating the possibility of whether or not a subassembly spontaneously falls apart, due to gravity, other external forces (such as vibrations), or internal forces (such as due to springs). As gravity is directed (usually in the $-z$ direction by convention), re-orientation of the product might be useful and is considered as well. Fasteners are applied to the product for guaranteeing its stability. Temporarily, fixtures and gauges can be required in the course of the operations. Stability analysis is particularly required if assembly processes are considered. Stability analysis will only be discussed concisely as further detail falls within domain of mechanical and design engineering and is beyond the scope of this book.

*Modularity analysis* deals with methods for distinguishing sets of components as distinct entities that can be manipulated as such. Although there is no precise way to define modules unambiguously, this can be done *a priori*, via intuitive rules, or automatically, via predefined heuristic rules. Stability usually is a prerequisite for a module. Apart from this, modules can be determined according to their functionality.

Before we further discuss the above analyses, a brief introduction of product classes and product representations are given.

## 6.2 PRODUCT REPRESENTATION AND CLASSES

Products traditionally represented by such mechanisms as assembly drawings and exploded views are amicable to human inspection rather than automated analysis. The introduction of computer aided design/computer aided manufacturing (CAD/CAM) enabled the application of virtual prototypes of increasing complexity that are accessible to computerized analysis as well. In both drawings and virtual prototypes, there are different approaches for product representation.

In the simplest form, all components are approximated by polygons or *polyhedra*, with the coordinates of the vertices as the principal data. An extension is the use of *nameable objects* such as cylinders and boxes, including intersections and unions of such shapes for representing a realistic component. Still more sophisticated is *boundary representation* (B-rep) in which unnameable shapes such as Bézier surfaces are applied (see Hoffman, 1990ab). We will restrict ourselves to the first two categories.

The two-dimensional (2D) case of Figure 5.6 is a typical example of a product that consists of polygons. Bourjault's ballpoint (Figure 5.1) consists of nameable surfaces, although these arise from rotation of a polyhedron on an axis. The single-piston engine of Figure 5.10 is represented via an exploded view, and its virtual prototype consists of a more or less complex set of unions and intersections of nameable surfaces. In calculations, these nameable surfaces can be approximated by polyhedra in a virtual model.

Automated movability and detachability analyses can be carried out via a trial and error method of simulating motion using virtual prototypes. This, however, usually requires lots of CPU time and is not a flexible method. In particular, if modifications are required or if online calculations have to be carried out, such a method is not appropriate. Apart from this, such a method generates suboptimum results, because not every possible motion can be analyzed.

These drawbacks can be avoided by the application of more abstract methods, which place restrictions on the class of product considered. In this case, it is stressed that a greater part of the products is characterized by a clearly defined and relatively simple geometry, which is the result of proper design and cost reduction. Therefore, products usually have a restricted set of directions of motion. If these are in a line, such as in case of Bourjault's ballpoint (Figure 5.1), one has a *motional one-dimensional* product. If these are in a plane, one has a *motional two-dimensional product*, etc. Not only planar products (such as the one in Figure 5.6), but also seemingly three-dimensional products (such as the single-piston engine of Figure 5.10) are motional two-dimensional.

Products that consist of one single type of component are analyzed by Garcia et al. (2000), who study various products that are built from LEGO-bricks. A frequently occurring category of products is the *isothetic products*, which have the mating faces in mutually orthogonal planes (see Dutta and Woo, 1995). An example of such a product is in Figure 5.9(h). The 2D-example of Figure 5.6 is not strictly isothetic, because it possesses some slanting surfaces. But even in this case, the product can be considered *motional isothetic*, i.e., the directions of motion are mutually orthogonal. The engine of Figure 5.10 is also motional isothetic, but the simple example of Figure 5.4 is not. However, a discrete set of three motional directions can be discerned here.

In the preceding chapters and subsections, such as in subsection 4.3.3, some other special categories were also presented. In the framework of this chapter, 1-removability is often assumed, which means that the detachment of a component or subassembly will take place in a single direction.

The example of Figure 5.6 is not strictly 1-removable. For example, the detachment of L from G with all the other components detached in advance, still requires a combination of two translational motions, namely, one in the $+x$ direction and one in the $-y$ direction. This implies that the detachment is 2-removable.

Many other examples of products have been discussed in the literature for illustrating various approaches in assembly and disassembly studies. Many of these products are motional one-dimensional, such as Bourjault's ballpoint (Figure 5.1) and the assembly from industry (Figure 5.7). Three-dimensional products that are presented in the literature appear motional two-dimensional, such as the single-piston engine (Figure 5.10), and a relay (see De Lit, 2001; De Lit et al., 2001). Interesting examples from the literature that are motional 2-dimensional are: a vise (Lee, 1989), a coffee maker (Knight, 1996), a toaster (Jovane et al., 1997), a hair dryer (Kroll et al., 1996), a camera (Lee and Bailey-Van Kuren, 1997; 2000), and an electric drill (Kroll and Carver, 1999). The number of components in these products range between 20 and 30.

## 6.3 SURFACE-ORIENTED ANALYSIS

### 6.3.1 General

In surface-oriented analysis, the components and subassemblies are considered in detail. If the components are represented by polyhedrons, a given component is confined by a discrete set of surfaces. If the components are confined by nameable surfaces, which can be other than planar ones, a similar reasoning can be applied. Surfaces can be either *mated surfaces*, which are in contact with other components, or *free surfaces*. Free surfaces can either be in contact with the environment, or with a cavity inside the product. Surface-oriented analysis considers the individual surfaces of the components.

### 6.3.2 Movability Analysis

A component is movable if it can be transported to a new position or orientation independently of all other components (Eng et al., 1999). Movability analysis will be illustrated here using the 2D example of Figure 5.6 where motions are restricted to the $x,y$-plane. Figure 6.1(b) depicts a subassembly of the product that consists of four components. A completely unconstrained component is able to move in all directions in the $x,y$-plane, which is indicated by a *unit circle* or *Gaussian circle*. Actually, every infinitesimal element of the circumference of this circle is perpendicular to a possible direction of motion.

The component N in Figure 6.1(b) is constrained in its motion by three adjacent components: H, L, and M. Each of these components shares a different plane-mating surface with component N. If the motion is constrained by a plane surface of a mating component, say M, only half of the unit circle is available, including the ends, which is indicated by dots (Figure 6.1(a)). Because motion is also constrained by component L, the direction is restricted according to a second unit circle. The same is true if component H is considered. If one considers movability of component N with respect to the remaining components of the subassembly HLMN, only those directions that result from the intersection of the three constrained unit circles are valid directions of motion. This intersection includes only one possible direction, the $-y$ direction, see Figure 6.1(c). If one adds component J to the subassembly, as in Figure 5.6, an additional constraint is introduced and the intersection of the four relevant circle segments becomes empty, thus implying that component N is not movable.

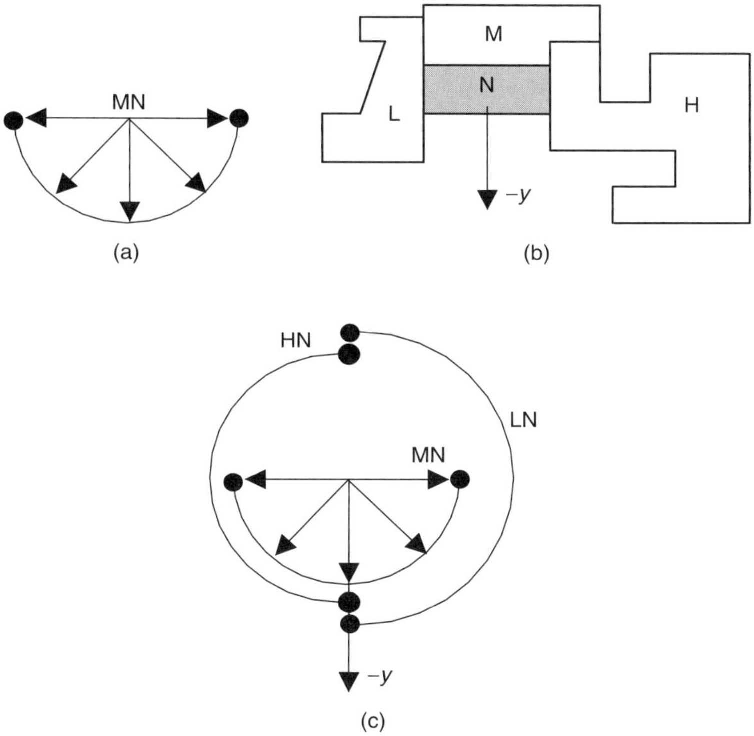


**FIGURE 6.1** Movability of component N in: (a) subassembly MN and: (c) subassembly HLMN.

When considering parallel disassembly, the subassembly that has to be detached is assumed to be a single supercomponent. Considering, for example, the subassembly JN in HJLMN (see Figure 5.6), it is evident that only motion along the $-y$ direction is possible here.

If there are multiple mating surfaces between components, such as between components C and D in Figure 5.6, the intersection of the two semicircles, each corresponding to a planar surface, represents the permitted directions of motion. If curved surfaces are present, such as the one between the components A and D in Figure 5.4, an analogous reasoning can be followed. Here also, the intersection of the semicircles represents the permitted directions of motion (see Figure 6.2).

In a 3D example a *unit sphere* (*Gaussian sphere*) instead of a unit circle is applied. Every surface element is orthogonal to a possible direction of motion here. A planar surface thus corresponds to a hemisphere instead of a semicircle. Here also motion of components can be restricted, apart from surfaces, by linear and point-like constraints. Examples are spline contacts, pin-and hole contacts, etc. In this case, the direction of motion is constrained to, for example, a single direction, or two opposite directions.

If a set of planes is considered, the resulting hemispheres are described by

$$x_i \cdot x + y_i \cdot y + z_i \cdot z \geq 0 \tag{6.1}$$

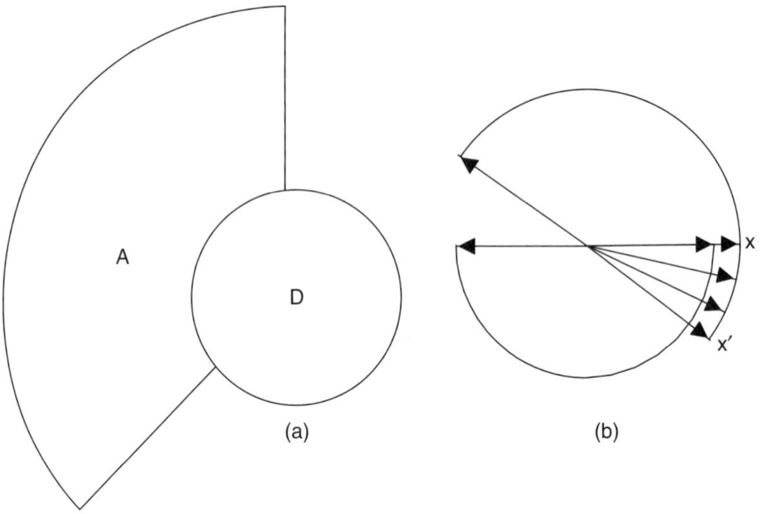


**FIGURE 6.2** Movability of a curved surface.

with

$$\sqrt{x^2 + y^2 + z^2} = 1 \tag{6.2}$$

The $x_i$, $y_i$, and $z_i$ define the directions that are perpendicular to the plane, see Siddique and Rosen (1997).

### 6.3.3 Disassembly Direction

In detachability analysis, infinity is usually approximated by the minimum distance of separation where the convex hulls of the two child subassemblies do not intersect (Oliver and Huang, 1994; Srinivasan et al., 1997). Movability analysis is only engaged with the surfaces that restrict infinitesimal motion. It does not consider obstruction at any other distance, which can be caused by interference of the static and the dynamic subassembly at some finite distance from the initial position.

Algorithms for determining the disassembly direction are most frequently used in *disassembly path generation*, which involves the calculation of the exact motion of the dynamic subassembly away from the static subassembly. Such studies stem from the domain of automated assembly planning.

Essential in disassembly path generation is the determination of the *separation direction vector*, which is the direction along which the dynamic subassembly is initially moved away from the static one. This can be done by the use of Gaussian spheres (see Oliver and Huang (1994)). If the components consist of intersecting plane surfaces, which is always true if the components are approximated by polyhedrons, the set of

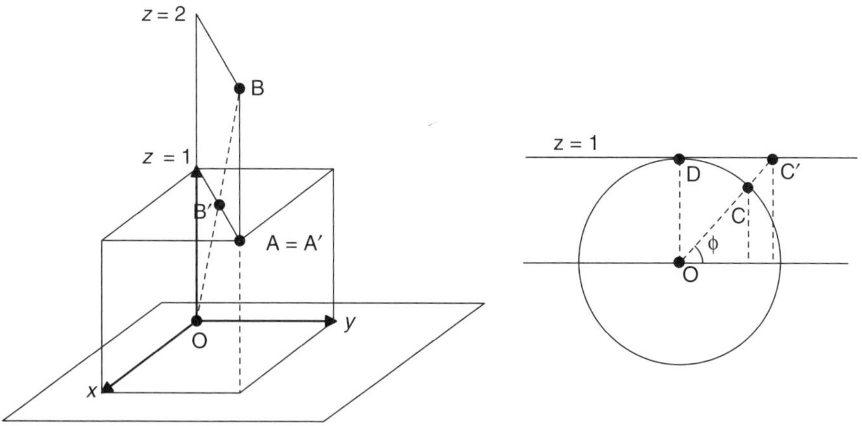


**FIGURE 6.3** Central projection of A and B in the $z = 1$ plane; central projection of points C and D on the Gaussian sphere in the $z = 1$ plane.

feasible directions of motion is represented by a *spherical polygon* that is confined by sectors of great circles. This polygon can be mapped as a polygon on the $z = 1$ plane via *central projection*. If the origin is connected to the edges of the polygon, a *polyhedral convex cone* is created, which represents all the possible directions of motion. This cone is thus confined to those vectors that point from the origin to the edges of the polygon. These vectors represent the *extreme directions*. Any linear combination of these vectors then corresponds to a potential direction of motion.

The method of *central projection* proceeds as follows: Consider the point $(x, y, z)$ in Cartesian coordinates, which has to be mapped on the $z = 1$ plane. Therefore, the coordinates of every point on the spherical polygon are divided by the value of its z coordinate, which results in $(x/z, y/z, 1)$. This implies that the antipode of the original point, i.e., $(-x, -y, -z)$, is projected on the same point of the plane. The central projection of points on the equator, with $z = 0$, thus has infinite $x$ and $y$ coordinates.

The basic features of central projection are illustrated in Figure 6.3. Point $C$ with the Cartesian coordinates $(0, \cos\phi, \sin\phi)$ has its central projection on point $C$' with coordinates $(0, \cot\phi, 1)$. Great circles on the Gaussian sphere with its center at the origin, are mapped as straight lines, except of the equator, which is at infinity. Consequently, a spherical polygon that is confined by great circles is mapped as a polygon. Connecting the edges of this polygon with the origin defines the polyhedral convex cone, which confines the possible directions of motion. The separation direction vector thus can be in any direction that is pointing from the origin to an arbitrary point on the polygon in the $z = 1$ plane. A desirable point can be, for example, the centroid of the polygon, which is determined via the arithmetic mean of its vertices.

Determination of the extreme directions and the associated polyhedral convex cone can proceed in several alternative ways.

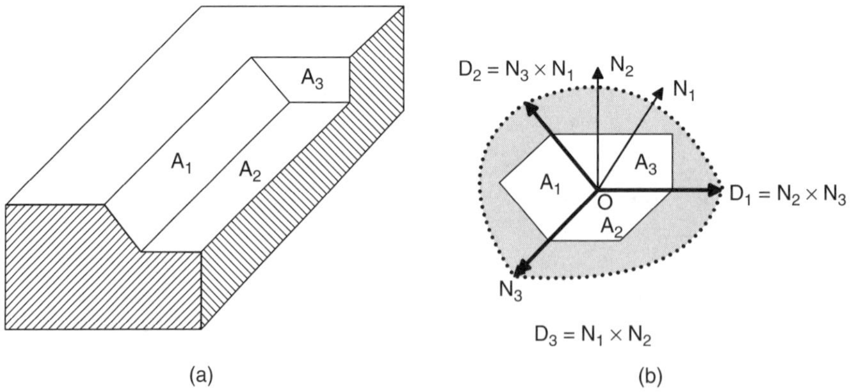


**FIGURE 6.4** Directions of motion with the polyhedral convex cone method (Nnaji, 1993).

Nnaji (1993) describes a method of the configuration in Figure 6.4. Three surfaces $A_1$ through $A_3$ of a particular component are considered (Figure 6.4(a)). The origin $O$ is chosen at the intersection of this set of surfaces. The normal vectors $\mathbf{N}_1$ through $\mathbf{N}_3$ of these surfaces are considered. If all of these are in a half space that is confined by at least one of the planes through $O$, a polyhedral convex cone exists. Next, this cone is constructed by defining its extreme directions via the products $\mathbf{N}_i \times \mathbf{N}_j$ (see Figure 6.4(b)). Its extreme directions are $\mathbf{D}_1$ through $\mathbf{D}_3$, which indeed span the set of directions in which motion is possible. Venuvinod (1993) presents an alternative algorithm for finding the principal disassembly directions if the mating surface consists of a set of multiple intersecting surfaces. For the intersection of any pair of surfaces, say $A_1$ and $A_2$ with normal vectors $\mathbf{N}_1$ and $\mathbf{N}_2$, there are four principal directions that have to be determined (see Figure 6.5):

$$\begin{aligned}
\mathbf{E}_1 &= \mathbf{N}_1 \times \mathbf{N}_2 \\
\mathbf{E}_2 &= -\mathbf{E}_1 \\
\mathbf{E}_3 &= \pm\, \mathbf{N}_1 \times \mathbf{E}_1 \text{ such that } \mathbf{E}_3 \cdot \mathbf{N}_2 \geq 0 \\
\mathbf{E}_4 &= \pm\, \mathbf{N}_2 \times \mathbf{E}_1 \text{ such that } \mathbf{E}_4 \cdot \mathbf{N}_1 \geq 0
\end{aligned} \tag{6.3}$$

Any of these vectors has to be normalized. After this, some vectors are deleted:

- From those vectors that point in the same direction, only one is left.
- Those $\mathbf{E}$ vectors that are derived for the intersection of the $i$th and $j$th plane with $\mathbf{E} \times \mathbf{N}_k \leq 0$ ($k \neq i,j$) are deleted, for these point in infeasible directions (for example, $\mathbf{E}_2$ in Figure 6.5).
- Those vectors that are coplanar with respect to two other vectors but are in between are deleted. For example, a vector in the ($\mathbf{E}_3$, $\mathbf{E}_4$) plane that can be formed via a linear combination of these vectors.

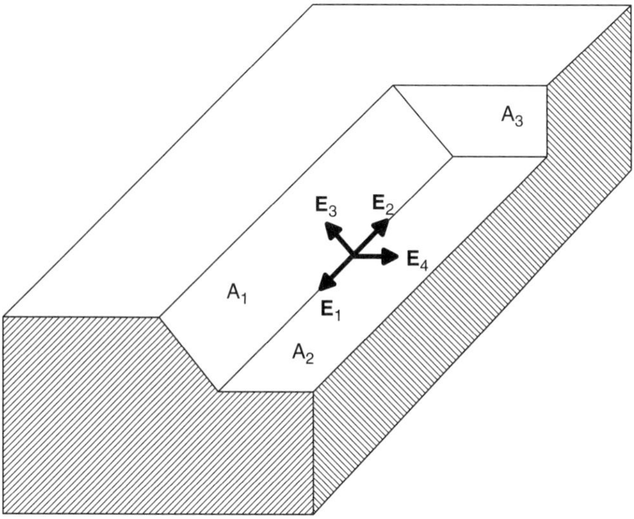


**FIGURE 6.5** Principal disassembly direction according to Venuvinod (1993).

Although Cartesian coordinates are frequently used, spherical coordinates are often applied, see, for example, Shin and Cho (1994). The relationships between spherical coordinates ($r$, $\varphi$, $\theta$) and Cartesian coordinates ($x$, $y$, $z$) are given by

$$\begin{aligned} x &= r\cos\theta\cos\varphi \\ y &= r\cos\theta\sin\varphi \\ z &= r\sin\theta \\ r &= \sqrt{x^2+y^2+z^2} \\ 0 &\le \varphi \le 2\pi \\ -\pi/2 &\le \theta \le \pi/2 \end{aligned} \tag{6.4}$$

In contrast with the central projection that uses Cartesian coordinates, the boundaries of the intersecting Gaussian spheres are mapped on the ($\varphi$, $\theta$)-plane, giving a set of intersecting sinusoids. This way, a specific area on this plane represents the separable range, which is the set of permitted directions of motion. The *separability index* can be defined as the ratio between this surface and the surface of the complete ($\varphi$, $\theta$)-plane. This index is not suitable if the separable range is given by an arc or a point, which results in zero separability index. Alternatively, an approximate index can be found via discretization of the plane by defining unit squares or circles, which are expressed in fractions of degrees or radians (see also Shyamsundhar and Gadh (1996) and Shyamsundhar et al. (1998)).

### 6.3.4 Detachability Analysis

Even if a component is movable, it might be possible that the motion to infinity of this component is obstructed by surfaces that are not mating with it. This can be demonstrated by the subassembly ACDEFGIJKLMN of the product in Figure 5.6. Although this is an infeasible subassembly from a disassembly point of view, it can nevertheless be obtained via disassembly and reassembly. The disassembly of a component is assumed here to proceed along a straight path, which is called a *ray* (see Rajan and Nof (1996)). Note that this is a restrictive assumption compared to the visual inspections on disassemblability that enables the detection of every feasible path (see chapter 5).

The set of separation direction vectors is in one plane. Consequently, it is represented by a circular sector with the angle $\varphi$, see Figure 6.6. However, motion along these directions, which were obtained via movability analysis only, does not guarantee the absence of collisions if the dynamic subassembly is moved. In this particular example, motion analysis reveals that motion of component I with respect to the remaining subassembly is permitted in the sector that is enclosed between the directions of $a_1$ and $a_2$. In part of this domain, detachment is interfered via the components A or C, which causes component I to collide. Parts of both components are shaded in Figure 6.6 to indicate this constraint. The figure also shows how the angle is constructed that corresponds to the permitted collision-free directions. This angle is represented by the symbol $\phi$. The range of directions that enable the detachment of component I in a single translational motion is represented by the intersection of $\phi$ and $\varphi$.

Although queries on detachability of given components or subassemblies can be answered automatically by this method, assumptions such as the motion along a straight line might be too rigid. Detachment of component I can also proceed via two subsequent translational motions, one in the $+x$ direction and one in the $-y$ direction. Apart from this, component I can also be detached via a combination of translational and rotational motion. Such alternative methods of detachment might remain possible even if no straight path is permitted. Detecting these

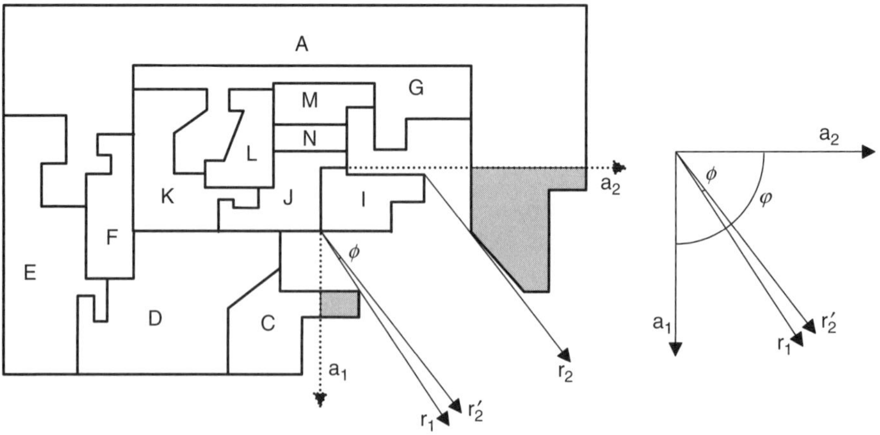


**FIGURE 6.6** Movability and detachability of component I.

composed paths requires either time-consuming trial and error calculations, or additional visual inspection.

For the subassembly of Figure 5.6, it can be concluded that many queries that previously needed visual inspection, can be automatically answered via this method. If motion analysis gives a negative result, it does not make sense to proceed with detachability analysis.

**Example 1**

The answer to the query: *is BC detachable from ABCDEFGHIJKLMN?* is YES, because a ray can be detected, which is in the negative *y* direction.

It is possible, however, that movability analysis returns a positive result, and detachability analysis returns a NO answer, because all the rays are obstructed by the remaining subassembly. This alone, however, does not guarantee that detachment is not possible.

**Example 2**

The answer to the query: *is L detachable from GL?* is NO, because no direction of detachment is detected automatically. Visual inspection, however, reveals that a composed disassembly path, consisting of two translative motions is possible.

**Example 3**

The query: *is component A detachable from ABCDEFGHIJKLMN?* is answered with NO by movability analysis only. This can be concluded automatically without further assumptions. It is even possible here to find automatically an appropriate selection rule. Therefore one selects those components that are mated with A (these are E, F, K, G, H, and B). Next one applies movability analysis with only two of those components present, e.g., E and F. As these obstruct the motion of A, the appropriate selection rule is *EF not A* (see table 5.10).

Not all the possible selection rules can be revealed via this method, because many relationships are caused by complex interactions between the components involved.

**Example 4**

The selection rule *DE not F* cannot be revealed directly from simple motion analysis, because motion of F is only indirectly obstructed by D and E, because component A is also immobilized, as long as both D, E, and F are present. A selection rule of this type is called an *indirect selection rule*.

### 6.3.5 Automated Derivation of Selection Rules with Arbitrary Surfaces

The automated generation of selection rules for axially symmetric products and for products with an orthogonal set of permitted directions of motion, will be discussed in subsection 6.4. There exists a method for dealing with products that have arbitrary

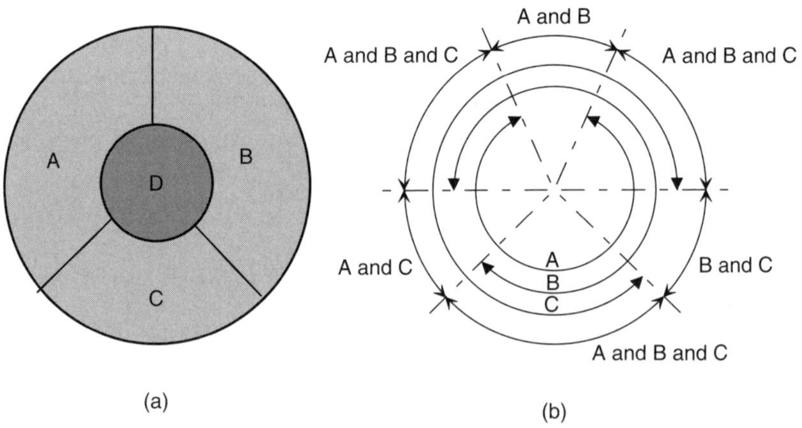


**FIGURE 6.7** Automatic determination of selection rules: (a) 2D example from Figure 5.4; (b) directions of motion of component D.

directions of motion permitted, such as the product of Figure 5.4. With this example in mind, the automatic detection of precedence relationships, as proposed by Rajan and Nof (1996), can be demonstrated (see Figure 6.7).

Movability analysis is applied first with respect to each of the peripheral components, A through C. The innermost concentric circle elements in Figure 6.7(b) represent the *obstructions* of motion of D by the components A, B, and C, respectively. These are complementary to the circle segments that represent the directions of motion as in Figure 6.1. The ends of those segments are therefore provided with arrows instead of dots, thus depicting that the end points are not in the segment. The outer circle is divided in different sectors, each corresponding to a combination of components that impede the motion in that direction.

The Boolean expressions represent the components that obstruct the motion in the corresponding direction. The expression A and B and C thus means that all three components A, B, and C have to be detached in order to enable the motion of D in that direction. The determination of the condition for motion in an arbitrary direction thus proceeds by adding the conditions via a complex AND/OR relationship, viz.:

(ABC or AB or BC or AC) not D

Since ABC not D is stricter than all of those selection rules with two components at the left, this can be relaxed into:

(AB or BC or AC) not D,

which corresponds to the three combined selection rules that were found in analyzing case (d) in Figure 5.9.

In this simple example, movability guarantees detachability, but in general products, this is not true. If one considers slightly more complex cases, such as that in

Figure 6.6, one has to account for detachability analysis as well. Therefore, any component that potentially obstructs the motion in different directions, has to be considered. This might result in the generation of lots of redundant precedence relationships. Fortunately, simpler methods can be applied in the frequently occurring class of products with a restricted set of disassembly directions (see subsection 6.4).

### 6.3.6 Destructive Disassembly

Although disassembly is normally assumed to be a nondestructive process, destructive disassembly might be beneficial in some specific cases. In the design phase, it is even possible to shape the components in such a way that the disassembly is made easier, by studying the virtual prototypes of the products. For instance, if some components obstruct the disassembly path, such as the shaded areas in Figure 6.6, the design can be reconsidered, which is sometimes called *virtual destruction* if this results in the redesign of a component, resulting in its replacement by two separate components. If the design is predefined, a similar analysis can be carried out for determining whether or not a component should be physically separated via a destructive operation aimed at facilitating the disassembly process.

In real life cases, there might be irreversible fasteners such as welds that make nondestructive disassembly impossible. In such cases, partial destruction of one or more components might be required for accomplishing the desired disassembly operations.

Figure 6.8 depicts the subassembly GL of the product in Figure 5.6. Considering the constraints that are imposed by the different surfaces, and proceeding from $S_1$ through $S_7$, it is observed that freedom of motion is gradually constrained to the $+x$ direction by $S_1$ through $S_6$. This can be detected via the application of the Gaussian spheres (circles). Since $S_7$ interferes with the motion to infinity in this direction, it can be advisable to make a planar cut in G to enable the motion in the $+x$ direction (see the dashed line in Figure 6.8). Algorithms that support destructive disassembly in this way are due to Lee and Gadh (1998).

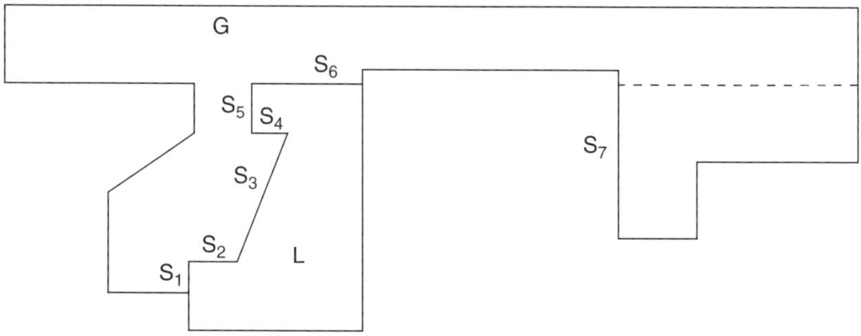


**FIGURE 6.8** Constraining surfaces for detachment of L in the $+x$ direction.

## 6.4 DIRECTION-ORIENTED APPROACH

### 6.4.1 Obstruction Diagrams

If the motion proceeds in a discrete set of directions only, the above-mentioned procedures can be considerably simplified. For this purpose, *obstruction diagrams* are used. An obstruction diagram is associated with every component. This diagram depicts those components that obstruct the detachment of a specific component in the indicated directions. It is assumed that detachment takes place in one translational direction only.

This is most clearly illustrated for axially symmetric products, which have only two principal directions of motions: $+x$ and $-x$. A multitude of example products in the assembly and disassembly sequence literature are axially symmetric, including Bourjault's ballpoint (Figure 5.1) and the AFI (Figure 5.7). We introduce obstruction diagrams with the help of Bourjault's ballpoint.

The complete set of selection rules for Bourjault's ballpoint (see Figure 5.1) is derived as follows: Consider the components sequentially, starting with component A. List the components that sequentially obstruct the motion in any of the permitted directions, i.e., the $-x$ direction and the $+x$ direction. Component A, for instance, is obstructed first by B, next by F, if moved in the $-x$ direction. The component is obstructed by the component E if moved in the $+x$ direction. This is indicated in Figure 6.9.

Obviously, component A can neither be detached if both B and E are present, nor if both E and F are present in the subassembly. This is expressed via the selection rules, which are written here as a complex AND/OR relation:

(BE or EF) not A

Along the same lines, the following additional selection rules can be derived:

(AF or CF or DF or EF) not B
(BE or EF) not C
(BE or EF) not D

Consequently, 10 selection rules are found. Only two of these are required for finding the complete set of feasible subassemblies, namely BE not D, and AF not B, see the Table 5.3(a) and Table 5.3(b). The remaining selection rules are redundant.

The obstruction diagram of the AFI of Figure 5.7 is shown in Figure 6.10. There are 130 selection rules here (see Figure 6.10). In Table 5.9, only 12 of these are used. The remaining selection rules are redundant.

If multiple directions of motion are possible, the obstruction diagram has to be extended. Assuming that only simple translational motions are permitted, the 2D example of Figure 5.6 can be disassembled via four principal directions of motion, which are parallel to two orthogonal axes: $\pm x$ and $\pm y$. An obstruction diagram that includes all permitted directions of motion for every component has to be constructed. This is depicted in Figure 6.11.

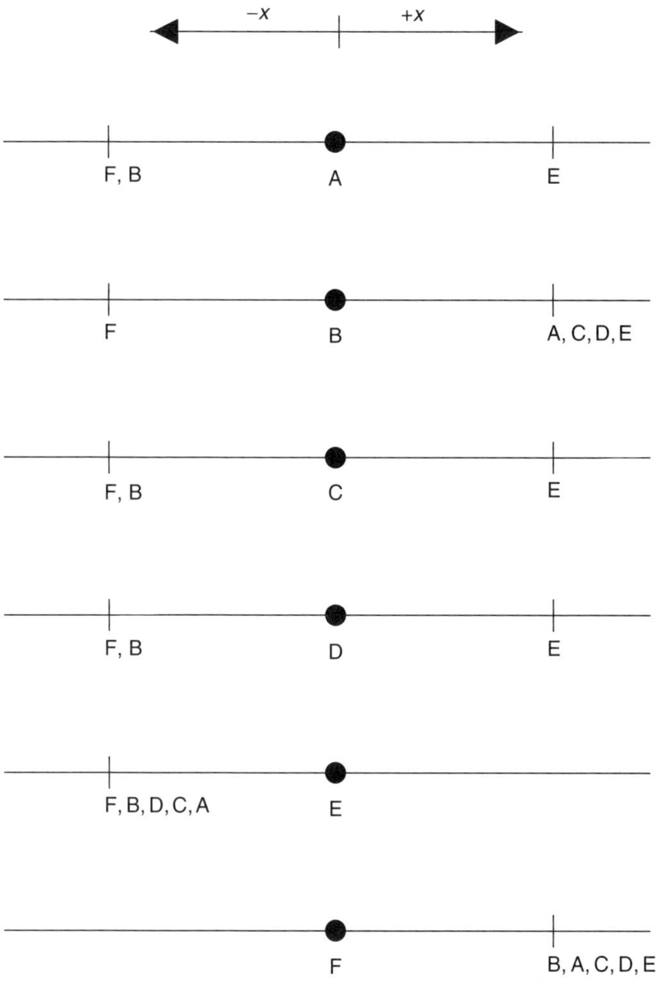


**FIGURE 6.9** Obstruction diagram for Bourjault's ballpoint of Figure 5.1.

The generation of this diagram can be automated if information on component surfaces is available. One can conclude from these diagrams that the product is not 1-disassemblable, because there is no component that is unconstrained in at least one direction.

### 6.4.2 Support of the Complete Cut-Set Method

The obstruction diagrams can deal with parallel disassemblies. Therefore, they can be used for the selection of feasible cut-sets, thus revealing all possibilities of parallel disassembly. We present a systematic algorithm that selects these cut-sets, even for a $k$-disassemblable product. The steps for the algorithm are presented with the help of the 2D example of Figure 5.6 as follows, see Figure 6.11:

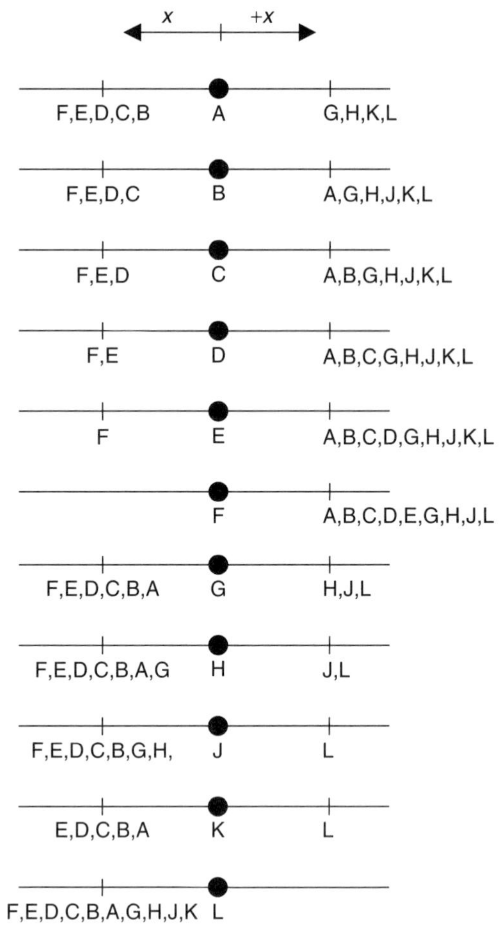


**FIGURE 6.10** Obstruction diagram for the AFI example of Figure 5.7.

*Step 1:* Start with component A. Here, only the $+y$ direction has to be considered. The motion of A is blocked by E and F, thus E and F have to be considered. Motion of E is obstructed by A and F, and motion of F is obstructed by A. Thus subassembly AEF can be detached in the $+y$ direction.

*Step 2:* Proceed with component B. Motion in the $-y$ direction is obstructed by C, motion in the $+x$ direction is obstructed by A. Motion of A is obstructed in the $+x$ direction by many components, thus this direction should not be considered. Considering the motion of component C in the $-y$ direction reveals that it is obstructed by B only. Thus subassembly BC can be detached from the product by motion in the $-y$ direction.

*Step 3:* Proceed with component C. This can be disassembled via motion in the $-y$ direction of BC. In the $-x$ direction one meets DEF. Thus F needs be detached simultaneously with C in this direction. Motion of F, however,

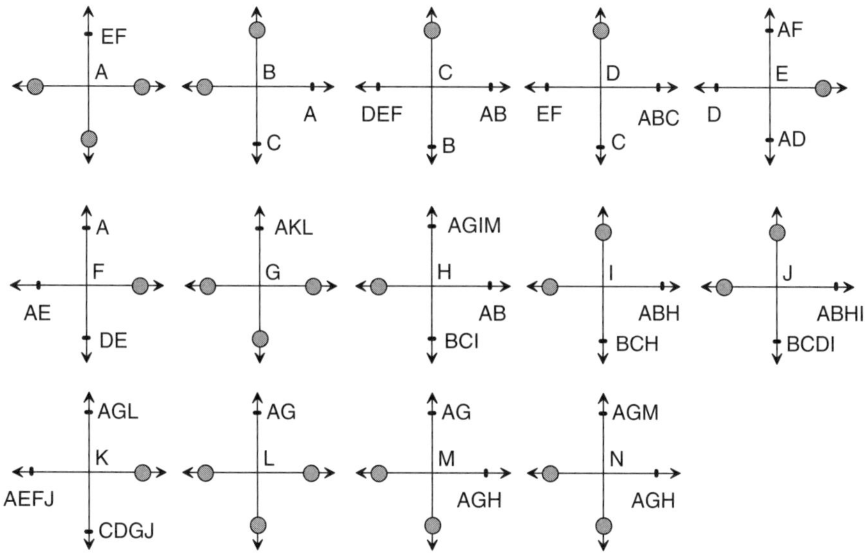


**FIGURE 6.11** Obstruction diagrams for the example of Figure 5.6. Shaded circles represent the obstruction of motion in the associated direction by five or more components.

is obstructed by component A, which cannot be detached in that direction, thus there is no subassembly in which C can be detached in the – *x* direction.

*Step 4:* Proceed with component D. It can be detached in the –*y* direction via subassembly BCD.

*Step 5:* Both the components E and F can be detached in the +*y* direction via the subassembly AEF.

*Step 6:* Study of the detachment of the component G is interesting because it reveals a nontrivial cut-set. The only reasonable direction of motion is +*y*. However, this motion is obstructed by A, K, and L. Motion of component A is obstructed by E and F; motion of component L is obstructed by A and G. Motion of component K is obstructed by A, G and L. Thus the detachment of AEFGKL in the +*y* direction is a possibility for the detachment of G from the complete product. This corresponds to the cut-set:

$$\text{ABCDEFGHIJKLMN} \rightarrow \text{AEFGKL, BCDHIJMN}$$

*Step 7:* Components H and I can be detached in the –*y* direction via subassembly BCHI.

*Step 8:* Component J can be detached in the –*y* direction via BCDHIJ.

*Step 9:* Component K can be detached in the +*y* direction via AEFGKL.

*Step 10:* Component L can be detached in the +*y* direction via AEFGL.

*Step 11:* Component M can be detached in the +*y* direction via AEFGLM.

*Step 12:* Component N can be detached in the +*y* direction via AEFGLMN.

It is evident that *symmetry* occurs. If, for example, AEFGKL can be detached via motion in the $+y$ direction, then the subassembly BCDHIJMN, which is a combination of BCD and AEFGKL, can be detached via motion in the $-y$ direction.

It should be noted that, once the obstruction diagrams are available, conclusions on detachability of subassemblies could be drawn automatically, without any user query.

Evidently, subassemblies can be studied along the same lines. If some components have been detached beforehand, the set of obstruction diagrams is relaxed. Those that correspond to an already detached component are rejected. From the remaining ones, the components that are no longer present are removed as constraints.

This way, it is possible to automatically find the possible modes of disassembly, without going into laborious cut-set methods.

A serious drawback of using obstruction diagrams is that it does not support $m$-removability, such as in the removal of L from G in Figure 5.6. This problem can be circumvented here somewhat artificially by removing the G constraint in the $+x$ direction: if J, M, and N are detached, L can be detached from G.

### 6.4.3 Motion Analysis

This subsection deals with direction-oriented motion analysis. It starts with *motion diagrams*, which are similar to obstruction diagrams, but with only those components included that restrict the freedom of motion. The motion diagram for Bourjault's ballpoint (Figure 5.1) is presented in Figure 6.12.

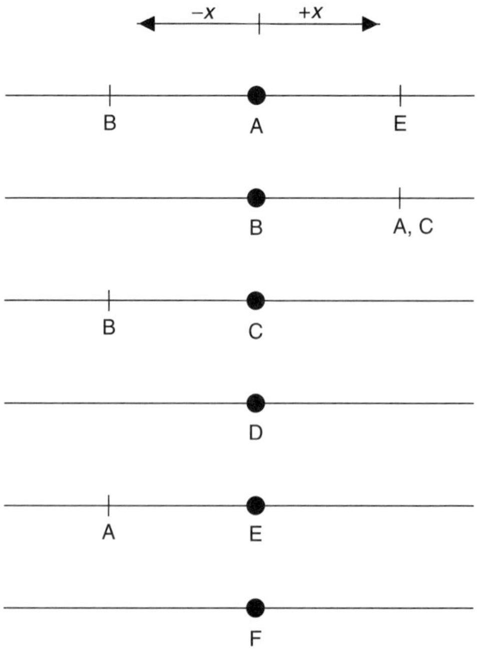


**FIGURE 6.12** Motion diagram of Bourjault's ballpoint of Figure 5.1.

From Figure 6.12, it clear that both components D and F can move freely along some distance with respect to the remaining subassembly (see Figure 5.1). We would like to remind the readers that the forces are neglected here.

Surface analysis is based on the construction of motion diagrams, enabling automated construction of motion diagrams by considering individual surfaces. The works of Dutta and Woo (1995), and Chen et al. (1997) deal with surface-oriented analysis that is based on detecting free surfaces and automatically determining disassembly sequences by a heuristic called *onion peeling*. It is based on systematically working through the product inward from outside. This work is mainly confined to isothetic 2D examples, see subsection 6.2. A main focus here is the search for appropriate combinations of components that can be detached via parallel disassembly. In the preceding subsection, it was revealed that the technique of obstruction diagrams offers an alternative method in detecting these combinations. The works of these authors go into more details, as it is aimed at automatically deriving the motion diagrams using the data on individual surfaces.

Surface-oriented analysis is indispensable for accessibility analysis, which distinguishes between mating surfaces and free surfaces. In a general case, a surface may also be partly free. Free surfaces may be either enclosed, or accessible, i.e., mating with the exterior space of the product. This is applied in the automatic search for accessible surfaces straight from a virtual prototype, aimed at finding attachments for grippers and fixtures (see Séré et al., 1995). In the course of the disassembly process, mating surfaces might transform into accessible surfaces.

Surfaces may be planar, such as the ones that occur in polygonal representations, or curved, such as the ones that occur in cylindrical contacts. In surface analysis, all the individual surfaces (also called *faces*) are labeled, which correspond to the surface-oriented approach.

The relationships between surfaces are represented by a *surface adjacency graph*, which indicates the mating between different surfaces. The mating can involve a *point contact* (0D), a *line contact* (1D) or a *face contact* (2D) (see Woo and Dutta (1991)). Apart from the components, a *virtual component* that reflects the outer space and, possibly, some virtual components that reflect the inner spaces or cavities, are added. From this, the *surface adjacency graph* is derived, which is an undirected graph with surfaces as nodes and contacts between surfaces as arcs (see Figure 6.13).

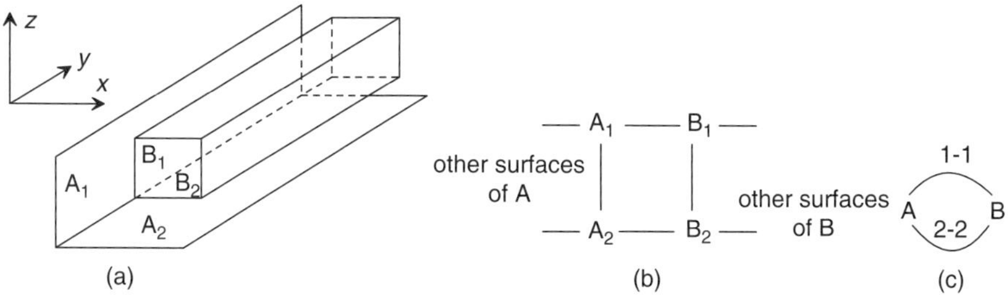


**FIGURE 6.13** (a) Example of mating components; (b) Surface adjacency graph; (c) Reduced surface adjacency graph.

The surfaces $A_1$ and $B_1$ are mating surfaces of the components A and B respectively; the same is true for $A_2$ and $B_2$. It is clear that such a diagram might be more complex in real life cases. For such applications, deleting the contacts between different surfaces of the same component can perform a reduction in the surface adjacency graph (see Eng et al. (1999)). This way, a *reduced adjacency graph*, which is an *extended connection diagram*, is constructed (see Figure 6.13(c)). Joining all the mating surfaces between the distinct components results in the conventional connection diagram (see subsection 3.2.9).

A surface-oriented mating matrix or *freedom matrix* can be determined for each pair of mating surfaces, which can be described by three rows, each representing motion along an orthogonal axis. The columns represent the positive and negative translational and rotational directions, which makes up a 3 ×4 matrix as follows:

$$F = \begin{vmatrix} +x & -x & \omega_x & \omega_{-x} \\ +y & -y & \omega_y & \omega_{-y} \\ +z & -z & \omega_z & \omega_{-z} \end{vmatrix} \tag{6.5}$$

This type of notation has also been discussed by Mascle and Figour (1990); Mascle (1992, 1995, 1998, 1999); Mascle and Balasoiu (2001), who call it a *connection pseudomatrix*. This kind of matrix can be used to determine the degrees of freedom of some component or group of components in the environment of other components.

If only translational motions are considered, the rotations can be removed and a 3 × 2 matrix results. For the case of Figure 6.13, the following matrices appear:

$$F_{AB,1-1} = \begin{vmatrix} 0 & 1 \\ 0 & 0 \\ 0 & 0 \end{vmatrix} \qquad F_{AB,2-2} = \begin{vmatrix} 0 & 0 \\ 0 & 0 \\ 0 & 1 \end{vmatrix} \tag{6.6}$$

In this notation, A is the static component and B is the dynamic component. It is read such that the motion of B is obstructed in the $-x$ direction by the mating surfaces $A_1 - B_1$ and in the $-z$ direction by the mating surfaces $A_2 - B_2$. This is indicated with the matrix element equaling 1. If no restriction is present, the corresponding matrix element equals 0. For representing the constraints in motion by the mating of the components A and B, all the matrices that correspond to the individual mating surfaces are added in such a way that the element that corresponds to a particular direction of motion equals 1 if at least one of the corresponding elements in a single surface matrix is 1, thus

$$F_{AB} = F_{AB,1-1} \vee F_{AB,2-2} = \begin{vmatrix} 0 & 1 \\ 0 & 0 \\ 0 & 1 \end{vmatrix} \tag{6.7a}$$

which means that motion of B relative to A is obstructed in both the $-x$ and $-z$ directions. These are called *matrices of half degrees* or *component-oriented interference matrices*. The symbol $\vee$ is the disjunctive relationship. In Mascle's original approach, additional information on the occurrence of zero and 1-elements in the disjunctive procedure is added, according to

$$\begin{aligned} 1^m \vee 1^n &= 1^{m+n} \\ 1^m \vee 0^n &= 1^m \\ 0^m \vee 0^n &= 0^n (n \geq m) \end{aligned} \tag{6.8}$$

with $m$ an $n$ as integers.

The following rule is valid: For any pair of components, say A and B, and for any direction, say $x$, the obstruction of the motion of component B by component A in the $-x$ direction is equivalent to the obstruction of the motion of A by B in the $+x$ direction. Thus

$$F_{BA} = \begin{vmatrix} 1 & 0 \\ 0 & 0 \\ 1 & 0 \end{vmatrix} \tag{6.7b}$$

Alternatively, *directed connectivity matrices C* are composed from the set of component-oriented interference matrices. These represent the restriction of the motion of a component by other components in a specific direction. From Expression 6.7a the following set of interference matrices is derived:

$C_{+x} =$

| | A | B |
|---|---|---|
| A | 0 | 1 |
| B | 0 | 0 |

$C_{+y} =$

| | A | B |
|---|---|---|
| A | 0 | 0 |
| B | 0 | 0 |

$C_{+z} =$

| | A | B |
|---|---|---|
| A | 0 | 1 |
| B | 0 | 0 |

(6.9a)

These can be interpreted as follows: $C_{+x}$ means that the motion of component A is constrained in the $+x$ direction by component B; motion of the component B is unrestricted by component A in the $+x$ direction. This set contains complete information, because the interference matrix that refers to the opposite direction is the transpose of the original one, thus

$$C_{-x} = C_{+x}^{T} \tag{6.9b}$$

The transpose means that columns and rows are reverted such that for any element of the matrix the following relationship holds:

$$C_{ij} = C_{ji}^{T} \tag{6.10}$$

The directed connectivity matrix contains all the information as the connectivity matrices (see Figures 3.6 and Figure 4.16). However, because this type of matrix contains additional information, a square matrix (and not a triangular matrix) is required here. Dini and Santochi (1992) introduced the directed connectivity matrices in disassembly studies. The following gives the directed connectivity matrix $C_{+x}$ for the Bourjault's ballpoint, which is derived from the motion diagrams that are presented in Figure 6.12.

| | A | B | C | D | E | F |
|---|---|---|---|---|---|---|
| A | 0 | 0 | 0 | 0 | 1 | 0 |
| B | 1 | 0 | 1 | 0 | 0 | 0 |
| C | 0 | 0 | 0 | 0 | 0 | 0 |
| D | 0 | 0 | 0 | 0 | 0 | 0 |
| E | 0 | 0 | 0 | 0 | 0 | 0 |
| F | 0 | 0 | 0 | 0 | 0 | 0 |

(6.11)

If the motion of a component is constrained by all the permitted directions, which is indicated by "1" in any of the directed connectivity matrices, the component is immovable. For instance, component N in the product of Figure 5.6 is immovable. These might not only include the six translational directions, but also rotational directions are included, such as $\omega_{\pm x}$, which represent rotation in the positive or negative $x$-directions. In cases such as Bourjault's ballpoint, rotational motion in the $x$ direction is possible for every component as long as no forces are considered and only geometric constraints are included.

Note that movability does not guarantee detachability, as the desired direction of motion may be obstructed. Similarly, in the case of rotational motions only, it may not result in detachment of the component.

**Example 5**

The motion diagram of the product in Figure 5.6 is derived using Figure 6.11 and is presented in Figure 6.14. The directed connectivity matrices can be derived from this.

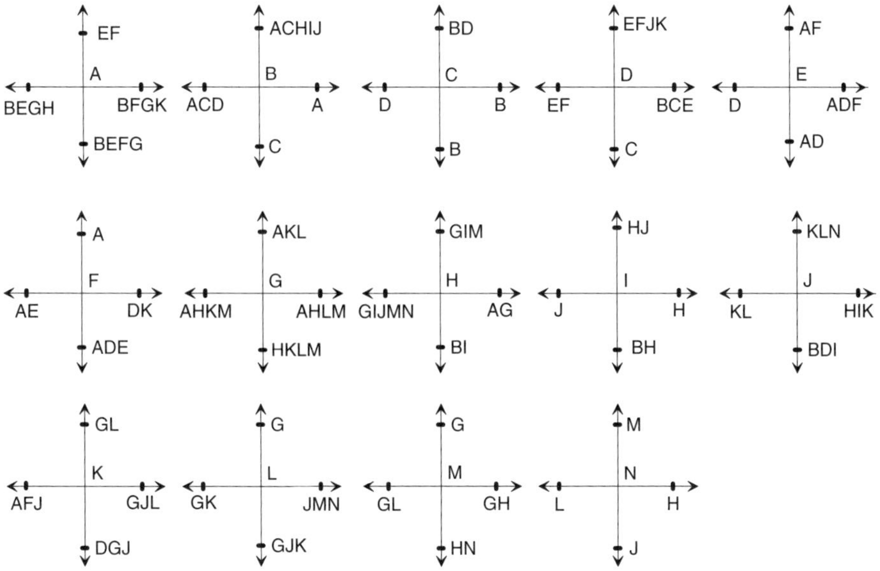


**FIGURE 6.14** Motion diagram for the product of Figure 5.6.

The related directed connectivity matrices are

$C_{+x} =$

| | A | B | C | D | E | F | G | H | I | J | K | L | M | N |
|---|---|---|---|---|---|---|---|---|---|---|---|---|---|---|
| A | 0 | 1 | 0 | 0 | 0 | 1 | 1 | 0 | 0 | 0 | 1 | 0 | 0 | 0 |
| B | 1 | 0 | 0 | 0 | 0 | 0 | 0 | 0 | 0 | 0 | 0 | 0 | 0 | 0 |
| C | 0 | 1 | 0 | 0 | 0 | 0 | 0 | 0 | 0 | 0 | 0 | 0 | 0 | 0 |
| D | 0 | 1 | 1 | 0 | 1 | 0 | 0 | 0 | 0 | 0 | 0 | 0 | 0 | 0 |
| E | 1 | 0 | 0 | 1 | 0 | 1 | 0 | 0 | 0 | 0 | 0 | 0 | 0 | 0 |
| F | 0 | 0 | 0 | 1 | 0 | 0 | 0 | 0 | 0 | 0 | 1 | 0 | 0 | 0 |
| G | 1 | 0 | 0 | 0 | 0 | 0 | 0 | 1 | 0 | 0 | 0 | 1 | 1 | 0 |
| H | 1 | 0 | 0 | 0 | 0 | 0 | 1 | 0 | 0 | 0 | 0 | 0 | 0 | 0 |
| I | 0 | 0 | 0 | 0 | 0 | 0 | 0 | 1 | 0 | 0 | 0 | 0 | 0 | 0 |
| J | 0 | 0 | 0 | 0 | 0 | 0 | 0 | 1 | 1 | 0 | 1 | 0 | 0 | 0 |
| K | 0 | 0 | 0 | 0 | 0 | 0 | 1 | 0 | 0 | 1 | 0 | 1 | 0 | 0 |
| L | 0 | 0 | 0 | 0 | 0 | 0 | 0 | 0 | 0 | 1 | 0 | 0 | 1 | 1 |
| M | 0 | 0 | 0 | 0 | 0 | 0 | 1 | 1 | 0 | 0 | 0 | 0 | 0 | 0 |
| N | 0 | 0 | 0 | 0 | 0 | 0 | 0 | 1 | 0 | 0 | 0 | 0 | 0 | 0 |

(6.12a)

$C_{+y} =$

| | A | B | C | D | E | F | G | H | I | J | K | L | M | N |
|---|---|---|---|---|---|---|---|---|---|---|---|---|---|---|
| A | 0 | 0 | 0 | 0 | 1 | 1 | 0 | 0 | 0 | 0 | 0 | 0 | 0 | 0 |
| B | 1 | 0 | 1 | 0 | 0 | 0 | 0 | 1 | 1 | 1 | 0 | 0 | 0 | 0 |
| C | 0 | 1 | 0 | 1 | 0 | 0 | 0 | 0 | 0 | 0 | 0 | 0 | 0 | 0 |
| D | 0 | 0 | 0 | 0 | 1 | 1 | 0 | 0 | 0 | 1 | 1 | 0 | 0 | 0 |
| E | 1 | 0 | 0 | 0 | 0 | 1 | 0 | 0 | 0 | 0 | 0 | 0 | 0 | 0 |
| F | 1 | 0 | 0 | 0 | 0 | 0 | 0 | 0 | 0 | 0 | 0 | 0 | 0 | 0 |
| G | 1 | 0 | 0 | 0 | 0 | 0 | 0 | 0 | 0 | 0 | 1 | 1 | 0 | 0 |
| H | 0 | 0 | 0 | 0 | 0 | 0 | 1 | 0 | 1 | 0 | 0 | 0 | 1 | 0 |
| I | 0 | 0 | 0 | 0 | 0 | 0 | 0 | 1 | 0 | 1 | 0 | 0 | 0 | 0 |
| J | 0 | 0 | 0 | 0 | 0 | 0 | 0 | 0 | 0 | 0 | 1 | 1 | 0 | 1 |
| K | 0 | 0 | 0 | 0 | 0 | 0 | 1 | 0 | 0 | 0 | 0 | 1 | 0 | 0 |
| L | 0 | 0 | 0 | 0 | 0 | 0 | 1 | 0 | 0 | 0 | 0 | 0 | 0 | 0 |
| M | 0 | 0 | 0 | 0 | 0 | 0 | 1 | 0 | 0 | 0 | 0 | 0 | 0 | 0 |
| N | 0 | 0 | 0 | 0 | 0 | 0 | 0 | 0 | 0 | 0 | 0 | 0 | 1 | 0 |

(6.12b)

The undirected connectivity matrix is found via

$$C_{+x} \vee C_{+x}^{T} \vee C_{+y} \vee C_{+y}^{T} \tag{6.13}$$

*Precedence relationships* can be found from the directed connectivity matrices as follows:

Consider an arbitrary subset of components, for example, ABC. The interference matrices that belong to this subset are as follows:

$C_{+x} =$

| | A | B | C |
|---|---|---|---|
| A | 0 | 1 | 0 |
| B | 1 | 0 | 0 |
| C | 0 | 1 | 0 |

(6.14a)

$C_{+z} =$

| | A | B | *C* |
|---|---|---|---|
| A | 0 | 0 | 0 |
| B | 1 | 0 | 1 |
| *C* | 0 | 0 | 0 |

(6.14b)

The column in $C_{+x}$, that corresponds to component C has only zeros, which means that C can be detached in the $-x$ direction. The row that corresponds to component A in $C_{+z}$ also has only zeros. Thus A can be detached in the $+y$ direction. B cannot be detached if both A and C are present, thus AC not B is a selection rule. This is an example of how selection rules can be automatically derived using subsets of directed connectivity matrices.

### 6.4.4 Detachability Analysis and Disassembly Sequence Generation

Obstruction diagrams are more restrictive than motion diagrams, for these also include the components that obstruct the path that is required for detachment even though they are not mutually connected. Similarly, an interference matrix is more restrictive than the corresponding directed connectivity matrix as it has additional elements with "1" representing the components that only obstruct the path but are not connected. For the AFI example (Figure 5.7), the obstruction diagram is presented in Figure 6.10 while the *interference matrix* is as follows:

$C_{+x} =$

| | A | B | C | D | E | F | G | H | J | K | L |
|---|---|---|---|---|---|---|---|---|---|---|---|
| A | 0 | 0 | 0 | 0 | 0 | 0 | 1 | 1 | 0 | 1 | 1 |
| B | 1 | 0 | 0 | 0 | 0 | 0 | 1 | 1 | 1 | 1 | 1 |
| C | 1 | 1 | 0 | 0 | 0 | 0 | 1 | 1 | 1 | 1 | 1 |
| D | 1 | 1 | 1 | 0 | 0 | 0 | 1 | 1 | 1 | 1 | 1 |
| E | 1 | 1 | 1 | 1 | 0 | 0 | 1 | 1 | 1 | 1 | 1 |
| F | 1 | 1 | 1 | 1 | 1 | 0 | 1 | 1 | 1 | 0 | 1 |
| G | 0 | 0 | 0 | 0 | 0 | 0 | 0 | 1 | 1 | 0 | 1 |
| H | 0 | 0 | 0 | 0 | 0 | 0 | 0 | 0 | 1 | 0 | 1 |
| J | 0 | 0 | 0 | 0 | 0 | 0 | 0 | 0 | 0 | 0 | 1 |
| K | 0 | 0 | 0 | 0 | 0 | 0 | 0 | 0 | 0 | 0 | 1 |
| L | 0 | 0 | 0 | 0 | 0 | 0 | 0 | 0 | 0 | 0 | 0 |

(6.15)

This matrix can also be used for automated disassembly sequence generation. Here the directed connectivity matrix does not result in any sequential sequence, for there are neither rows nor columns with only zeros. This is because the product considered is *m*-disassemblable.

The automated disassembly sequence generation starting from the interference matrix Expression (6.15) for the AFI example (Figure 5.7) proceeds as follows:

Search either rows or columns that contain "0" elements only. The row that corresponds to L and the column that corresponds with F have zeros only. Thus L can be detached in the $+x$ direction and L can be detached in the $-x$ direction.

Reduce the matrix by removing the rows and columns related to F and L as follows:

$C_{+x} =$ (6.15a)

| | A | B | C | D | E | G | H | J | K |
|---|---|---|---|---|---|---|---|---|---|
| A | 0 | 0 | 0 | 0 | 0 | 1 | 1 | 0 | 1 |
| B | 1 | 0 | 0 | 0 | 0 | 1 | 1 | 1 | 1 |
| C | 1 | 1 | 0 | 0 | 0 | 1 | 1 | 1 | 1 |
| D | 1 | 1 | 1 | 0 | 0 | 1 | 1 | 1 | 1 |
| E | 1 | 1 | 1 | 1 | 0 | 1 | 1 | 1 | 1 |
| G | 0 | 0 | 0 | 0 | 0 | 0 | 1 | 1 | 0 |
| H | 0 | 0 | 0 | 0 | 0 | 0 | 0 | 1 | 0 |
| J | 0 | 0 | 0 | 0 | 0 | 0 | 0 | 0 | 0 |
| K | 0 | 0 | 0 | 0 | 0 | 0 | 0 | 0 | 0 |

Search for new rows and columns with "0" elements only. The column that corresponds to E and the rows that correspond to J and K have zeros only. This implies that J and K can be detached in the $+x$ direction and component E can be detached in the $-x$ direction. Continuing in a similar fashion would lead to a complete disassembly sequence.

### 6.4.5 Aggregation of Components

Parallel disassembly can be analyzed via the combination of multiple components. The combined components can act as a single supercomponent in an interference matrix. Such matrix is called a *contracted interference matrix*. This concept is demonstrated here for the AFI example of Figure 5.7 (see Expression 6.15), by considering DEF as a subassembly. Here, the rows and columns that correspond to the individual components of DEF are combined. The diagonal element is put to zero, because the detachment of a subassembly cannot be obstructed by itself. If detachment of a component that does not belong to the subassembly is obstructed in the $+x$ direction by any of the components that belong to the subassembly, the corresponding row element is given a value of 1, else the value of 0 is given. If detachment of the component is obstructed by any of the components of the subassembly in the $-x$ direction, the corresponding column element is given a value of 1, else the value of 0 is given. The result, which is derived from Expression 6.15, is given below:

$C_{+x} =$ (6.15b)

| | A | B | C | G | H | J | K | L | DEF |
|---|---|---|---|---|---|---|---|---|---|
| A | 0 | 0 | 0 | 1 | 1 | 0 | 1 | 1 | 0 |
| B | 1 | 0 | 0 | 1 | 1 | 1 | 1 | 1 | 0 |
| C | 1 | 1 | 0 | 1 | 1 | 1 | 1 | 1 | 0 |
| G | 0 | 0 | 0 | 0 | 1 | 1 | 0 | 1 | 0 |
| H | 0 | 0 | 0 | 0 | 0 | 1 | 0 | 1 | 0 |
| J | 0 | 0 | 0 | 0 | 0 | 0 | 0 | 1 | 0 |
| K | 0 | 0 | 0 | 0 | 0 | 0 | 0 | 1 | 0 |
| L | 0 | 0 | 0 | 0 | 0 | 0 | 0 | 0 | 0 |
| DEF | 1 | 1 | 1 | 1 | 1 | 1 | 1 | 1 | 0 |

From Expression (6.15b), it is clear that the subassembly DEF can be detached as a whole in the $-x$ direction, because its column elements are all zeroes. Because detachment of component C in the $-x$ direction is only obstructed by DEF, this component can be added via an additional contraction step, thus resulting in a subassembly CDEF, with zero column elements implying that this can also be disassembled as a whole. Using similar iterations, one detects that subassemblies BCDEF or ABCDEF can also be disassembled as a whole. This illustrates a valuable tool for automatic detection of parallel disassembly operations.

## 6.4.6 Alternative Matrix Representations

### 6.4.6.1 Disassembly Precedence Matrices

Approaches that are closely related to the previously described representations can be found throughout the literature. Although different authors use different notations, the matrices nevertheless contain similar kinds of information. Laperriere and ElMaraghy (1992, 1994, 1996), and Huang and Huang (2002) use the *obstruction component matrix* notation, in which the components are represented by rows and the directions by columns. For the sake of generality, we will use a similar notation here. The obstruction component matrix for the Bourjault's ballpoint (Figure 5.1) can be derived from the obstruction diagram of Figure 6.9 as follows:

| | $x$ | $-x$ |
|---|---|---|
| A | E | BF |
| B | ACDE | F |
| C | E | BF |
| D | E | BF |
| E | ∅ | ABCDF |
| F | ABCDE | ∅ |

(6.16)

For products involving multidimensional directions of detachment, additional columns can be added to the obstruction matrix representing the additional directions.

From the obstruction matrix, an $N \times N$ the *disassembly precedence matrix* (DPM) (see Gungor and Gupta, 2001a) can be derived that contains the directions in which a given component (represented in the row) is obstructed. This matrix represents the same information as is in the obstruction component matrix. The DPM derived from the obstruction matrix in Expression 6.16 is as follows:

| | A | B | C | D | E | F |
|---|---|---|---|---|---|---|
| A | 0 | $-x$ | 0 | 0 | $x$ | $-x$ |
| B | $x$ | 0 | $x$ | $x$ | $x$ | $-x$ |
| C | 0 | $-x$ | 0 | 0 | $x$ | $-x$ |
| D | 0 | $-x$ | 0 | 0 | $x$ | $-x$ |
| E | $-x$ | $-x$ | $-x$ | $-x$ | 0 | $-x$ |
| F | $x$ | $x$ | $x$ | $x$ | $x$ | 0 |

(6.17)

This formalism has been used by Zhou et al. (1998) as a tool for studies in the design for disassembly.

Selection rules can be derived from the disassembly precedence matrix using the following method that is illustrated using Expression 6.17. Consider the row that corresponds to A. It is clear that if both components, B and E are present in this row, detachment of A is constrained in both the $+x$ and the $-x$ directions. Consequently, component A cannot be detached prior to the detachment of either B or E. Following similar reasoning, the following selection rules are obtained:

$$\begin{gathered}\text{BE not A; EF not A} \\ \text{AF not B; CF not B; DF not B; EF not B} \\ \text{BE not C; EF not C} \\ \text{BE not D; EF not D}\end{gathered} \tag{6.18}$$

When multiple directions are involved, detachment is inhibited if a component is constrained in every permitted direction.

In studies by Moore et al. (1998abc; 2001), the authors assign a value of "1" to elements that refer to fasteners, such as bolts, which constrain detachment in every direction. This represents an AND relationship, because all the components that are referred to with a "1" in a row, have to be detached prior to the detachment of the component that is assigned to this specific row.

### 6.4.6.2 Reduced Disassembly Precedence Matrices

Moore et al. (2001) present an algorithm for deriving reduced disassembly precedence matrices. The rationale behind this is to remove those constraints that make detachment in a particular direction unlikely. For instance, the detachment of the head B of Bourjault's ballpoint (Figure 5.1) in the $+x$ direction is not likely because too many components are obstructing this. Unlikeliness of detachment in a particular direction reflects itself by the appearance of many components in an element of the obstruction component matrix (see Expression 6.16). These are replaced by the $\infty$ symbol, which is comparable to the shaded circles in obstruction diagrams, such as in Figure 6.11. The *reduced obstruction component matrix* that replaces Expression 6.16 is as follows:

$$\begin{array}{c|cc} & x & -x \\ \hline A & E & BF \\ B & \infty & F \\ C & E & BF \\ D & E & BF \\ E & \varnothing & \infty \\ F & \infty & \varnothing \end{array} \tag{6.19}$$

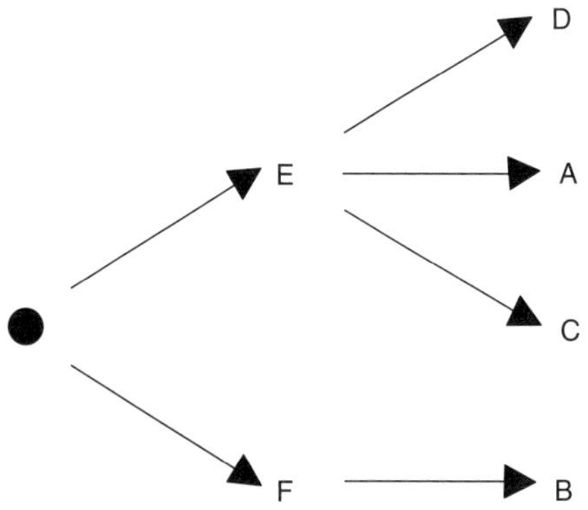


**FIGURE 6.15** Disassembly precedence graph of Bourjault's ballpoint.

The already derived corresponding selection rules can be derived from this once again, with the selection rule *F not B* replacing the four rules in Expression (6.18) with B on the right. It can be demonstrated that the reduced obstruction component matrix Expression (6.19) corresponds to the disassembly precedence graph in Figure 6.15.

### 6.4.7 Parallel Disassembly Aimed at Component Recovery

The matrix formalism is frequently applied in repair and maintenance optimization. The basic problem here is to remove a specific component at minimum disassembly costs. This can be done by minimizing the number of disassembly operations, which might include maximization of parallelism. A similar approach is useful in optimizing the detachment of given components, aimed at complying with the regulation on the environment or safety, meeting the customer's demand, or enabling the recovery of valuable components.

The problem of detaching a specific component with minimum disassembly actions has been addressed by Lee and Kumara (1992). These authors state that different components can be detached in one action as a complete subassembly if:

- They have the same disassembly direction
- They are connected

A search is carried out for a maximum subassembly that can be detached as a whole, which does not include the target component. This is called the *carpenter's approach*. The subassembly that is left is called the minimum subassembly. The procedure is carried out iteratively.

### Example 6

This method is illustrated with a product that consists of the components GHIJKLMN in the 2D-example of Figure 5.6. Let the target component be N. The obstruction component matrix is as follows:

| | $+x$ | $-x$ | $+y$ | $-y$ |
|---|---|---|---|---|
| G | *HLMN* | *HKLMN* | KL | *HIJKLMN* |
| H | G | *GIJKLMN* | GIM | I |
| I | H | JKL | *GHJMN* | H |
| J | HI | GKL | *GKLMN* | I |
| K | *GHIJLMN* | J | GL | GJ |
| L | *GHIJMN* | GK | G | GJK |
| M | GH | GKL | G | *HIJN* |
| *N* | *GH* | *GKL* | *GM* | *IJ* |

(6.20)

The subassemblies that correspond to the target component N are excluded; the matrix elements that contain component N are in italics. The search for subassemblies starts with considering those groups that are obstructed by N, which proceeds by columns. For instance: G, K, and L are obstructed by N in the $+x$ direction. Consequently, these components cannot be included in a subassembly that is detached in the $+x$ direction. Next, the components that are obstructed in the $+x$ direction by G, K, or L are considered. These are H and M. Subsequently, those additional components that are obstructed by H and M are considered. These are I and J. It can therefore be concluded that no subassembly exists that can be detached in the $+x$ direction.

Analogous reasoning reveals that no subassembly exists that can be detached in the $-x$ direction. Considering the $+y$ direction, it is observed that I and J are obstructed by N. However, no additional component exists that is obstructed by I or J. As a result, the subassembly GKLM can be detached as a whole in the $+y$ direction. Similarly, the subassembly HIJ can be detached in the $-y$ direction. Consequently, GKLM is the maximum subassembly. Once the maximum subassembly is detached, one is left with HIJN. This subassembly is investigated according to the same rationale. The obstruction component matrix is as follows:

| | $+x$ | $-x$ | $+y$ | $-y$ |
|---|---|---|---|---|
| H | ∅ | *IJN* | I | I |
| I | H | J | *HJN* | H |
| J | HI | ∅ | *N* | I |
| *N* | *H* | ∅ | ∅ | *IJ* |

(6.21)

It can be concluded that HIJ is the maximum subassembly that can be detached in either the $+x$ or the $-y$ direction. The subassembly IJ can be detached in the $-x$ direction. No subassembly can be detached in the $+y$ direction.

From this it follows that component N can be obtained by the subsequent detachment of GKLM and HIJ. Obviously, detaching HIJ from HIJN corresponds to detaching N from HIJN in the reverse direction.

### 6.4.8 Removing Redundancy

The disassembly precedence matrix that corresponds to Expression 6.20 is as follows:

| | G | H | I | J | K | L | M | N |
|---|---|---|---|---|---|---|---|---|
| G | $0$ | $\pm x,-y$ | $-y$ | $-y$ | $-x,\pm y$ | $\pm x,\pm y$ | $\pm x,-y$ | $\pm x,-y$ |
| H | $\pm x,y$ | $0$ | $-x,\pm y$ | $-x$ | $-x$ | $-x$ | $-x,y$ | $-x$ |
| I | $y$ | $x,\pm y$ | $0$ | $-x,y$ | $-x$ | $-x$ | $y$ | $y$ |
| J | $-x,y$ | $x$ | $x,-y$ | $0$ | $\pm x,y$ | $-x,y$ | $y$ | $y$ |
| K | $x,\pm y$ | $x$ | $x$ | $\pm x,-y$ | $0$ | $x,y$ | $x$ | $x$ |
| L | $\pm x,\pm y$ | $x$ | $x$ | $x,-y$ | $-x,-y$ | $0$ | $x$ | $x$ |
| M | $\pm x,y$ | $x,y$ | $-y$ | $-y$ | $-x$ | $-x$ | $0$ | $-y$ |
| N | $\pm x,y$ | $x$ | $-y$ | $-y$ | $-x$ | $-x$ | $y$ | $0$ |

(6.22)

Obviously, this matrix contains redundant information, because if an element contains a specific direction, for instance $-y$ in the (G,H) element, its transpose, the (H,G) element, contains the opposite direction $+y$. Therefore, only positive directions need to be listed without any loss of information, see Expression 6.23:

| | G | H | I | J | K | L | M | N |
|---|---|---|---|---|---|---|---|---|
| G | $0$ | $x$ | $0$ | $0$ | $y$ | $x,y$ | $x,y$ | $x$ |
| H | $x,y$ | $0$ | $y$ | $0$ | $0$ | $0$ | $y$ | $0$ |
| I | $y$ | $x,y$ | $0$ | $y$ | $0$ | $0$ | $y$ | $y$ |
| J | $y$ | $x$ | $x$ | $0$ | $x,y$ | y | y | y |
| K | $x,y$ | $x$ | $x$ | $x$ | $0$ | $x,y$ | $x$ | $x$ |
| L | $x,y$ | $x$ | $x$ | $x$ | $0$ | $0$ | $x$ | $x$ |
| M | $x,y$ | $x$ | $0$ | $0$ | $0$ | $0$ | $0$ | $0$ |
| N | $x,y$ | $x$ | $0$ | $0$ | $0$ | $0$ | $y$ | $0$ |

(6.23)

Now the detachability analysis can proceed as follows. First, the matrix is searched for rows and columns that do not contain all the directions, which implies that the corresponding component is free to be detached in the missing direction if it belongs to a row, or in the opposite of the missing direction, if it belongs to a column. Because no such rows or columns exist in Expression 6.23, all the components have a few constraints in every direction. Consider column H. A $y$ appears in the position that corresponds to I, which means that H is constrained in the $-y$ direction by I only. Consider column I. Detachment of I is constrained in the $-y$ direction by H only. Thus it can be concluded that HI together can be detached from GHIJKLMN in the $-y$ direction, and so on.

In the original approach due to Gungor and Gupta (2001ab), this method is applied together with a heuristic that confines the disassembly-sequencing problem to a predetermined disassembly depth. This heuristic offers an opportunity to find a set of near optimum disassembly sequences, without generating the set of all possible sequences.

A method for data storage in the disassembly precedence matrix is presented by Huang and Huang (2002). A binary number represents the possible directions here. For the case where there are only two principal directions of motion, viz., $x$ and $y$, they can be represented in binary notation, as $x = 10$; $y = 01$; $x,y = 11$. These are represented in decimal notation as 2, 1, and 3, respectively. If z direction is also allowed, the corresponding binary number will have three positions and might run from 0 through 7. Using these notations, the matrix in Expression 6.23 can be represented by the following matrices, which are called the *disassembly matrices* in the *binary system* and the *decimal system,* respectively.

| | G | H | I | J | K | L | M | N | sum |
|---|---|---|---|---|---|---|---|---|---|
| G | 00 | 10 | 00 | 00 | 01 | 11 | 11 | 10 | 11 |
| H | 11 | 00 | 01 | 00 | 00 | 00 | 01 | 00 | 11 |
| I | 01 | 11 | 00 | 01 | 00 | 00 | 01 | 01 | 11 |
| J | 01 | 10 | 10 | 00 | 11 | 01 | 01 | 01 | 11 |
| K | 11 | 10 | 10 | 10 | 00 | 11 | 10 | 10 | 11 |
| L | 11 | 10 | 10 | 10 | 00 | 00 | 10 | 10 | 11 |
| M | 11 | 10 | 00 | 00 | 00 | 00 | 00 | 00 | 11 |
| N | 11 | 10 | 00 | 00 | 00 | 00 | 01 | 00 | 11 |
| sum | 11 | 11 | 11 | 11 | 11 | 11 | 11 | 11 | |

(6.24a)

| | G | H | I | J | K | L | M | N |
|---|---|---|---|---|---|---|---|---|
| G | 0 | 2 | 0 | 0 | 1 | 3 | 3 | 2 |
| H | 3 | 0 | 1 | 0 | 0 | 0 | 1 | 0 |
| I | 1 | 3 | 0 | 1 | 0 | 0 | 1 | 1 |
| J | 1 | 2 | 2 | 0 | 3 | 1 | 1 | 1 |
| K | 3 | 2 | 2 | 2 | 0 | 3 | 2 | 2 |
| L | 3 | 2 | 2 | 2 | 0 | 0 | 2 | 2 |
| M | 3 | 2 | 0 | 0 | 0 | 0 | 0 | 0 |
| N | 3 | 2 | 0 | 0 | 0 | 0 | 1 | 0 |

(6.24b)

## 6.5 INTERFERENCE GRAPHS

Interference matrices can be visually presented by *interference graphs*. It should be noted that interference graphs do not provide information on connections, which is contained in connection diagrams or connectivity matrices.

Interference graphs are directed graphs with components as nodes. Directed arcs point from each node to those nodes that correspond to components that obstruct the motion in a specific direction. Examples of interference graphs can be found in Ong and Wong (1999).

An example of such a graph in the $+x$ direction, which is derived from the obstruction diagram of Figure 6.9, which corresponds to Bourjault's ballpoint, is presented in Figure 6.16. We see, for example, that motion of component A in the $+x$ direction is obstructed by component E. Consequently, a directed arc from A to E is drawn. It is evident that those components with no departing arcs can be detached. In this case, E has no departing arcs. Consequently, it can be detached in the $+x$ direction.

Components with no incoming arcs, on the other hand, can be detached in the opposite direction. Thus F, which has outgoing arcs only, is detachable in the $-x$ direction. Once a component is detached, the incoming and outgoing arcs of the corresponding node are deleted and the interference diagram is relaxed, possibly revealing additional components that can subsequently be detached.

Note that the interference graph in the $-x$ direction is similar to that of Figure 6.16, but with the arcs pointing in the opposite direction.

If *closed directed paths* are present in an interference graph, sequential disassembly in that specific direction to its final state (i.e., consisting of individual components) is not possible. Therefore, if detachment along a specific direction is required, those components that are connected via closed, directed paths can be combined resulting in a contracted interference graph. This is demonstrated with the graph in Figure 6.17, which exhibits a closed directed path (or *cycle*) running from B via C and D and back to B. Considering BCD as a single component results in a contracted interference graph that is also depicted in Figure 6.17.

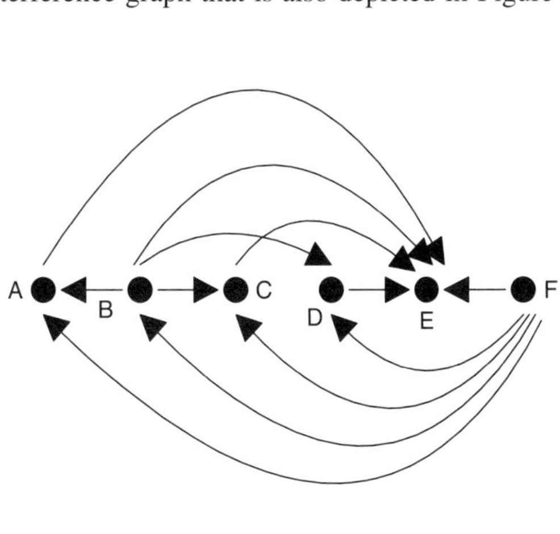


**FIGURE 6.16** Interference graph, corresponding to the obstruction diagram of Figure 6.9.

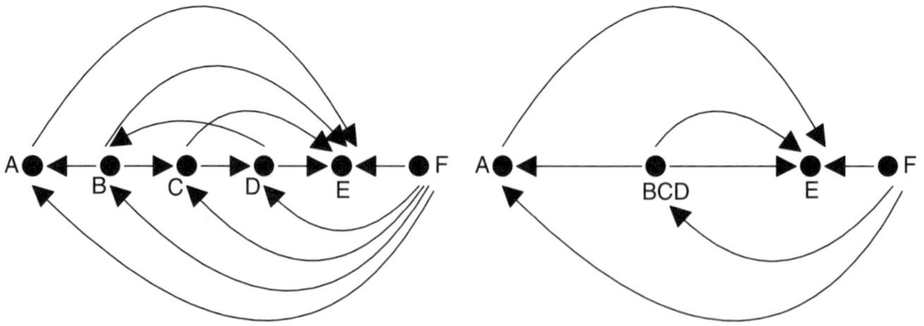


**FIGURE 6.17** Contraction of an interference graph if closed directed paths exist.

Possible sequences are the subsequent detachment of F, BCD, and A in one direction; or the subsequent detachment of E, A, and BCD in the other direction.

Apart from the occurrence of closed directed paths, combination of connected components aimed at simultaneous detachment can also be studied with this type of graph.

The occurrence of *meshes* in interference graphs can also be used as a tool for case-based reasoning (Swaminathan and Barber, 1996, and Swaminathan et al., 1998). *Leaves* are distinguished, which refer to components that are not a part of any mesh. As an example, the three components, viz., A, B, and F in Bourjault's ballpoint (Figure 5.1) are considered. The arcs that are connecting these components form a mesh, which is depicted in Figure 6.18. From this, disassembly plans can be derived. The corresponding disassembly precedence graphs can also be found in Figure 6.18.

The basic philosophy in *case-based reasoning* is that, if a similar pattern is encountered in a new problem, the set of plans in the library of cases is used instead of recalculating the plans (Veerakamolmal and Gupta, 2002).

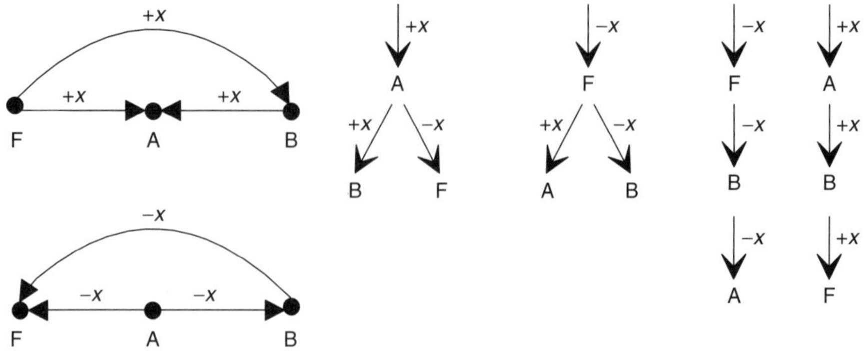


**FIGURE 6.18** Directed paths and associated disassembly precedence graphs.

## 6.6 MODULARITY ANALYSIS

### 6.6.1 Introduction

Disassembly problems can be reduced in size by defining modules, which are subassemblies that are characterized by specific internal coherence such as functionality and stability. There might also be a hierarchy of modules where larger modules constitute smaller modules incorporating subfunctionalities. For example, an engine may be considered a module of a car. In turn, a fuel pump may be considered a module of the engine. The systematic analysis of a product aimed at defining modules is called *modularity analysis*. In many cases, unfortunately, the definition of a module is ambiguous, which implies that exact methods for defining modules cannot be given. However, there exist heuristic methods that might be helpful in determination of practically useful modules.

Modules are frequently predefined (*a priori* modules). This may be market-forced, because demands often exist for specific modules, such as printed circuit boards, engines, and other groups of components that usually provide a desired functionality. This is also relevant in repair and remanufacturing processes. In production, on the other hand, modularization is also discernible, because the structure of a product often has a modular characteristic. In present practice, modules are delivered by suppliers and subcontractors. Apart from this, product designs often use *a priori* modules aimed at rapid exchange of modules, or for providing the customer with a broad range of product configurations.

Modules may also be defined as groups of components that are stable, e.g., because the components are secured by fasteners. The often intuitive character of module detection can be illustrated by the product that is depicted in Figure 5.6. For example, the subassembly GHIJKLMN can be considered a module here, although this does not follow from any rigid rule. In assembly, where parallelism is frequently desired, the subassemblies that are assembled as a whole have to be stable to some extent. These subassemblies can be considered modules. It is also possible to define modules as structures that repeatedly appear in products, which is related to case-based reasoning (Veerakamolmal and Gupta, 2002).

Despite the ambiguous character of modularization, multiple studies have been carried out aimed at formalizing the search for modules. By breaking up the product into modules, the size of disassembly sequencing problems can be considerably reduced. PC systems, for example, have exploited the use of modularization to a considerable extent by making use of multiple exchangeable units, such as drives and PWBs.

Although formal methods for modularity analysis exist, they are, unfortunately, not appropriate for unambiguously separating a product into subassemblies. Therefore, formal methods complemented with intuitive methods are applied for determining modules. This can proceed using one or more of the following approaches:

1. *Market-oriented approach*, which accounts for the demand on functional units, such as engines or power supply units.
2. *Structure-based approach*, emphasizing the occurrence of specific structures that appear multiple times in a product, such as in case-based reasoning.

3. *Subassembly-oriented approach*, which is based on the existence of large subassemblies that can be detached as a whole.
4. *Stability approach*, aimed at the stability of modules.
5. *Functional approach*, aimed at detecting modules that have similar functionality in common.
6. *Disassembly-oriented approach*, focuses on sets of equivalent disassembly operations, such as disassembly direction and the use of specific tools.
7. *Service-oriented approach,* focuses on grouping components with comparable breakdown time. This approach appears in maintenance and repair studies.

What follows is a review of the results on modularity analysis that have been reported by several authors.

### 6.6.2 Weighted Connection Diagram

Lee and Shin (1990ab) perform modularity analysis on the basis of assigning weight factors to connections. Subsequently, those connections with the highest weight factor are considered, and the components that are attached by these connections are merged successively, up to some minimum value of the weight factor. Consequently, a hierarchy of *tentative modules* is generated, which starts with single components up to the complete product (see Figure 6.19).

First, the connection with the highest weight is established, which is assumed to be the one between B and C. Therefore, B, C, and BC are considered tentative modules. Next, the establishment of the connection with the second-highest weight factor is carried out, which is the one between BC and D. The subassemblies D and BCD, are the new tentative modules. This procedure is repeated till the complete product is obtained. In practice, however, the procedure stops if the weight of a new connection drops below a predetermined threshold value. The possible tentative subassemblies are depicted in Figure 6.19(b).

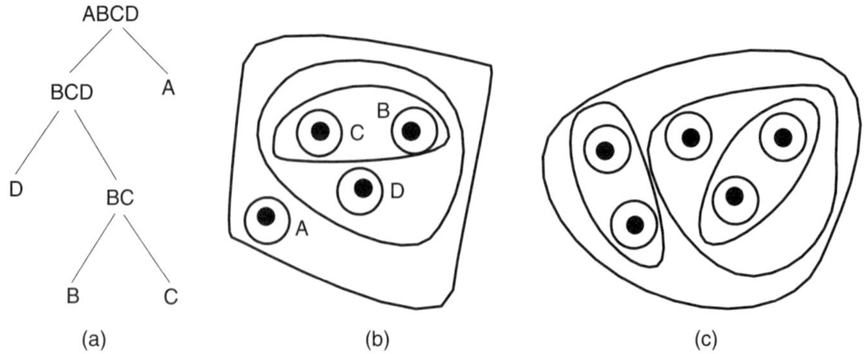


**FIGURE 6.19** The maximum number of tentative modules. (a) Binary tree with 4 terminal nodes; (b) Increase of 2 with each new component; (c) Alternative arrangement for $N = 5$.

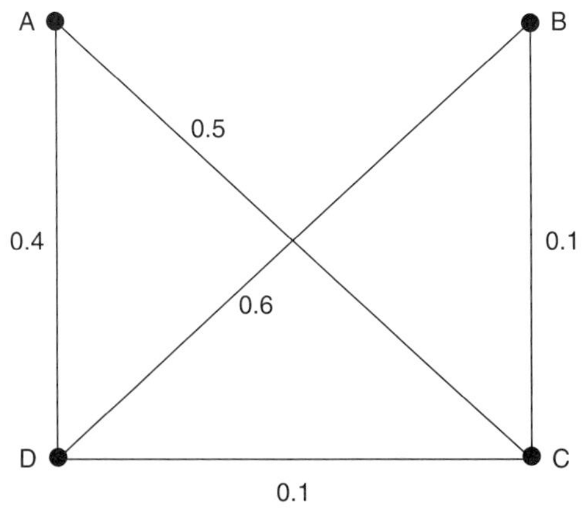


**FIGURE 6.20** Example of modularity analysis according to Lee and Shin (1990ab).

For every possible sequence, there are $2N - 1$ tentative subassemblies, consisting of $N$ components and $N - 1$ composed subassemblies with the number of components running from 2 through $N$. The example in Figure 6.19(a) and Figure 6.19(b) is confined to sequential disassembly. Apart from this, it is assumed that the original product is weakly connected. A similar reasoning holds for cases in which parallel disassembly is applied (see Figure 6.19(c) for an example with $N = 5$). From Figure 6.19(b) it directly follows that, if an additional component is added, the number of tentative subassemblies increases by 2.

#### Example 7

See Figure 6.20. It is called a *weighted connection diagram*. The first tentative subassemblies are the four individual components A, B, C, and D. The strongest connection is between B and D. Thus BD is a tentative subassembly. Next, AC is selected. Next, ABCD, which is the complete product. If the threshold is at 0.55, BD is the only module with more than one component.

In general, the weight of a connection is determined not only by the connection itself, but also by the surroundings of the component. Therefore, the weight factors have to be recalculated after every disassembly operation.

Once the tentative subassemblies have been determined, these have to be further evaluated with respect to stability and the coherence of their structures. A module should have a strong internal coherence. Apart from this, it should be easily detachable from the remaining product. Lee and Shin (1990ab) and Lee (1992) derive formulae for determining the *structural preference index*, which is a single figure that is composed of the *stability index* and the *directionality index*. The value of the structural preference index is decisive whether or not a tentative module is selected as a module. The formulae of the cited authors are rather detailed, as these are combination of many different mechanical properties, each provided with an appropriate weight factor. The aim of this work has been the formulation of an objective method for module detection.

### 6.6.3 Graph Decomposition

#### 6.6.3.1 Complete Separation

Instead of assigning weight factors to connections, one could make use of the structure of the connection diagram (Zhang and Kuo, 1996; Zhang and Yu, 1997; Kuo et al., 2000). The following features of a connection diagram are considered (Gu and Yan, 1995; Yan and Gu, 1995; Kuo et al., 2000):

1. *Cut-node:* It is a node, if removed, breaks the connection diagram into two or more disjoint graphs. The nodes A, B, and C in the connection graph of Bourjault's ballpoint (Figure 5.1(b)) correspond to cut-nodes. Node E is the only cut-node in the connection graph of the AFI (see Figure 5.7(b)). The nodes F, M, and N are cut-nodes in the connection diagram of the single-piston engine (see Figure 5.11(a)).
2. *Pendent node* or *satellite node:* It is a node that is connected to the remaining components of the product by only one arc. The components D, E, and F are pendent nodes of Bourjault's ballpoint (Figure 5.1(b)). Component F is a pendent node of the AFI (Figure 5.7(b)). The components G and O are pendent nodes of the single-piston engine (see Figure 5.11(a)).
3. *Biconnected graph:* It is a connection diagram without cut-nodes. The connection-diagram of the 2D-example in Figure 5.6(b) is a biconnected graph.
4. *Degrees of a node:* It is the number of arcs that are incident to the node.

**Example 8**

The degrees of the nodes in the connection diagram in Figure 5.6(b) are as follows:

Nodes H, and J have seven degrees
Nodes A, B, D, and K have six degrees
Node G has five degrees
Nodes F, L, and M have four degrees
Nodes E, I, and N have three degrees
Node C has two degrees

The degrees of a node equal the number of the nonzero elements in the corresponding row or column of the connectivity matrix. A pendent node has a degree of 1.

According to Kuo et al. (2000) and Kuo (2000), the connection diagram is subdivided in disconnected subgraphs by removing the cut-node of the highest degree. As a result, at least two disconnected subgraphs arise. Next, this cut-node is assigned to that subgraph which has the most arcs connected with this cut-node. If this number equals 1, the cut-node is treated as a separate module.

**Example 9**

In Bourjault's ballpoint (Figure 5.1), component A corresponds to a cut-node of degree 3, which is the highest degree of a cut-node. Detachment of A results in the disconnected subgraphs: E, F, and BCD. A is not assigned to any of those subgraphs, because it is connected with each of these subgraphs by one arc. Next one considers BCD. C is a cut-node. It leaves us with the subgraphs B and D, and C is not assigned to any of those subgraphs. These are not modules, however, but components. Subassembly BCD can be considered a module, which is apparently a realistic choice.

An important drawback of this graphical method is its restricted applicability. It can hardly be applied in typical products such as the ones in Figure 5.6, Figure 5.7, and Figure 5.11.

#### 6.6.3.2 Merging

In papers due to Yan and Gu (1995) and Gu and Yan (1995), a modified method for modularity analysis is presented. It is based on cut-node detection and the decomposition of the connection graphs into biconnected graphs. However, the cut-nodes are assigned to all the subgraphs that arise from graph decomposition.

**Example 10**

Component E is a cut-node in the AFI of Figure 5.7(b). Consequently, the related connection graph is decomposed in two subgraphs: EFQ and ABCDEGHJKLMNOP, which have component E in common. Next, it is searched for additional cut-nodes in both subgraphs. Because no cut-nodes are present, both subgraphs are biconnected.

Using some heuristic, disassembly sequences are generated for the subassemblies that correspond to the biconnected subgraphs thus found. Next, these sequences are merged so that a distinction is made between those disassembly operations of all sequences that have to be carried out prior to the detachment of the component that corresponds to the cut-node, and those operations that have to be carried out after the detachment of the cut-node.

Although algorithms like this might be helpful in reducing the problem complexity by subdividing it into smaller subproblems, there is no guarantee that the optimum sequence will be found. Furthermore, many connection diagrams are either completely biconnected or the cut-node separates the connection diagram in a principal subassembly and one or more satellite nodes or small subassemblies.

### 6.6.4 Biconnected Graph Analysis

#### 6.6.4.1 Core Component Method

Tanaka et al. (1996) have applied the graph structure, particularly the degree of the nodes, for automating modularity analysis. The authors search for the component with the largest number of connections. This component might be chosen as the *core component*. Together with its surrounding components, a core module is defined. Next, one looks at the noncore or peripheral components. Distinction is made between the connections in the core module, the peripheral module, and the connections between both modules. If a peripheral component has relatively many connections with the core, it might be transferred to the core module. As a criterion, the ratio between the number of connections of the component with the core module and the rest of the peripheral modules can be used.

**Example 11**

In Bourjault's ballpoint (Figure 5.1), A is the core component and ABEF is the preliminary core module. CD is the peripheral module. C is the connecting component that might be subordinated to the core. Component D is left as the peripheral component.

In the AFI example (Figure 5.7), component A is the core component. The core module is ACDGKLMNP. The peripheral module is BEFHJOQ. Component H has three connections with the core, and two connections with the periphery. Consequently, it is transferred to the core.

In the 2D-example of Figure 5.6 there are two candidates for core component, viz., H and J. Assuming H is selected, the core module is ABGHIJMN. The peripheral component L has three connections with the core, and one with the periphery. Consequently, it may be transferred from the periphery to the core.

#### 6.6.4.2 Biconnected Graph Disconnection

Biconnected graphs can also be subjected to modularity analysis. Therefore one has to determine the degree of each node. Next, one searches for a *pair* of nodes of high degree that, if removed, causes disconnection of the graph.

**Example 12**

The nodes A, B, and H, in the AFI of Figure 5.7, all have five degrees, which are the highest possible degrees. Detachment of both A and B results in separation of the graph in two disconnected subgraphs: CDEF and GHJKL (see Figure 6.21). If Kuo's rules are implemented, A and B should be assigned to GHJKL. This results in the modules CDEF and ABGHJKL. However, intuitively, it can be advocated that separating the product in the modules AK, BCDEF, and GHJL might be more appropriate.

#### 6.6.4.3 Fastener-Oriented Method

A similar method is proposed by Ong and Wong (1999). Modularity analysis is carried out by first grouping the components that are connected by fasteners. Next the base component is defined. Tentative modules that have the base component as a member are excluded. Next, those components, which correspond to nodes with the highest degrees, are considered. Weight factors that account for the relative importance of connections might also be introduced.

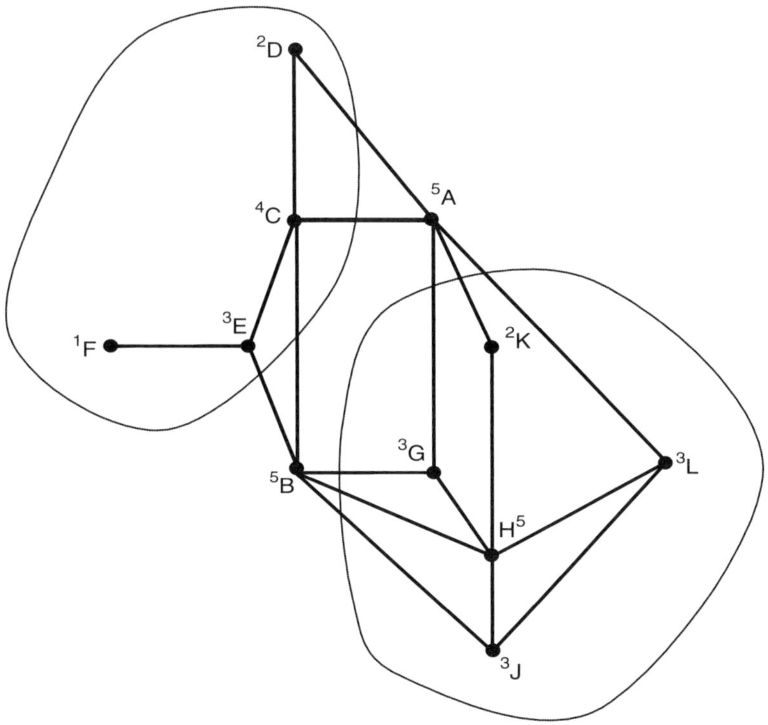


**FIGURE 6.21** Separation of the connection diagram of the AFI in two subassemblies, see text.

### 6.6.5 Functional Analysis

Functionality consideration is discussed by Lambert (1999) who analyzed the toaster that was introduced by Jovane et al. (1997). The exploded view of the toaster is given in Figure 6.22.

The connection diagram of the toaster, presented as a spaghetti diagram, is shown in Figure 6.23.

Detailed and systematic analysis of this graph proceeds as follows. First, the components of the toaster are introduced. These include: (1) Back front (2) Slot (3) Forefront (4,5) Strip (6) Bottom (7) Housing (8) Back support (9) Front support (10) Heat element support (11) Heat elements (12,13,15,16) Cables (14) Switch (17) Plug (18) Dial (19) Button (20) Clamp. Next, a modified connection graph is constructed (see Figure 6.24), which accounts for the symmetry properties of the product. Distinction is made between mechanical joints such as the strips 4 and 5, and electrical connections.

Distinction has also been made in firm connections, which are established via fasteners, and loose connections, such as mating. This corresponds to assigning weight

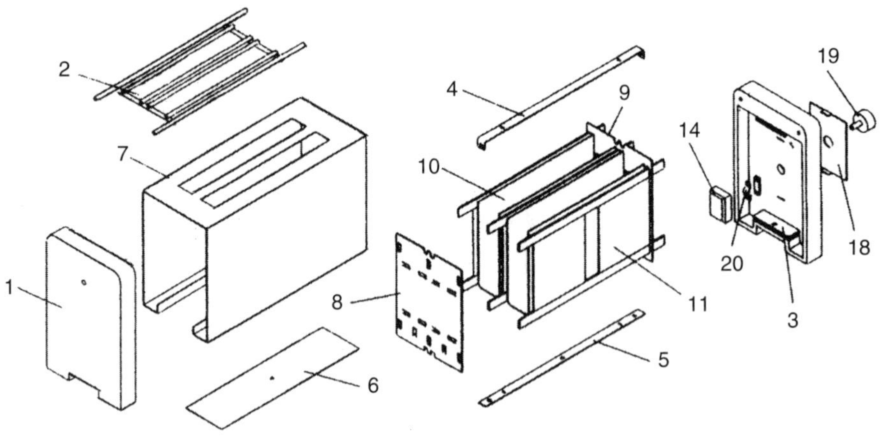


**FIGURE 6.22** Exploded view of a toaster (see Jovane et al., 1997).

factors to connections. It should be noted that the cables are not rigid, but flexible, components. In the disassemblability analysis in chapter 5, all the components were assumed to be rigid bodies. The presence of flexible components relaxes this constraint. Component 3 appears to be a cut-node. Evidently, the product can be broken up via removal of the module 3-14-16-17-18-19-20, which is the forefront module. If removed, the back front (component 1) appears as a cut-node in the remaining subassembly. Its removal leaves us with the module 4-5-6-8-9-10-11-12. Since its

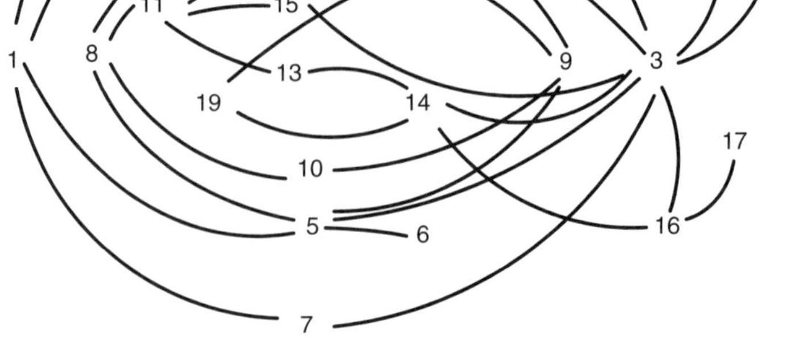

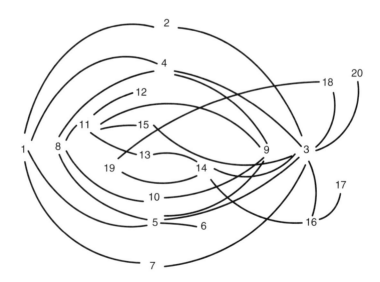


**FIGURE 6.23** Connection diagram of the toaster, in provisional format.

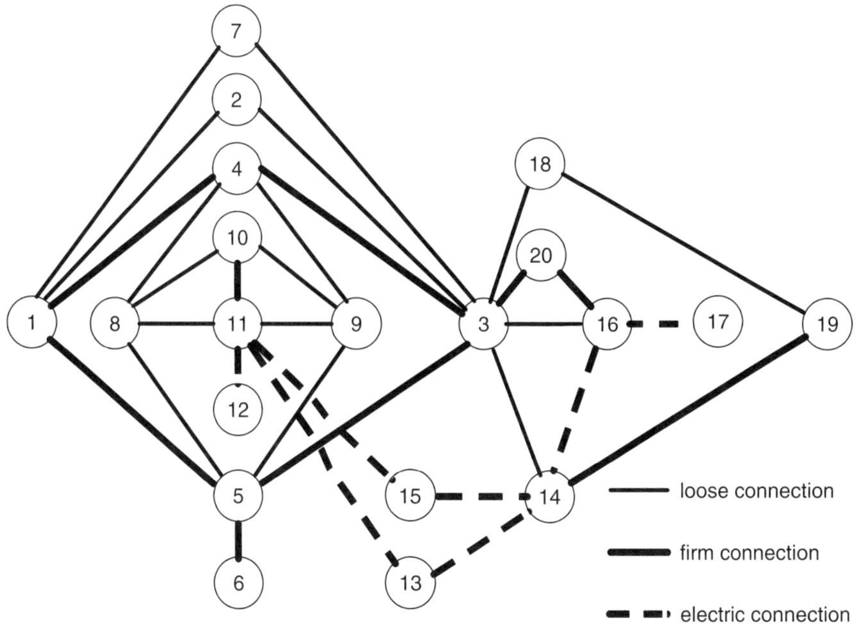


**FIGURE 6.24** Modified connection diagram of the toaster.

components (except the heat elements and the cable) are not connected with fasteners, the nonelectrical components can be easily separated. By doing so, a disassembly sequence is generated along heuristic lines.

### 6.6.6 Stability-Oriented Analysis

The stability-oriented approach is used by Kroll (1993). Distinction between different types of fasteners is made here (see subsection 3.2). Groups of components are considered that are held together by reversible fasteners such as bolts and nuts. Also groups of components are considered that are connected by irreversible connections such as soldering, press-fitting, and adhesive bonding. This author also includes the merging of specific modules.

#### Example 13

If applied to the AFI, with fasteners included (see Figure 5.7), it appears that, according to this definition, EF and BE are modules that are held together with bolts Q and O, respectively. Similarly, AD and AC can be considered modules with bolts P and N as fasteners. Thus, due to merging, ACD is a module as well. We also have to keep the precedence constraints in mind. For example, module BEF cannot be obtained via disassembly, because the precedence constraints require that C and D have to be present in the module. By merging, which is partly based on intuition, the module ABCDEF is obtained.

In Bourjault's ballpoint (Figure 5.1), components C and D are connected by adhesive bonding; B and C are connected by press fitting. Here, BCD can be considered as a module. The connections between A and E, A and B, and A and F are also due to press fitting, but the forces that are included here are smaller.

### 6.6.7 Matrix-Based Modularity Analysis

Mascle (1998) has carried out modularity analysis in a systematic way. It is based on the calculation of freedom matrices. This method starts with the analysis of the connection diagram of the product. Fasteners are usually connected with two mating components. This is visible for the case of fastener M in Figure 5.7, which *bridges* the connection between A and L. After the detachment of the fastener, the components A and L are considered two *poles*, which need further research to see whether or not these poles can be detached individually. This is done by considering the freedom matrix of the related components (see Expression 6.5). The heuristic starts with the pole that has the least number of connections (in this case, component L), which has three connections in contrast to component A, which has five connections. With M detached, L can be detached individually in the $+x$ direction. Therefore, it will not be incorporated in a module. This follows from

$$F_{L,ABCDEFGHJK} = F_{L,A} \vee F_{L,H} \vee F_{L,J} \tag{6.25a}$$

With only translations in the $\pm x$ permitted, we have:

$$F_{L,ABCDEFGHJK} = |0\ 1| \vee |0\ 1| \vee |0\ 1| = |0\ 1| \tag{6.25b}$$

This means that L is detachable in the $+x$ direction. Consequently, L is not in a module.

However, if L is connected to H with a fastener, individual detachment of L is impeded, because

$$F_{L,ABCDEFGHJK} = |0\ 1| \vee |1\ 1| \vee |1\ 1| = |1\ 1| \tag{6.26}$$

Consequently, L is not detachable, which means that further investigation is required. Since H immobilizes component L, the combination of L and H is explored. The following matrix is studied:

$$F_{HL,ABCDEFGJK} = F_{HL,A} \vee F_{HL,J} \vee F_{HL,B} \vee F_{HL,G} \vee F_{HL,K} \tag{6.27}$$

This results in

$$F_{HL,ABCDEFGJK} = |0\ 1| \vee |1\ 1| \vee |0\ 1| \vee |0\ 1| \vee |0\ 1| = |1\ 1| \tag{6.28}$$

HL is not detachable, because it is constrained by J. Therefore, HL and J are combined. Consequently

$$F_{\text{HJL,ABCDEFGK}} = F_{\text{HJL,A}} \vee F_{\text{HJL,K}} \vee F_{\text{HJL,G}} \vee F_{\text{HJL,B}} \tag{6.29}$$

This results in

$$F_{\text{HJL,ABCDEFGK}} = |0\ \ 1| \vee |0\ \ 0| \vee |0\ \ 1| \vee |0\ \ 1| \vee |0\ \ 1| = |0\ \ 1| \tag{6.30}$$

From this, it follows that HJL is detachable in the $+x$ direction. Therefore, HJL can be considered a module, provided it is internally stable. This can be checked by carrying out the stability analysis of HJL (see subsection 6.7). It appears that HJL is stable and, consequently, subassembly HJL is considered a module.

This illustrates how modules can be detected automatically. This has to take place in combination with *a priori* and intuitive methods.

### 6.6.8 Cluster Graph Analysis

O'Shea et al. (2000) have also discussed modularity analysis. They use *modularity graphs* (*cluster graphs*), which are hierarchically ordered interference graphs (see Figure 6.25), with the directional information not included.

The components that can be detached from the product without the need of previously detaching other components are on the first level of the cluster graph. Those components that can be detached once the components on the first line are detached are at the second level, etc. Figure 6.25 shows the cluster graph of the AFI from Figure 5.7, with fasteners included. Defining a module starts with a fastener, which is component Q here. From this, a downward search is performed, which proceeds till an end node with no exiting arcs is reached. The following potential modules can be discerned: BCDEFNOPQ; AGHJ; KLM. It is noticed that modules need not be strictly connected, unless stability requirements have to be dealt with.

Further work on modularity analysis is due to Quian and Pagello (1994), who distinguish between *reliable* and *unreliable* connections. The latter might spontaneously disconnect, for example, due to vibrations, or gravity, and thus might not guarantee stability. Examples of reliable connections may be irreversible, such as welds, or reversible, such as bolts. Mating is a notorious example of an unreliable connection. Quian and Pagello also distinguish between *fastened groups* and *units*, both of them being modules that cannot be disassembled without detachment of one or more reliable connections, thus having no internal degrees of freedom that might disrupt the coherence of the module. However, fastened groups are connected with the product with one or more reliable connections. Once these connections are detached, such a group transforms into a unit, which can be detached as a whole from the rest of the product.

Cluster graphs are also applied by Tseng and Liou (2000), who discuss the integration of assembly planning and machining planning, which is beyond the scope of this book.

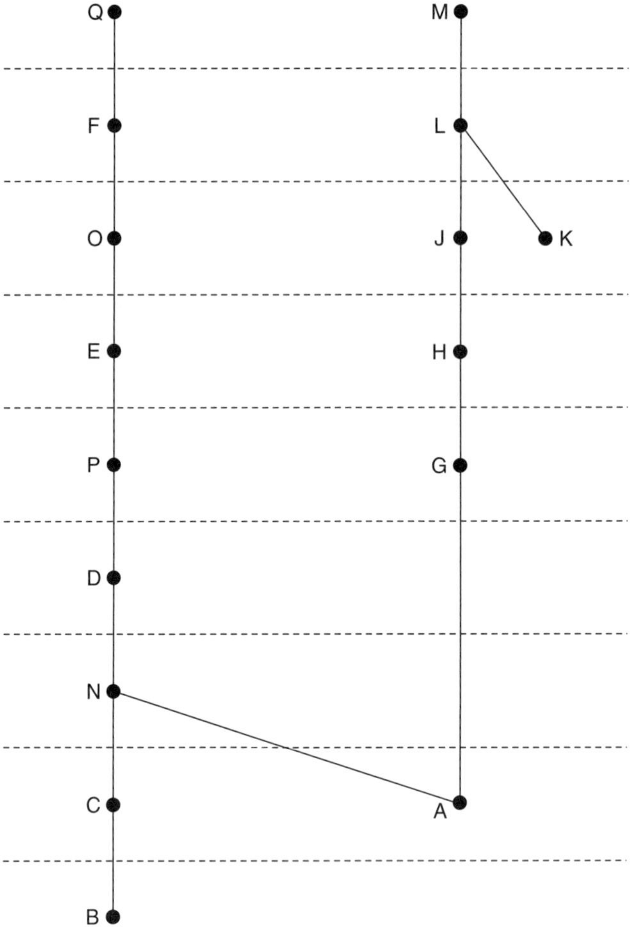


**FIGURE 6.25** Modularity graph of the AFI (see Figure 5.7).

### 6.6.9 Functionality Metrics

Modularity analysis is also discussed by Gu and Sosale (1999). In an attempt to quantify modularity, they design a metric. With a module defined as a subassembly, which is a physically connected set of components, with a specific functionality, two types of relationships are discerned, viz., functional and physical. The functional relationship refers to the exchange of information, energy, and matter, such as in electronic circuitry, electric wires, gears, and ducts. The physical relationship refers strictly to constraints in degrees of freedom, the kind of fastener, and the alignment. A laser, which throws a bundle of light on a mirror, for example, has a functional relationship with the mirror. The only physical relationship is the alignment, because the mirror and the laser are not physically connected. Therefore, there exists a moderately strong functional relationship, but a weak physical relationship. Modularization is carried out for different purposes. First, a metric for the intensity of relationship between two components has to be applied. Functional relationship differs depending on the purpose it is aimed for. For recycling, matching materials composition contributes to a strong relationship. For fasteners, the strongest relationship is formed with a permanent connection, such as weld, and the weakest relationship is when no connection exists.

With a weighted average of the different factors, an interaction matrix $I_{i,j}$ is calculated, which is symmetric and has the weighted average interaction between two components $i$ and $j$ as elements. The interactions are on a 0,..., 10 scale, with 10 for the strongest interaction and 0 for no interaction at all. Of course, some arbitrariness is present in determining the values of these elements. With a given number of modules, the optimum separation in modules can be determined.

### Example 14

For a simple product with four components, this proceeds as follows. Let the interaction matrix $I$ be:

$I =$

| | A | B | C | D |
|---|---|---|---|---|
| A | 10 | 7.2 | 0.4 | 2.1 |
| B | 7.2 | 10 | 3.8 | 2.1 |
| C | 0.4 | 3.8 | 10 | 1 |
| D | 2.1 | 2.1 | 1 | 10 |

(6.31)

If there were only one module ABCD the value for the internal interaction of this module should be $I_{AB} + I_{AC} + I_{AD} + I_{BC} + I_{BD} + I_{CD} = 16.6$, which is the sum of the elements of the extreme upper right triangular matrix. The possible separations in two modules are as follows: A,BCD; B,ACD; C,ABD; D,ABC; AB,CD; AC,BD; AD,BC.

For the selection of the appropriate module, the sum of the values of the elements that belong to the still established connections has to be maximized. If a module with three components is selected, triplets such as: $I_{BC} + I_{BD} + I_{CD}$ have to be considered. If two modules with two components each are considered, the sum of the elements that corresponds to pairs of elements, such as: $I_{AB} + I_{CD}$, has to be considered. The values that correspond to the above-mentioned seven possible separations are, respectively 6.9, 3.5, 11.4, 11.4, 8.2, 2.5, 5.9. From this, it follows that both C,ABD and D,ABC are optimum separations in modules. The connections that hold the module together have a relatively high value with respect to the inter-modular interactions.

Since the number of possible separations increases exponentially with increasing number of components, Gu and Sosale (1999) advocate using metaheuristic methods, such as simulated annealing algorithm.

### 6.6.10 Concluding Remark

A plethora of methods aimed at module detection, as presented by various authors, have been presented in this subsection. It can be concluded that, since modularity analysis is performed for different purposes, no unequivocal definition for a module exists. Therefore, no exact method can be formulated aimed at modularity analysis. Even a seemingly exact method, such as functionality metrics, will not *a priori* result in a rational subdivision in modules. Intuition, and external factors such as the market demand, play a significant role here. Most objective is simply the method of searching for maximum parallelism, which results in detecting those subassemblies that can be detached as a whole, and considering these as modules. Functionality metrics is also exact, but here the problem is shifted toward the determination of the interaction matrix elements, which also suffers from ambiguity. The other methods together, although not generally applicable, offer a versatile tool that might be valuable in the analysis of many product configurations. This tool can support intuitive reasoning.

## 6.7 STABILITY ANALYSIS

### 6.7.1 Introduction

Stability is an important aspect of modularity and it also plays a significant role in the feasibility of parallel assembly. In end-of life disassembly, stability might be of lesser importance. Stability is considered a technical feature. Constraints that originate from stability issues are technical constraints. It should be noted that topological and geometric constraints are frequently studied under the assumption that no forces are present. Stability refers to the coherence of a subassembly if forces are applied, which might be external forces, such as gravity and vibrations, and internal forces such as the ones caused by deformation, magnetism, etc. In particular, if disassembly studies are carried out aimed at assembly, it is of crucial importance that the subassemblies involved are stable. If these are not quite stable, temporary support has to be utilized. For this purpose, use can be made of gauges and related tools. This, however, complicates the operation and enhances the costs.

Stability analysis is not comprehensively discussed here, because it belongs to the structural and design engineering discipline, which is beyond the scope of this book. It requires detailed analysis including statics, the incorporation of friction, and even dynamics if vibrations are considered. However, the general idea, particularly involving the mutual immobilization of the components that are part of the subassembly, is considered. Much statics is in fasteners such as screws. The details of this are beyond the scope of this book. It is assumed that connections such as screw-connections are completely stable, but that the fasteners can be released subjected to specific rules and with appropriate tools.

### 6.7.2 Gravitation

Since gravitation is directed, product orientation plays a crucial role in gravitational stability. Analysis based on movability analysis is due to Mattikali et al. (1994, 1995, 1996). These authors use the Gaussian sphere formalism, which was discussed in subsection 6.3. A linear programming method is presented that reveals all the gravitationally stable orientations of subassemblies. If friction is neglected, this reveals an exact solution; if friction is included, a heuristic is introduced.

Stability analysis including gravity has also been addressed by Gu and Yan (1995). Gravity is assumed to be pointed in the $-z$ direction. A heuristic is presented that searches for detachable components that mate with other components in the $+z$ direction. Those components are called virtually instable and should be checked if it would remain stable when the detachable component is removed. If it appears to be unstable, then it is considered a suspended candidate, and another detachable component is searched for. If no further candidate can be found, a suspended candidate has to be detached, but temporary support needs to be applied in such case. It is evident, that such a consideration is significant in repair and maintenance, or in reverse assembly studies, but is less important in end-of-life disassembly.

Abe et al. (1999) describe a search for the existence of directions of motion that have an angle of less than $\pi/2$ with the $-z$ axis. If motion is possible in this direction, the component or subassembly will spontaneously slip, unless the product is reoriented or some fixture is applied.

Onozuka et al. (1994) use a heuristic that is based on distinguishing three types of stability of components in a subassembly: Stable, partly stable, and unstable components. Stable components are connected with fasteners and cannot move in any direction. *Partly stable* components are not fixed but they are stable under current product orientation. Unstable components cannot remain in their position. According to this, a set of heuristic priority rules is proposed for the disassembly of a product. This includes the following hierarchy in detachment:

1. The unstable components
2. The partly stable components
3. The detected modules
4. The fasteners

### 6.7.3 Connections

Stability analysis needs information on connections. Subsection 3.2 presented a typology of connections. Aimed at stability, the following three types of connection are considered (Lee and Shin, 1990ab; Lee, 1992):

*Floating connections*, which are not held together by internal forces, e.g., mating surfaces,

*Firm connections*, which are self-stable but result in relative motion if appropriate external forces are applied, e.g., a screw connection,

*Rigid connections*, which are stable, even under external force, e.g., a weld.

In addition, a metric for determining the strength of a connection on a continuous scale can be applied. This might include both the constraints in freedom of motion, and the force that is required for the disestablishment of the connection.

Besides this, structural considerations have to be added, because freedom of motion is not only constrained by the connection itself, but also by the remaining components in the subassembly that is going to be separated. This kind of analysis is related to movability analysis.

**EXAMPLE 15**

In the product of Figure 5.6, motion of the subassembly MN is completely constrained if GHIJL is present, even if the connection between M and N is a floating connection.

## 6.7.4 MATRIX-BASED STABILITY ANALYSIS

### 6.7.4.1 Stability Matrices and Stability-Directed Graphs

Since there is a strong relationship between movability and stability analysis, the use of matrices in stability analysis, analogous to movability analysis, is evident. Laperriere and Lavoie (1997) discuss the use of matrices. One starts with a component that is constrained by a fixture. Next, one adds components to it and freedom matrices are determined. Different forces, such as gravitation, vibrations, and internal forces, can be considered separately. Since gravitational force points in the $-z$ direction, the motion of components in the $-z$-direction should be inhibited, which can be accomplished by the presence of a fixture or a constraining component such as a table. On the other hand, rotational motion in both the $\pm x$ and the $\pm y$ directions should also be inhibited, as this would mean that the added component is falling on a surface of the fixed component. Note that rotation here means the spontaneous rotation of a component that has a point or line contact with the fixed component. Axial rotation without translation of the center of gravity, which does not spontaneously occur, such as if axles, screws, etc. are considered, is not what is meant here.

A *stability matrix* is introduced, which is a modified freedom matrix with the above-mentioned forbidden directions of motion excluded (see Expression 6.5). For example

$$F = \begin{vmatrix} +x & -x & 1 & 1 \\ +y & -y & 1 & 1 \\ +z & 1 & \omega_z & \omega_{-z} \end{vmatrix} \tag{6.32}$$

This provides an inventory of the constrained and unconstrained directions of motion. The constrained directions are indicated with elements equal to 1. If stability with respect to vibration is considered, motion in the $\pm x$ and in the $\pm y$ direction should also be avoided. Only translational motion in the $+z$ direction and rotational motion in the $\pm z$ direction is permitted in this case.

The method is complemented with a heuristic graphical representation. This is based on the directed connection diagram. Directed arcs here point from a component to those components that grant stability to the specific component. This type of graph is called a *stability directed graph*. The determination of such a graph starts with a fixed component. Next, those components are considered that are stabilized by the fixed component. A disassembly operation, which is the detachment of a specific subassembly from the product, is represented by a cut-set in this graph. If any arc points from a component of the static subassembly to the dynamic subassembly, which means that this component is supported by one or more components of the dynamic subassembly, the remaining static subassembly is unstable. If, on the other hand, the stability of the dynamic subassembly depends on components of the static subassembly, arcs pointing from the dynamic to the static subassembly represent it.

Metrics such as the *stability index* proposed by Shin and Cho (1994) might further support stability analysis.

### Example 16

The stability-directed graph of the components A, G, H, J, K, and L of the AFI from Figure 5.7 is presented in Figure 6.26. Component A is the fixed one. It supports G, K, and L in the $-z$ direction, which guarantees stability. If the dynamic subassembly represents all the components other than the static one (geometric constraints are neglected here), which corresponds to cut-set 1, the dynamic subassembly appears to be unstable. If, however, the cut-set 2 is considered, and H is the component that is temporarily fixed (via a gripper, for example), the dynamic subassembly appears to be stable.

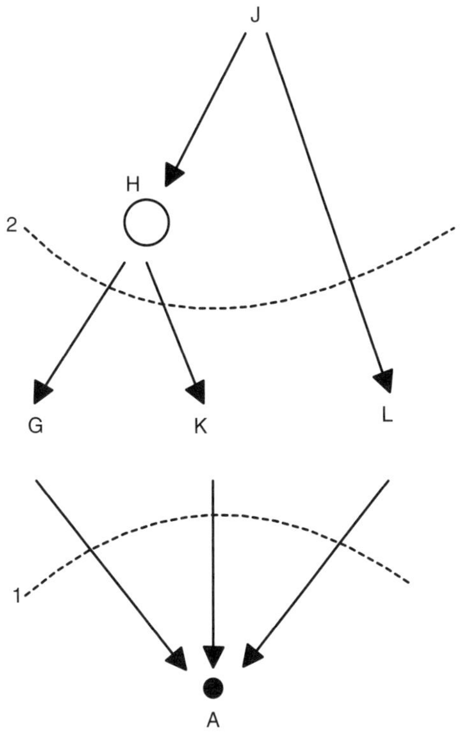


**FIGURE 6.26** Stability-directed graph of some AFI components (see Figure 5.7).

It should be mentioned here that the direction of the arcs in the stability directed graph strongly depends on (1) the fixed component and (2) the orientation of the product. Because these are decision variables, adaptation of the problem to the desires of the operator is possible to some extent. For the AFI case, the situation changes drastically if the product is reoriented such that the direction of gravity is parallel to the $x$-axis of the product.

### 6.7.4.2 Modified Interference Matrices

An alternative matrix approach, due to De Lit et al. (2001), is based on the interference matrix (see Expression 6.15). The concept of *indirect constraint* is introduced, which implies that if a component A is constrained by component B, and component B by component C, then component A is constrained by component C. With this in mind, *transitive closure* of the interference matrices is applied. Stability of a subassembly in a particular direction (and its opposite direction) is reached if the motion of every component is constrained by the remaining components of that subassembly.

#### Example 17

The transitive closure of the interference matrix for the AFI, Expression 6.15 is given in Expression 6.33. The underlined elements have been switched from 0 to 1. For instance, motion of F is not obstructed by K, but motion is obstructed by A. The motion of A is obstructed by K. Consequently, the motion of F is indirectly obstructed by K in the $+x$ direction.

$$NC_{+x} = \begin{array}{c|ccccccccccc} & A & B & C & D & E & F & G & H & J & K & L \\ \hline A & 0 & 0 & 0 & 0 & 0 & 0 & 1 & 1 & \underline{1} & 1 & 1 \\ B & 1 & 0 & 0 & 0 & 0 & 0 & 1 & 1 & 1 & 1 & 1 \\ C & 1 & 1 & 0 & 0 & 0 & 0 & 1 & 1 & 1 & 1 & 1 \\ D & 1 & 1 & 1 & 0 & 0 & 0 & 1 & 1 & 1 & 1 & 1 \\ E & 1 & 1 & 1 & 1 & 0 & 0 & 1 & 1 & 1 & 1 & 1 \\ F & 1 & 1 & 1 & 1 & 1 & 0 & 1 & 1 & 1 & \underline{1} & 1 \\ G & 0 & 0 & 0 & 0 & 0 & 0 & 0 & 1 & 1 & 0 & 1 \\ H & 0 & 0 & 0 & 0 & 0 & 0 & 0 & 0 & 1 & 0 & 1 \\ J & 0 & 0 & 0 & 0 & 0 & 0 & 0 & 0 & 0 & 0 & 1 \\ K & 0 & 0 & 0 & 0 & 0 & 0 & 0 & 0 & 0 & 0 & 1 \\ L & 0 & 0 & 0 & 0 & 0 & 0 & 0 & 0 & 0 & 0 & 0 \end{array} \qquad (6.33)$$

From this it follows that no stability in the $+x$ direction exists, because all the row elements of component L in the transient closure are zeros.

Stability can be obtained with the use of fasteners. In the AFI, stability in the $+x$ direction is obtained via the fastener M, which combines the components A and L and immobilizes motion of these components with respect to each other. This is incorporated in the matrix by contraction.

$NC_{+x} =$ (6.34)

| | AL | B | C | D | E | F | G | H | J | K |
|---|---|---|---|---|---|---|---|---|---|---|
| AL | 1 | 0 | 0 | 0 | 0 | 0 | 1 | 1 | 1 | 1 |
| B | 1 | 0 | 0 | 0 | 0 | 0 | 1 | 1 | 1 | 1 |
| C | 1 | 1 | 0 | 0 | 0 | 0 | 1 | 1 | 1 | 1 |
| D | 1 | 1 | 1 | 0 | 0 | 0 | 1 | 1 | 1 | 1 |
| E | 1 | 1 | 1 | 1 | 0 | 0 | 1 | 1 | 1 | 1 |
| F | 1 | 1 | 1 | 1 | 1 | 0 | 1 | 1 | 1 | 1 |
| G | 1 | 0 | 0 | 0 | 0 | 0 | 1 | 1 | 1 | 1 |
| H | 1 | 0 | 0 | 0 | 0 | 0 | 1 | 1 | 1 | 1 |
| J | 1 | 0 | 0 | 0 | 0 | 0 | 1 | 1 | 1 | 1 |
| K | 1 | 0 | 0 | 0 | 0 | 0 | 1 | 1 | 1 | 1 |

In this case, the motion of the components G, H, J, and K is obstructed by AL, which causes the corresponding elements in the AL column to become equal to 1. Furthermore, motion of AL is obstructed by G, but motion of G is obstructed by AL, which puts the diagonal element for AL equal to 1. The same is true for the diagonal elements that correspond to G, H, J, and K. The transitive closure thus results in Expression 6.34. From this it follows that any element of the $5 \times 5$ matrix with the components AL, G, H, J, K equals 1. Consequently, the subassembly AGHJKL is stable with respect to the $\pm x$ directions.

### 6.7.5 Conclusion

Several valuable tools on stability analysis are available. This subsection gave a survey of such tools without going too far into details. It is demonstrated that the matrix approach of movability and detachability analyses is also useful in stability analysis. An approach that accounts for mechanical detail has also been described. This is not exhaustively described here, because detailed stability analysis is beyond the scope of this book.

## 6.8 FORCE-FLOW ANALYSIS

Force-flow analysis deals with internal forces. These mainly result from the deformation of components. Elastic forces caused by press fitting and springs, and adhesive forces frequently occur. For the case of Bourjault's ballpoint (Figure 5.1) it was asserted that, once the button E is removed, the ink D could be detached. Evidently this is only valid if the components are assumed to be rigid bodies and forces are neglected. If F is immobilized with a fixture, the detachment of the body A, without dragging the head B with it, is also possible

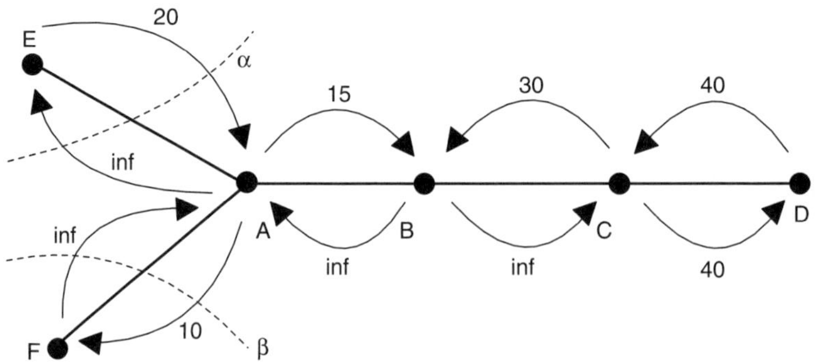


**FIGURE 6.27** Force-flow diagram for Bourjault's ballpoint. F is a static component and E is a dynamic component.

under this assumption. In practice, forces do exist. One has to apply a finite force to the component that should be detached, aimed at counteracting the forces that are rendering stability to the product.

If internal forces exist, one has to properly deal with the connection that one wants to disestablish. This feature has been investigated by Lee and Moradi (1999).

These authors use the connection diagram, with the forces that are required for detaching connections indicated for a particular direction. This is illustrated in Figure 6.27 for Bourjault's ballpoint in the $+x$ direction. If button E is moved with respect to the body A in this direction, a force of 20 N is supposed to be applied. If A is moved with respect to E in the $+x$ direction, motion is obstructed, thus the force is infinite. Analogously, all the forces are determined and added to the connection diagram. Such a diagram is called a *force-flow diagram*. Let component F be the one that is immobilized with a fixture. Assume that the gripper picks component E. All the possible cut-sets are considered. In this case, two cut-sets are possible:

- Only button E is detached, with ABCDF the static subassembly (cut-set $\alpha$)
- The subassembly ABCDE is detached, with F the static subassembly (cut-set $\beta$)

In the force-flow diagram, the static subassembly is considered a sink; the dynamic subassembly is considered a source. The *capacity* of the cut-set is the sum of the forces that cross the cut-set toward the sink. If the capacity is infinite, the cut-set is infeasible. In this case, the capacity of cut-set $\alpha$ is 20; that of cut-set $\beta$ is 10. The value of the maximum flow equals that of the cut-set with minimum capacity. Consequently, E cannot be separately detached from the product in the $+x$ direction, if F is immobilized in the fixture. If component E needs to be detached, the body A has to be immobilized.

The analysis of internal forces belongs to a detailed approach. Therefore, only the principle of this method was outlined here.

## 6.9 CONCLUSION

This chapter dealt with additional information on product structure. Several main topics were discussed. Surface-oriented analysis deals with the different surfaces of the components. With additional information on the individual surfaces, automatic generation of selection rules appears possible. For products with a finite set of directions in which components can move with respect to each other, movability and detachability analysis can be carried out automatically, with the aid of obstruction diagrams and disassembly precedence matrices. These can be constructed from the assembly drawing or virtual prototype in a direct way.

Various matrix-oriented and graphical representations that support modularity analysis, stability analysis, and the influence of internal forces were also discussed. The different available methods, each with their pros and cons, have been concisely discussed. Many of the detailed methods are based on heuristic rules, that are only applicable on a restricted class of products and that do not return a near-optimum solution.

With this, the domain of disassembly sequence detection has been discussed together with some related topics. The following chapters, which are devoted to optimization problems, consider the set of possible disassembly sequences, presented as AND/OR graphs, state diagrams, or disassembly precedence graphs, as a prerequisite.

## REFERENCES

Abe, S., Murayama, T., Oba, F., and Narutaki, N., 1999, Stability check and reorientation of subassemblies in assembly planning. *Proceedings of the IEEE International Conference on Systems, Man, and Cybernetics*, **2**, 486–491.

Chen, S.-F., Oliver, J.H., Chou, S.-Y., and Chen, L.-L., 1997, Parallel disassembly by onion peeling. *Journal of Mechanical Design*, **119**, 267–274.

De Lit, P., 2001, *A comprehensive and integrated approach for the design of a product family and its assembly system*. Ph.D. Thesis, Université Libre de Bruxelles.

De Lit, P., Latinne, P., Rekiek, B., and Delchambre, A., 2001, Assembly planning with an ordering genetic algorithm. *International Journal of Production Research*, **39**(16), 3623–3640.

Dini, G. and Santochi, M., 1992, Automated sequencing and subassembly detection in assembly planning. *Annals of the CIRP*, **41**(1), 1–4.

Dutta, D. and Woo, T.C., 1995, Algorithm for multiple disassembly and parallel assemblies. *Journal of Engineering for Industry*, **117**, 102–109.

Eng, T.H., Ling, Z.K., Olson, W., and Mclean, C., 1999, Feature-based assembly modeling and sequence generation. *Computers and Industrial Engineering*, **36**(1), 17–33.

Garcia, M.A., Larré, A., López, B., and Oller, A., 2000, Reducing the complexity of geometric selective disassembly. *Proceedings of IEEE/RSJ International Conference on Intelligent Robots and Systems*, 1474–1479.

Gu, P. and Yan, X., 1995, CAD-directed automatic assembly sequence planning. *International Journal of Production Research*, **33**(11), 3069–3100.

Gu, P. and Sosale, S., 1999, Product modularization for life cycle engineering. *Robotics and Computer Integrated Manufacturing*, **15**, 387–401.

Gungor, A. and Gupta, S.M., 2001a, Disassembly sequence plan generation using a branch-and-bound algorithm. *International Journal of Production Research*, **39**(3), 481–509.

Gungor, A. and Gupta, S.M., 2001b, A solution approach to the disassembly line balancing problem in the presence of task failures. *International Journal of Production Research*, **39**(7), 1427–1467.

Hoffman, R., 1990a, Assembly planning for B-rep objects. *Proceedings of IEEE Rensselaer's 2nd International Conference on Computer Integrated Manufacturing*, 314–321.

Hoffman, R., 1990b, Automated assembly planning for B-rep objects. *Proceedings of IEEE International Conference on Systems Engineering*, 391–334.

Huang, Y.M. and Huang, C.T., 2002, Disassembly matrix for disassembly processes of products. *International Journal of Production Research*, **40**(2), 255–273.

Jovane, F., Semeraro, Q., and Armillotta, A., 1997, Computer-aided disassembly planning as a support to product redesign. In: Krause, F.L. and Seliger, G., (eds.), Life-cycle networks: *Proceedings of 4th CIRP International Seminar on Life-Cycle Engineering*, 388–399. London: Chapman and Hall.

Knight, W.A., 1996, *Software tools to evaluate cost benefits and environmental impact of designing for disassembly, recycling and the environment*. Design for the 21st Century Technical Papers, Regional Technical Conference, Society of Plastics Engineers, Brookfield CT, 27–38.

Kroll, E., 1993, Modelling and reasoning for computer-based assembly planning. In: A. Kusiak (ed), *Concurrent engineering: automation, tools, and techniques*, New York: Wiley-Interscience, chapter 8, pp. 177–205.

Kroll, E., Beardsley, B., and Parulian, A., 1996, A methodology to evaluate ease of disassembly for product recycling. *IIE Transactions*, **28**, 837–845.

Kroll, E. and Carver, B.S., 1999, Disassembly analysis through time estimation and other metrics. *Robotics and Computer Integrated Manufacturing*, **15**, 191–200.

Kuo, T.C., 2000, Disassembly sequence and cost analysis for electromechanical products. *Robotics and Computer Integrated Manufacturing*, **16**, 43–54.

Kuo, T.C., Zhang, H.C., and Huang, S.H., 2000, Disassembly analysis for electromechanical products: a graph-based heuristic approach. *International Journal of Production Research*, **38**(5), 993–1007.

Lambert, A.J.D., 1999, Disassembly process modelling: problems and solutions concerning their adaptation to practice, *Proceedings of 15th International Conference on Computer-Aided Production Engineering* CAPE'99, Durham, UK, 485–492.

Laperriere, L. and ElMaraghy, H.A., 1992, Planning of products assembly and disassembly. *Annals of the CIRP*, **41**(1), 5–9.

Laperriere, L. and ElMaraghy, H.A., 1994, Assembly sequences planning for simultaneous engineering applications. *International Journal of Advanced Manufacturing Technology*, **9**, 231–244.

Laperriere, L. and ElMaraghy, H.A., 1996, GAPP: a generative assembly process planner. *Journal of Manufacturing Systems*, **15**(4), 282–293.

Laperriere, L. and Lavoie, A., 1997, Evaluating subassembly stability using stability directed subgraphs. *International Journal of Computer Applications in Technology*, **10**(5/6), 348–360.

Lee, S., 1989, Disassembly planning based on subassembly extraction. *Proceedings of 3rd ORSA/TIMS Conference on Flexible Manufacturing Systems: Operations Research Models and Applications*, 383–388.

Lee, S. and Shin, Y.G., 1990a, A cooperative planning system for flexible assembly. *Proceedings of Rensselaer's 2nd International Conference on Computers and Intelligent Manufacturing*, 306–313.

Lee, S. and Shin, Y.G., 1990b, Assembly planning based on subassembly extraction. *Proceedings of 1990 IEEE Conference on Robotics and Automation*, 1606–1611.

Lee, Y.Q. and Kumara, S.R.T., 1992, Individual and group disassembly sequence generation through freedom and interference spaces. *Journal of Design and Manufacturing*, **2**, 143–154.

Lee, S., 1992, Backward assembly planning with assembly cost analysis. *Proceedings of 1992 IEEE International Conference on Robotics and Automation*, 2382–2391.

Lee, K.M. and Bailey-Van Kuren, M.M., 1997, Supervisory control of an automated disassembly workcell based on blocking topology. *Proceedings of 1997 IEEE International Conference on Robotics and Automation*, 1523–1528.

Lee, K. and Gadh, R., 1998, Destructive disassembly to support virtual prototyping. *IIE Transactions*, **30**(10), 959–972.

Lee, S. and Moradi, H., 1999, Disassembly sequencing and assembly sequence verification using force flow networks. *Proceedings of 1999 IEEE International Conference on Robotics and Automation*, 2762–2767.

Lee, K.M. and Bailey-Van Kuren, M.M., 2000, Modeling and supervisory control of a disassembly automation workcell based on blocking topology. *IEEE Transactions on Robotics and Automation*, **16**(1), 67–77.

Mascle, C. and Figour, J., 1990, Methodological approach of sequences determination using the disassembly method. *Proceedings of IEEE Rensselaer's 2nd International Conference on Computer Integrated Manufacturing*, 483–490.

Mascle, C., 1992, Détermination automatique à propos des sous-assemblages (automatic determination of subassemblies). *Automatique Productique Informatique Industrielle*, **26**(2), 167–192. (in French).

Mascle, C., 1995, Features modeling in assembly sequence and resource planning. *Proceedings of IEEE International Symposium on Assembly and Task Planning*, 232–237.

Mascle, C., 1998, Automatic apriori, a posteriori or appropriate determination of subassemblies. *International Journal of Production Research*, **36**(4), 1001–1021.

Mascle, C., 1999, Feature-based assembly model and multi-agents system structure for computer-aided assembly. *Proceedings of 1999 IEEE International Symposium on Assembly and Task Planning*, 8–13.

Mascle, C. and Balasoiu, B.A., 2001, Disassembly-assembly sequencing using feature-based life-cycle model. *Proceedings of 2001 IEEE International Symposium on Assembly and Task Planning*, 31–36.

Mattikali, R., Baraff, D., and Khosla, P., 1994, Finding all gravitationally stable orientations of assemblies. *Proceedings of 1994 IEEE International Conference on Robotics and Automation*, 251–257.

Mattikali, R., Baraff, D., and Khosla, P., 1995, Gravitational stability of frictionless assemblies. *IEEE Transactions on Robotics and Automation*, **11**(3) 374–388.

Mattikali, R., Baraff, D., and Khosla, P., 1996, Finding all stable orientations of assemblies with friction. *IEEE Transactions on Robotics and Automation*, **12**(2) 290–301.

Moore, K.E., Gungor, A., and Gupta, S.M., 1998a, Disassembly process planning using Petri nets. *Proceedings of 1998 IEEE Conference on Electronics and the Environment*, 88–93.

Moore, K.E., Gungor, A., and Gupta, S.M., 1998b, A Petri net approach to disassembly process planning. *Computers and Industrial Engineering*, **35**(1–2), 165–168.

Moore, K.E., Gungor, A., and Gupta, S.M., 1998c, Disassembly Petri net generation in the presence of XOR precedence relationships. *Proceedings of 1998 IEEE International Conference on Systems, Man, and Cybernetics*, 13–18.

Moore K.E., Gungor, A., and Gupta, S.M., 2001, Petri net approach to disassembly process planning for products with complex AND/OR precedence relationships. *European Journal of Operations Research*, **135**, 428–449.

Nnaji, B.O., 1993, Program synthesis and other planners, *Theory of Automatic Robot Assembly and Programming*. London: Chapman and Hall. Chapter 11, 267–299.

Oliver, J.H. and Huang, H.T., 1994, Automated path planning for integrated assembly design. *Computer-Aided Design*, **26**(9), 658–666.

Ong, N.S. and Wong, Y.C., 1999, Automatic subassembly detection from a product model for disassembly sequence generation. *International Journal of Manufacturing Technology*, **15**, 425–431.

Onozuka, M., Kanai, S., and Takahashi, H., 1994, Computer aided evaluation of assembly sequences based on the analysis of methods time measurement and design for assembly. *Advances in Intelligent Production*, 707–712.

O'Shea B., Kaebernick H., and Grewal, S.S., 2000, Using a cluster graph representation of products for application in the disassembly planning process. *Concurrent Engineering*, **8**(3), 158–170.

Quian, W.H. and Pagello, E., 1994, On the scenario and heuristics of disassemblies. *Proceedings of 1994 IEEE International Conference on Robotics and Automation*, 264–271.

Rajan, V.N. and Nof, S.Y., 1996, Minimal precedence constraints for integrated assembly and execution planning. *IEEE Transactions on Robotics and Automation*, **12**(2), 175–186.

Séré, A.T., Laperriere, L., and Mascle, C., 1995, Automatic identification of available grasping and fixturing surfaces in assembly resources planning, *Proceedings of INRIA/IEEE Symposium on Emerging Technologies and Factory Automation*, 69–77.

Shin, C.K. and Cho, H.S., 1994, On the generation of robotic assembly sequences based on separability and assembly motion stability. *Robotica*, **12**, 7–15.

Shyamsundar, N. and Gadh, R., 1996, Selective disassembly of virtual prototypes. *Proceedings of 1996 IEEE International Conference on Systems, Man, and Cybernetics*, 3159–3164.

Shyamsundar, N., Ashai, Z., and Gadh, R., 1998, Geometry-based metric formulation and methodology to support virtual design for disassembly. *Engineering Design and Automation*, **4**(1), 13–26.

Siddique, Z. and Rosen, D.W., 1997, A virtual prototyping approach to product disassembly reasoning. *Computer-Aided Design*, **29**(12), 847–860.

Srinivasan, H., Shyamsundar, N., and Gadh, R., 1997, A framework for virtual disassembly analysis. *Journal of Intelligent Manufacturing*, **8**(4), 277–295.

Swaminathan, A. and Barber, K.S., 1996, An experience-based assembly sequence planner for mechanical assemblies. *IEEE Transactions on Robotics and Automation*, **12**(2), 252–267.

Swaminathan, A., Shaikh, S.A., and Barber, K.S., 1998, Design of an experience-based assembly sequence planner for mechanical assemblies. *Robotica*, **16**, 265–283.

Tanaka, M., Iwama, K., and Watanabe, T., 1996, A method of generating assembly plans by assembly matrices. *Proceedings of ASME Japan/USA Symposium on Flexible Automation*, **2**, 803–806.

Tseng, Y.J. and Liou, L.C., 2000, Integrating assembly and machining planning using graph-based representation models. *International Journal of Production Research*, **38**(12), 2619–2641.

Veerakamolmal, P. and Gupta, S.M., 2002, A case-based reasoning approach for automating disassembly process planning. *Journal of Intelligent Manufacturing*, **13**(1), 47–60.

Venuvinod, P.K., 1993, Automated analysis of 3-D polyhedral assemblies: assembly directions and sequences. *Journal of Manufacturing Systems*, **12**(3), 246–252.

Woo, T.C. and Dutta, D., 1991, Automatic disassembly and total ordering in three dimensions. *Journal of Engineering for Industry*, **113**, 207–213.

Yan, X. and Gu P., 1995, Assembly/disassembly sequence planning for life-cycle cost estimation. *ASME Manufacturing Science and Engineering*, 935–956.

Zhang, H.C. and Kuo, T.C., 1996, A graph-based approach to disassembly model for end-of-life product recycling. *Proceedings of IEEE/CPMT International Electronic Manufacturing Technology (IEMT) Symposium*, 247–254.

Zhang, H.C. and Yu, S.Y., 1997, An environmentally conscious evaluation/design support tool for personal computers. *Proceedings of IEEE International Symposium on Electronics and the Environment*, **5**, 131–136.

Zhou, M.C., Caudill, R.J., and He, X., 1998, Evaluation of environmentally conscious product designs. *Proceedings of 1998 IEEE International Conference on Systems, Man, and Cybernetics*, **4**, 4057–4062.

# 7 Selecting the Optimum Disassembly Sequence

## 7.1 INTRODUCTION

The optimum disassembly sequence is the best out of the many possible sequences that may exist in fulfilling specific criteria. If these criteria are quantifiable, often methods can be identified for automatic selection of such sequence. On the other hand, if such methods require too much effort or if no linear models can be formulated, then one could use metaheuristic and heuristic methods, many of which are available. However, they often return suboptimal solutions. Methods for determining the maximum possible number of disassembly sequences and, consequently, the upper limit of the size of the search space, were discussed in chapter 4. It was observed there, that this number tends to increase exponentially with the number of components $N$, as long as no additional constraints are present. However, in real world products, a set of both topological and geometric constraints is present, which considerably reduces the maximum size of the problem. Even so, the search space remains computationally large, even if it is reduced by a couple of orders of magnitude. The semiautomated incorporation of these constraints in the model was discussed in chapter 5. We discussed three basic representation methods: the state diagram, the disassembly AND/OR graph, and the disassembly precedence graph. Of these, the disassembly AND/OR graph appear to be the most appropriate tool for representing all the possible sequences. It can also be conveniently transformed into a mathematical model. We, therefore, will focus on this representation methodology. We will mainly discuss exact methods as they are considered the benchmarks for evaluating heuristic methods. Heuristic methods can be designed to study arbitrary complex products, mainly by simplifying the problem by introducing additional assumptions and rules, thus returning possibly good, but suboptimal solutions.

If geometric constraints are present, the size of the search space can no longer be determined using combinatorial mathematics, as in chapter 4. The specific structure of the corresponding graph requires methods that account for these constraints, such as graphical and iterative methods. We encountered such methods for AND/OR graphs in subsection 4.6.3, and for disassembly precedence graphs in section 5.9.

The exact optimization problem is quite simple to solve provided the costs of a specific disassembly operation are considered sequence independent. Although rarely recognized, this can be done using linear programming, which gives access to efficient algorithms that return the optimum solution without the need for visiting the complete search space. This method directly follows from graphical determination of optimum sequences, but it also results from a Petri net approach.

Because linear programming is used, NP completeness is avoided. We will discuss this method and see that it is appropriate for many types of problems in disassembly sequencing, including adaptive planning, repair and maintenance, and design. The linear programming approach can also be extended to disassembly/clustering, which includes the adequate sorting of the subassemblies that result from the disassembly process.

In real world problems, sequence dependent operation costs are often encountered, which violates the assumption of sequence independent costs. In such a case, the problem has to be reformulated as a binary integer linear programming problem which can be solved via an iterative procedure or, if a rigorous model is required, as an integer programming problem. We will discuss the methods that are presently available to deal with this problem. We will also introduce some basic heuristic methods. If exact methods are applied, NP completeness cannot be avoided, which reflects itself in terms of central procesing unit (CPU) time that rapidly increases with product complexity. This can be mitigated to some extent if iterative methods are used. We, however, are left with a binary integer programming problem. Both the number of its binary integer variables and the number of required iterations will increase with product complexity.

A quite different class of problems that can be dealt with mathematical programming are the disassembly-to-order problems, which frequently occur in end-of-life disassembly aimed at remanufacturing. Here the product structure is simplified to a hierarchical tree similar to the one that appears in materials requirement planning (MRP) systems. This reflects itself in a disassembly precedence graph representation rather than an AND/OR graph. The model is quite a lot simpler than if AND/OR graphs are considered, which enables its extension to product families, including multiplicity and commonality of components. The subassembly revenue is replaced here with a finite demand on the number of items of different components, which can be considered as a scheduling problem as well. This results in an adaptive model that can be adequately used if both the supply and the demand of the various products change over time.

Since this chapter deals with optimization, we have to appreciate the fact that optimization takes place in the environment of multiple and partly subjective criteria. We have already seen in chapter 3 that both economic and environmental benefits play a role here, which, in turn, are both composed of multiple components that have to be weighed according to some criteria. It can also be observed that, even if we confine ourselves to financial criteria, it is impossible to exactly define the costs of every possible disassembly operation, even if considered sequence independent. In addition, the exactness of the costs and revenues parameters remains questionable. Therefore, the optimum that is returned is considered a guideline that should be further evaluated *a posteriori* by the decision-maker. It can be assessed on technical criteria and, when rejected, a second best solution can be generated. If multiple criteria play a role, one can consider a weighted sum of those as an objective. It is also possible to generate a set of near optimum solutions according to a single criterion, and next select the solution from this set that fits a second criterion best. Finally, sensitivity analyses with respect to parameters that are inaccurate will often be applied. Although the exact method thus shows shortcomings, it still has clear advantages in offering clear selection criteria and making rational choices automatically. Modifications can easily be carried through. Therefore, the software based on these methods can be considered a decision support system rather than a decision making machine!

## 7.2 SEQUENCE INDEPENDENT COSTS

### 7.2.1 Introduction

As a basis for optimization theory, models are discussed in which it is assumed that the costs of single disassembly operations are independent of the order in which the operations are carried out. In the subsequent section, the model will be extended to include sequence dependent costs.

### 7.2.2 Early Methods

Homem De Mello and Sanderson (1990, 1991) first addressed the optimization problem by representing the complete set of disassembly sequences using a single (AND/OR) graph. They suggested *pruning* the graph by judicially rejecting individual subassemblies and operations. Baldwin et al. (1991), who worked with the state diagrams, also suggested the editing of such diagrams via visual inspection of both the states and the individual disassembly sequences. These authors also noticed that a state diagram shows some redundancy compared to the AND/OR graph, which can be counteracted by purging the redundant sequences (see subsection 4.3.4). This redundancy originates because the state diagrams are sensitive to the ordering of the parallel operations. Pruning, which can proceed on an intuitive basis, can also be done automatically with a set of predefined heuristic rules. These can be based on stability and accessibility criteria, etc. Baldwin et al. realized the importance of sequence dependent costs issues such as refixturing and reorientation. These topics were originally dealt with by manual editing of the state diagram. Automatic heuristics and exact methods are discussed in section 7.3.

An exact algorithm for selecting the optimum disassembly sequence was not obtained by the above-mentioned authors, for their main task was to generate a complete set of disassembly sequences. However, their research made it possible to select optimal disassembly sequence using mathematical programming algorithms. Early attempts in the use of mathematical programming are due to Navin-Chandra (1994) and Penev and De Ron (1996).

Navin-Chandra starts with the *disassembly state diagram* of a product. Costs are assigned to disassembly operations, and revenues to the subassemblies and components. The detachment of a fastener is considered a disassembly unit operation. Both tangible fasteners (such as screws) and intangible fasteners (such as welds) are considered. Consequently, Navin-Chandra's approach is a connection-oriented approach, because the fasteners are assigned to connections. The method supports selective disassembly, for it is aimed at end-of-life disassembly. Profit maximization is aimed for. The problem is formulated on the basis of a modified traveling salesperson problem (TSP), which searches for the shortest path in a network. The nodes are considered cities and the arcs, which represent profits by carrying out disassembly operations, correspond to distances. Flow variables $x_{s,t}$ are defined here, which are binary integers that equal 1 only if a shift from state $s$ to state $t$ takes place. Unfortunately, this method requires excessive CPU times for products with even a modest number of components. It appears that the TSP approach is not appropriate for the case with sequence independent costs, which is discussed by Navin-Chandra. Simpler modeling methods are available, which can be solved via linear programming. We will show in the sequel of this section (see section 7.3) that a TSP approach is required only if both sequence dependent costs and parallel disassembly are incorporated in the model.

Penev and De Ron (1996) also use the state diagram. Their aim is to obtain a specific component with minimum number of partitions. In general, there are multiple sequences available to obtain this state. Consequently, the optimization problem is considered equivalent to the minimum distance problem from the initial state to this particular final state. This is usually done in a reverted way by starting with the final state and working in the reverse direction. This method has been called *dynamic programming*, because the authors used Bellman's principle, which had been originally used for solving problems that include variation over time. Bellman's principle is often formulated as the *Principle of Optimality*:

> *An optimal policy has the property that whatever the initial state and decision are, the remaining decisions must constitute an optimal policy with regard to the state resulting from these decisions.* (Shapiro, 1979)

Penev and De Ron (1996) describe this method only schematically, without a general reformulation as a linear programming problem. Since the original approach is confined to a defined final state, which is intrinsic to Bellman's principle, the main disadvantage of this method is that it can only be applied if the final state is identified (which implies complete disassembly). In disassembly problems, unfortunately, this is usually not the case: incomplete disassembly is often performed in a large number of cases. Even if selective disassembly of an *a priori* defined set of components is considered, it is not clear beforehand what the final state will be, i.e., which components and subassemblies have to be disassembled to reveal the desired set of components. If the final state is *a priori* undetermined, Bellman's principle cannot be applied. Unfortunately, the final state is usually a decision variable in end-of-life disassembly.

One intuitively prefers to follow a disassembly strategy where the most profitable component is detached first, and so on, till only those components are left that are not disassemblable in any profitable way. Because of geometric constraints, however, carrying out some unprofitable operations may be a prerequisite for enabling the detachment of profitable components, which essentially complicates the problem. Bellman's principle can be applied here by starting with the final state (single components) and considering the profit of the last operations that have to be carried out. If such an operation does not provide some added value, it is not carried out. This approach has been followed by Lambert (1994, 1997) via a graphical method, which is based on the AND/OR graph. The method applies

**TABLE 7.1**
**Revenues and Costs of Subassemblies and Operations (see Figure 5.5(b))**

| Subassembly | Revenue ($) | Operation | Cost ($) |
|---|---|---|---|
| ABCDEF | −10 | 1 | 5 |
| ABCDE | −9 | 2 | 4 |
| ABCDF | −8 | 3 | 3 |
| ABCD | −6 | 4 | 2 |
| ABF | −3 | 5 | 1 |
| BCD | −2 | 6 | 5 |
| AB | −1 | 7 | 4 |
| AE | 0 | 8 | 3 |
| CD | 1 | 9 | 2 |
| A | 3 | 10 | 1 |
| B | 4 | 11 | 5 |
| C | 5 | 12 | 4 |
| D | 6 | 13 | 3 |
| E | 7 | | |
| F | 8 | | |

Bellman's principle of optimality in a systematic way and includes incomplete disassembly quite naturally.

### Example 1

We will demonstrate the graphical method with the AND/OR graph of Figure 5.5(b). In the graph, the operations are enumerated. The revenues from the subassemblies and the costs of the disassembly operations are given in Table 7.1.

In practice, the revenues obtained from subassemblies may be high either because the subassemblies are modules and are useful as such, or because the subassemblies are homogeneous from a materials point of view, which may also include the case of mutually compatible materials (see subsection 3.5.2).

For every subassembly, the method determines whether or not further disassembly can be carried out in a profitable manner. It also evaluates the revenues for the best alternative. It starts with considering the disassembly of those subassemblies that consist of two components (see Figure 7.1).

The rectangles in the figure represent operations; the circles contain the profit of the optimum subsequent processing, which follows from preceding calculations. The subassembly at the left of each small graph represents the subassembly that is under consideration.

Subassembly AB is considered first. It is possible to leave this subassembly intact, which is indicated with the shunt. Alternatively, AB can be disassembled. This can proceed in one way only, by destroying AB, and creating the components A and B.

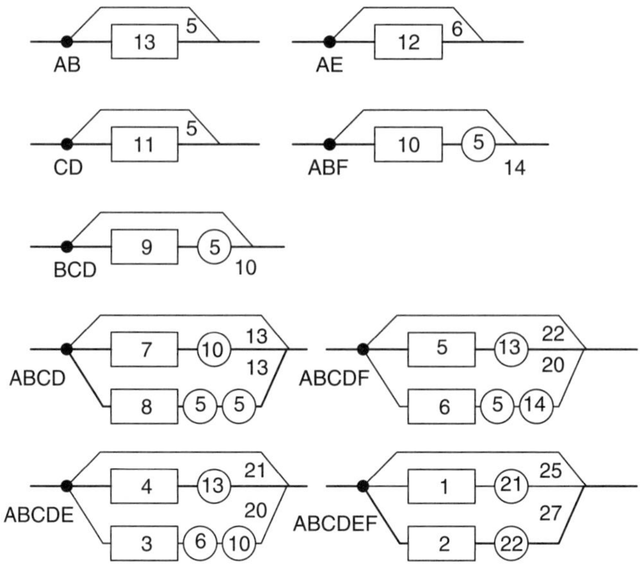


**FIGURE 7.1** Graphical method for disassembly sequence optimization.

This corresponds to operation 13. Therefore, the profit of operation 13 is established via the relation

$$P_{13} = -R_{AB} + R_A + R_B - C_{13}$$

This additional profit equals +5, which follows from the values given in Table 7.1. Leaving AB unprocessed, in contrast, creates an additional profit of zero. The profit that is assigned to AB is the maximum profit that is obtained via one of these two values. Hence:

$$P_{AB} = \max(0, 5) = 5$$

This means that, whenever AB is encountered, it should always be disassembled. Therefore, the value 5 is inserted in the circles that are assigned to the operations that create AB. These operations are 8 and 10 (see Figure 5.5(b)). If the profit of operation 13 were negative, the choice for the shunt should be made and a zero should be assigned to the respective circles.

The calculations for AE and CD proceed along the same lines. These reveal

$$P_{AE} = 6 \text{ and } P_{CD} = 5$$

Next, ABF is considered. It can either be separated in AB and F, or left unprocessed. Thus

$$P_{10} = -R_{ABF} + R_{AB} + R_F - C_{10} + P_{AB} = 14$$

Consequently, $P_{ABF} = 14$. Analogous reasoning reveals: $P_{BCD} = 10$. The subassembly ABCD can either be processed in two different ways, viz. via the operation 7 and operation 8 respectively, or it can be left unprocessed. The choice for each of the operation 7 or operation 8 reveals a profit of 13. If ABCD remains unprocessed, a profit of 0 is obtained. Consequently, $P_{ABCD} = 13$.

It should be noted that, if parallel disassembly occurs, two circles appear in the associated branch, each referring to one of two subassemblies that appear as a result of the operation. The calculations of the profits for ABCDF and ABCDE proceed analogously. These results in $P_{ABCDF} = 22$ via operation 5; $P_{ABCDE} = 21$ via operation 4. Finally, it appears that $P_{ABCDEF} = 27$ via operation 2. This is the maximum profit that can be obtained once the product is available.

The disassembly sequence follows from considering the graphs, starting with that of the complete product, and looking for the branches with optimum profit. Two sequences with equal maximum profit are revealed

Sequence 1: 2-5-7-9-11

Sequence 2: 2-5-8-11-13 (or its equivalent: 2-5-8-13-11)

Obviously, these sequences represent connected hypergraphs, which can be checked in Figure 5.5(b).

This method enables the selection of the optimum disassembly sequence, including those operations that are unprofitable if considered separately, but contribute to a sequence of operations with maximum profit.

### 7.2.3 Linear Programming (LP) Methods

#### 7.2.3.1 Early LP Approach

The graphical method described in the preceding subsection can be used for any given disassembly AND/OR graph. It is, however, a tedious method. Fortunately, directly from the basic reasoning of this graphical approach, a method based on linear programming can be derived that no longer explicitly makes use of the principle of optimality. We will derive a mathematical model that can be solved via standard software without the need of excessive CPU time, as the problem of NP-completeness is circumvented by transforming the problem into a linear programming problem (see Lambert, 1999, 2002).

Prior to discussing this formalism, however, we present the work of Kanehara et al. (1993), who were the first to discuss a linear programming method as a byproduct of the study of Petri nets. Unfortunately, their study has not got the credit it deserves based on its significance.

Kanehara et al. studied the disassembly process mainly aimed at assembly. Topological and geometric constraints were considered. The study considered the properties of *disassembly Petri nets* that can be derived from disassembly AND/OR graphs. The Petri net formalism is a modeling tool for the study of general discrete event systems. Its mathematical and practical foundations are well developed, and it supports both the qualitative and the quantitative aspects of a model. It is frequently used in the study of disassembly systems, although alternatives do exist.

Petri nets basically consist of four elements: *place*, *transition*, *state shift matrix*, and *arc* (see Figure 7.2). In the equivalent AND/OR graph, these elements correspond

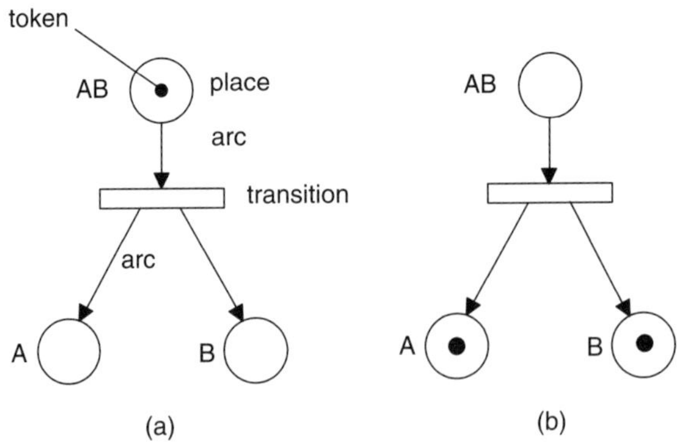


**FIGURE 7.2** An elementary disassembly operation in a Petri net representation: (a) before firing and (b) after firing.

to node (subassembly), AND relation (operation), transition matrix, and arc, respectively. Tokens can be present in the places. *Tokens* are conceptual entities that represent real-world objects, which are subjected to transformations (Moore and Gupta, 1996). The distribution of tokens over places represents the *state* of the system. The total number of tokens in the net is not necessarily constant over time. A state shift of the network is realized by the *firing* of tokens from one place to another, starting with an *initial configuration*. Transitions, which result from disassembly operations, are represented by *state shifts*. A state shift can take place only if at least one token is present in the input place. The state shift matrix is equivalent to the transition matrix in AND/OR graph theory, which will be explained in the next section. Firing a token from the place AB to the places A and B represents the separation of subassembly AB into A and B. It is evident that, before the state transition, only one token is present, and in the final state there are two tokens. The *state shift matrix* controls this firing. It contains the information that, if a disassembly operation acts upon AB, AB is destroyed and both A and B are created. Petri nets show a discrete event approach. In contrast, AND/OR networks are based on stationary flow variables. The model that is depicted in Figure 7.2 thus corresponds to a hyperarc pointing from a parent subassembly toward two child subassemblies. *State shift matrices* $T$ with elements $T_{i,j}$ are crucial in Petri net theory. In such a matrix, the index $i$ refers to the places, which correspond to the subassemblies; the index $j$ refers to the arcs, which correspond to the operations. The value of an element $T_{i,j}$ of the matrix equals the increase of the number of tokens in place $i$ due to operation $j$. In disassembly problems without component multiplicity and with binary operations only, those elements equal 0, −1, or +1. If multiplicity of components is permitted, such as in an action that implies the detachment of 4 identical bolts in one operation, the number might be larger than 1. This case is not considered here.

**Example 2**

The matrix in Expression 7.1 is the state shift matrix for Bourjault's ballpoint (Figure 5.1), from which the corresponding AND/OR graph is depicted (Figure 5.5(b)). All the elements that correspond to an operation that destroys a specific subassembly are equal to −1, those that correspond to operations that create the subassembly are equal to +1, and the other elements are 0. Consequently, every column contains one −1 and two +1 elements. The other elements are zeros.

| | 1 | 2 | 3 | 4 | 5 | 6 | 7 | 8 | 9 | 10 | 11 | 12 | 13 |
|---|---|---|---|---|---|---|---|---|---|---|---|---|---|
| ABCDEF | −1 | −1 | 0 | 0 | 0 | 0 | 0 | 0 | 0 | 0 | 0 | 0 | 0 |
| ABCDE | 1 | 0 | −1 | −1 | 0 | 0 | 0 | 0 | 0 | 0 | 0 | 0 | 0 |
| ABCDF | 0 | 1 | 0 | 0 | −1 | −1 | 0 | 0 | 0 | 0 | 0 | 0 | 0 |
| ABCD | 0 | 0 | 0 | 1 | 1 | 0 | −1 | −1 | 0 | 0 | 0 | 0 | 0 |
| ABF | 0 | 0 | 0 | 0 | 0 | 1 | 0 | 0 | 0 | −1 | 0 | 0 | 0 |
| BCD | 0 | 0 | 1 | 0 | 0 | 0 | 1 | 0 | −1 | 0 | 0 | 0 | 0 |
| AB | 0 | 0 | 0 | 0 | 0 | 0 | 0 | 1 | 0 | 1 | 0 | 0 | −1 |
| AE | 0 | 0 | 1 | 0 | 0 | 0 | 0 | 0 | 0 | 0 | 0 | −1 | 0 |
| CD | 0 | 0 | 0 | 0 | 0 | 1 | 0 | 1 | 1 | 0 | −1 | 0 | 0 |
| A | 0 | 0 | 0 | 0 | 0 | 0 | 1 | 0 | 0 | 0 | 0 | 1 | 1 |
| B | 0 | 0 | 0 | 0 | 0 | 0 | 0 | 0 | 1 | 0 | 0 | 0 | 1 |
| C | 0 | 0 | 0 | 0 | 0 | 0 | 0 | 0 | 0 | 0 | 1 | 0 | 0 |
| D | 0 | 0 | 0 | 0 | 0 | 0 | 0 | 0 | 0 | 0 | 1 | 0 | 0 |
| E | 0 | 1 | 0 | 1 | 0 | 0 | 0 | 0 | 0 | 0 | 0 | 1 | 0 |
| F | 1 | 0 | 0 | 0 | 1 | 0 | 0 | 0 | 0 | 1 | 0 | 0 | 0 |

(7.1)

A *firing vector* $x$ with components $x_j$ is defined, such that $x_j = 1$ if transition $j$ has fired. A state vector $M$ with components $M_i$ is defined. The component $M_i$ is 1 if the corresponding subassembly $i$ exists and 0 otherwise. The cost vector $C$ with components $C_j$ reflects the costs of performing operation $j$.

### Example 3

If complete disassembly is considered, the initial state of Bourjault's ballpoint is given by

$$M_I = (1\ 0\ 0\ 0\ 0\ 0\ 0\ 0\ 0\ 0\ 0\ 0\ 0\ 0\ 0)$$

The final state is represented by

$$M_F = (0\ 0\ 0\ 0\ 0\ 0\ 0\ 0\ 0\ 1\ 1\ 1\ 1\ 1\ 1)$$

Although the components of the firing vector are binary integers, the problem relaxes to a linear programming problem. However, a continuous time approach is applied, rather than the discrete time approach that is intrinsic to Petri nets. Its basic formulation, as given by Kanehara et al. (1993) is as follows:

$$\text{Minimize: } C = \sum_j C_j \cdot x_j \tag{7.2a}$$

$$\text{Subject to: } \sum_j T_{i,j} \cdot x_j = (M_F - M_I)_i \tag{7.2b}$$

$$x_j \geq 0 \tag{7.2c}$$

This formalism has a drawback in that the final state must be identified. In end-of-life disassembly, however, it is considered a decision variable. A related disadvantage is that subassembly revenues are not incorporated in this model. The problem formalism was extended to include *a priori* unknown final state via combining the firing vector with a final state vector. This way, problems from maintenance, such as detaching a specific component at minimum expense, can be solved. Kanehara et al. (1993) discussed three types of problems:

1. Attaining complete disassembly at minimum costs
2. Maximizing parallelism in complete disassembly
3. Minimizing costs for detaching a predefined set of components

### 7.2.3.2 Linear Programming Method, Network Approach

Incomplete disassembly can be included in a more comprehensive graphical based model, which has been discussed by Lambert (1999). This model proceeds directly from the AND/OR graph, and thus avoids the Petri nets formalism. It is a continuous model with respect to time, contrary to Petri nets that are intended for discrete events. It will be called *AND/OR graph based optimization.* This method proceeds as follows.

The starting point is a continuous flow of available products. A set of *flow variables* $x_j$ is then defined, each of which corresponds to a binary disassembly operation, which is represented by a hyperarc in the AND/OR graph. This is equivalent to the firing vector of the previous subsection. The flow variables are binary integer variables, which relax into continuous variables. These are equal to 1 if the corresponding operation is performed, and 0 otherwise. Each node in the AND/OR graph has one or more incoming arcs and one or more outgoing hyperarcs, apart from the single components that have incoming arcs only. Outgoing hyperarcs are related to an exclusive OR relationship, which means that only one of these can be performed. There can only be an outgoing flow if the sum of the incoming flows equals 1. This implies that at most one of the incoming flows can be equal to 1, because that is the condition for producing the subassembly that corresponds to the specific node.

For every node, the sum of the outgoing flow variables is equal to, or smaller than, the sum of the incoming flow variables. Formally, with the index $i$ referring to subassemblies and the index $j$ to operations, this can be written as follows:

$$
\begin{gathered}
x_j \in \{0,1\}, \forall j \\
\sum x_{i,in} \geq \sum x_{i,out}, \forall i \qquad (7.3) \\
\sum_i x_{i,in} \leq 1, \forall i
\end{gathered}
$$

**Example 4**

Consider node ABCD in figure 5.5(b). This subassembly is produced if the flow variable $x_4$ OR $x_5$ is equal to 1. Subassembly ABCD can be decomposed in either (AB AND CD),

**TABLE 7.2**
**Permitted Combinations of Flow Variables for the Node ABCD**

| $x_4$ | $x_5$ | $x_7$ | $x_8$ |
|---|---|---|---|
| 0 | 0 | 0 | 0 |
| 1 | 0 | 0 | 0 |
| 1 | 0 | 1 | 0 |
| 1 | 0 | 0 | 1 |
| 0 | 1 | 0 | 0 |
| 0 | 1 | 1 | 0 |
| 0 | 1 | 0 | 1 |

OR in (BCD AND A). This means that $x_7$ OR $x_8 = 1$. The product of the incoming flow variables should be 0; that of the outgoing flow variables should also be zero. This is equivalent to the *exclusive OR* (XOR) operator. Because incomplete disassembly is also permitted, both flow vectors might be zero. Therefore, the following inequality is valid for node ABCD:

$$x_4 + x_5 \geq x_7 + x_8$$

This can be expressed in the truth Table 7.2, which represents all the permitted combinations of the incoming and outgoing flows for node ABCD.

The transition matrix $T$ in the AND/OR graph is comparable to the state shift matrix. Consequently, nonzero elements are −1 if an operation causes the destruction of a subassembly, and +1 if an operation causes the creation of a subassembly.

The introduction of an initial flow variable $x_0$ is useful. This flow variable is set equal to 1. It adds a zeroth column to the transition matrix (Expression 7.1), which only has the element $T_{1,0}$ different from zero. The associated operation 0 is called the *pseudo operation*, which corresponds to initiating the availability of the product or the flow of products.

The relationship between the incoming and outgoing flow variables for every node is thus given by

$$\sum_j T_{i,j} \cdot x_j \geq 0 \tag{7.4a}$$

The presence of the ≥ sign instead of the = sign, which would imply node equilibrium, enables the subgraph to stop at any arbitrary node, thus dealing with both the final components, which have no outgoing hyperarcs, as well as incomplete disassembly.

**EXAMPLE 5**

The correctness of (Expression 7.4a) can be checked in Figure 5.5(b) for all the nodes.

For instance

$$x_0 - x_1 - x_2 \geq 0, \text{ for node ABCDEF}$$
$$x_1 - x_3 - x_4 \geq 0, \text{ for node ABCDE}$$
$$x_4 + x_5 - x_7 - x_8 \geq 0, \text{ for node ABCD}$$
$$x_1 + x_5 + x_{10} \geq 0, \text{ for node F}$$

We can observe that the transition matrix reflects the structure of the AND/OR graph. Additionally, an objective function has to be introduced which reflects the profit that is obtained via the disassembly process, which has to be maximized. The basic ideas to establish costs can be found in section 3.4. Costs are assigned to the disassembly operations, and this is represented via the cost vector $C$ with components $C_j$. Costs are only encountered if the action is actually performed, which is only true if the corresponding flow variable is different from zero. The complete disassembly costs thus are given by

$$C = \sum_j C_j \cdot x_j \tag{7.5a}$$

Revenues also have to be included. These are related to the availability of subassemblies. The revenues thus can be grouped in a revenue vector $R$ with components $R_i$. These might have both positive and negative values, because the availability of a specific subassembly may represent costs, for instance, when it cannot be reused or recycled and has to be discharged at some expense. The associated cost is considered negative revenue. Revenues are only present if the subassembly is available, which occurs if the sum of the incoming flow variables surmounts the sum of the outgoing flow variables. This can be expressed as

$$R = \sum_i \sum_j R_i \cdot T_{i,j} \cdot x_j \tag{7.5b}$$

Subtracting Expression (7.5a) from Expression 7.5b gives the profit, which is represented by $P_{\text{tot}}$. Thus

$$P_{\text{tot}} = R - C = \sum_j \left( \sum_i (R_i \cdot T_{i,j}) - C_j \right) \cdot x_j \tag{7.4b}$$

Consequently, the model is formulated with Expression 7.4ab and is aimed at maximizing the profit.

### Example 6

The validity of Expression 7.5b is demonstrated when $x_0 = x_1 = x_4 = 1$ and all the other flow variables are zero. In this case, only the 1st and the 4th columns of (Expression 7.1) are relevant. Expression 7.5b then results in

$$R = T_{0,1} \cdot R_1 + (T_{1,1} \cdot R_1 + T_{2,1} \cdot R_2 + T_{15,1} \cdot R_{15}) + (T_{2,4} \cdot R_2 + T_{4,4} \cdot R_4 + T_{14,4} \cdot R_4)$$

Here subassemblies 1, 2, 4, 14, and 15 correspond to ABCDEF, ABCDE, ABCD, E, and F, respectively. Inserting the values of the matrix elements results in

$$R = R_1 + (-R_1 + R_2 + R_{15}) + (-R_2 + R_4 + R_{14}) = R_4 + R_{14} + R_{15}$$

This corresponds to the combined revenue of the subassemblies ABCD, E, and F.

In the absence of capacity constraints, the integer linear programming problem relaxes into a linear programming problem. The only additional condition is that the variables $x_j$ be considered as *nonnegative* variables. This property of the model has already been recognized and proved by Kanehara et al. (1993). It is a consequence of the fact that a basic solution of a linear programming problem is an extreme point of a polyhedron. It turns out that the model is simple and flexible indeed. Changing the model only requires the modification of the transition matrix and new cost and revenue vectors.

### 7.2.4 Applications

#### 7.2.4.1 Optimum Solution

The optimum solution corresponds to a connected subgraph of the disassembly AND/OR graph that maximizes profit. This property can serve as a debugging tool in modeling.

**Example 7**

Consider the case of Figure 5.5(b) and the transition matrix T from Expression 7.1. Assume that the following values are assigned to the cost vector, with the index $j$ running from 0 through 13:

$$C = (0,.347,.286,.135,.621,.315,.108,.806,.438,.216,.811,.719,.571,.485)$$

The following values are assigned to the revenue vector $R$, with the index $i$ running from 1 through 15, corresponding to the rows of matrix (Expression 7.1):

$$R = (-12, -4, -2, 1, 3, 5, 7, 4, 3, 2, 1, 4, 6, 9, 5)$$

The optimum solution with the profit 41.242 is revealed, which corresponds to the sequence 0-2-5-8-11. This is indeed a *connected* subgraph of hypergraph in Figure 5.5(b). Revenues are those of AB, C, D, E, and F, and the inverse of ABCDEF, which is destroyed. Costs are those of the operations of the sequence.

**TABLE 7.3**
**The Ten Best Sequences for the Case of Figure 5.5(b)**

| No. | Sequence | Profit | Revenue/Cost |
|---|---|---|---|
| 1 | 0-2-5-8-11 | 29.242 | 17.634 |
| 2 | 0-2-6-10-11 | 29.076 | 16.112 |
| 3 | 0-1-4-8-11 | 28.875 | 14.588 |
| 4 | 0-1-3-9-11-12 | 25.012 | 13.581 |
| 5 | 0-2-5-7-9-11 | 24.658 | 11.529 |
| 6 | 0-1-4-7-9-11 | 24.291 | 9.967 |
| 7 | 0-2-5-8 | 22.961 | 23.099 |
| 8 | 0-2-6-10 | 22.795 | 19.917 |
| 9 | 0-1-4-8 | 22.594 | 17.070 |
| 10 | 0-2-6-11 | 20.887 | 19.766 |

#### 7.2.4.2 Suboptimum Solutions

If a reasonable disassembly sequence is sought out of the multitude of possible sequences, it is not always practical to rely on "the optimum" sequence only, because such a sequence may have disadvantageous properties that are not articulated in the model. That is, even though, multiple criteria play a role here, many criteria are not easily quantifiable. In addition, the data on costs and revenues are often fuzzy, because these are usually based on rough estimates only. Therefore, it is reasonable to search for a set of near optimum solutions, as these can be checked on other criteria. A set of reasonable (albeit suboptimum) solutions can be found either via additional constraints that inhibit the optimum solution, or by putting an upper limit on the profit. The latter method has a disadvantage that the model no longer relaxes into a linear programming problem. Consequently, the flow variables have to be explicitly declared binary integer variables.

**EXAMPLE 8**

The set of the ten best solutions for the already presented instance of the disassembly AND/OR graph (see example 7 above) is given in Table 7.3. These are all the solutions with a profit higher than an *a priori* chosen lower limit of 20. Also listed in the table are the ratios of revenue to cost corresponding to each solution. Note that in this table only the revenues due to the released subassemblies are considered, therefore, the optimum value is 29.242 instead of 41.242 (as reported in example 7). The difference is the additional revenue 12 that results from destructing ABCDEF.

This table clearly shows an alternative criterion that may be decisive, namely the ratio of revenue to cost, which is the maximum (23.099) for the seventh best solution based on profit.

Note that the rejection of some solutions cannot always take place due to extra constraints. This can be demonstrated by considering the optimum sequence 0-2-5-8-11.

The constraint

$$x_2 + x_5 + x_8 + x_{11} \leq 3$$

not only inhibits the optimum sequence, but also inhibits possible extended sequences, such as

0-2-5-8-11-13

Extended sequences have to be investigated by separately selecting the suboptimal solutions that are subjected to an additional constraint, in this case

$$x_2 + x_5 + x_8 + x_{11} = 4$$

and that have a profit that surpasses the lower limit previously set.

Those sequences that have the *maximum* number of partitions can be directly inhibited, because there are no extensions possible here. In this case, there is a maximum of $(N - 1) = 5$ partitions possible. For example, sequence 0-1-3-9-11-12 is explicitly inhibited via

$$x_1 + x_3 + x_9 + x_{11} + x_{12} \leq 4$$

Adding an upper limit to the profit and gradually lowering it, is another method for deriving a set of suboptimum sequences. In this case, one has to deal with a binary linear programming problem that does not relax into a linear programming problem.

#### 7.2.4.3 Maximum Parallelism

For both assembly and disassembly plans, maximum parallelism may be beneficial. Simultaneous disassembly of subassemblies is often helpful in line balancing problems. This was noticed by Kanehara et al. (1993), who carried out the calculation for obtaining the complete disassembly sequence of a product with maximum parallelism as an objective. This can be done by assigning reduced costs to a parallel disassembly operation. This was obtained in the original paper by defining the costs of an operation as the inverse of the number of components of the smallest child subassembly.

**Example 9**

For the Bourjault's ballpoint case (Figure 5.1 and Figure 5.5(b)), assume that the cost vector is as follows:

$$C = (0, 1, 1, 0.5, 1, 1, 0.5, 1, 0.5, 1, 1, 1, 1, 1)$$

Based on this, the following four optimum sequences can be obtained:

0-1-3-9-11-12
0-1-4-8-11-13
0-2-5-8-11-13
0-2-6-10-11-13

One can also include all other equivalent sequences here in which parallel executable operations, such as 11 and 13, are reverted.

If parallelism is not the only criterion, other techniques can be applied. The original cost vector is modified via assigning low costs to the parallel operations.

**Example 10**

If we assign reduced costs to the parallel operations 3, 6, and 8 (Figure 5.5(b)) in example 7, the following cost vector may be established:

$$C = (0, .347, .286, .01, .621, .315, .01, .806, .01, .216, .811, .719, .571, .485)$$

The corresponding set of near optimum results is listed in Table 7.4. It can be observed that the fifth best sequence does not involve any parallel operation.

In increasingly complex products, as in the AFI case of Figure 5.7, in which many possible parallel operations can be selected, the same method is easily applicable.

### 7.2.4.4 Detaching Selected Component(s)

In maintenance and repair operations, one often wants to determine the optimum way for detaching a specific component or set of components from the product, aimed at replacing or repairing them. In end-of-life disassembly also, similar situation may occur, such as retrieving a specific component that is needed to meet a certain demand or the compulsory removal of hazardous components due to government regulations.

**TABLE 7.4**
**Optimum Sequences with Parallel Disassembly**

| Sequence | Profit |
|---|---|
| 0-2-5-8-11 | 29.670 |
| 0-1-4-8-11 | 29.303 |
| 0-2-6-10-11 | 29.174 |
| 0-1-3-9-11-12 | 25.137 |
| 0-2-5-7-9-11 | 24.658 |

In general, this also requires the detachment of additional components because of geometrical constraints. Therefore, both the final state as well as the sequence of operations need to be determined.

**Example 11**

Assume that the ink D must be removed from Bourjault's ballpoint example of Figure 5.5(b) to comply with the government regulation, which requires the removal of hazardous working fluids. This can be obtained by extending the model with an additional constraint, i.e., one of the operations that have D as child subassembly should be 1. The only operation that creates component D is operation 11 (see the transition matrix (Expression 7.1)). Therefore, the constraint $x_{11} = 1$ has to be added to the model. If this is done, the optimum solution from Table 7.3 is returned as a solution, which is trivial in this specific case. If, on the other hand, component A is desired, it can be obtained via the operations 7, 12, or 13. Only one of these operations can be executed, thus the extra constraint should be as follows:

$$x_7 + x_{12} + x_{13} = 1$$

In this case, the following sequence is returned:

0-2-5-7-9-11

It should be noted that the operations 9 and 11 are not necessary if only the retrieval of component A is required. However, if A is obtained, which is compulsory, it is reasonable from an *economic* point of view to proceed with operations 9 and 11 as well.

#### 7.2.4.5 Detaching Selected Component(s) with Additional Constraints

It is often required that a specific component be detached with minimum number of collateral components. In this case, the cost of an operation corresponds to the number of disassembled components other than the one targeted. Here we have to distinguish between the static and the dynamic subassembly. We assume that the static subassembly is the one with the maximum number of components. The cost equals the number of the detached components, except the desired one.

**Example 12**

If we want to detach component A from Bourjault's ballpoint (Figure 5.1), costs may be assigned according to the above-mentioned rules as follows (see Figure 5.5(b)). The cost of operation 1 equals 1 because F is detached. This also holds true for operations 2, 3, 4, and 5. In operation 6, the subassembly CD is detached. Therefore, cost 2 is assigned to this operation. The desired component is detached in operation 7 at zero cost. Via operation 8 it is assumed that AB is the dynamic subassembly because it includes the desired component. Hence the cost equals 1. Operation 9 and operation 10, each has cost 1. Operation 11 is carried out at cost 2, because two additional components are released.

The costs of operation 12 and 13 are zero. Therefore

$$C = (0, 1, 1, 1, 1, 1, 2, 0, 1, 1, 1, 2, 0, 0)$$

For guaranteeing that component A is detached, one adds the following constraint:

$$\sum_j T_{i=A,j} \cdot x_j = 1$$

This corresponds to

$$x_7 + x_{12} + x_{13} = 1$$

The solution reveals sequence 0-2-5-7, which costs 2. This means that for minimum cost, two collateral components, viz., E and F, have to be detached to obtain the desired component A.

### 7.2.5 Clustering

The mathematical programming approach based on AND/OR graphs described so far can be adapted and extended to accommodate many other requirements. Apart from the modifications as described in the previous subsection, adaptive planning can be mentioned, which means that modifications in both parameter values and network structure can be smoothly adapted, which is useful in both design issues and in a rapidly changing environment. In addition, extensions to other processes in the end-of-life chain can be carried through. We already saw that many different processes in bulk recycling and disassembly sequencing can be modeled with integer linear programming models on the basis of a network. In particular, the combination of a disassembly sequence optimization problem together with a clustering problem can be easily accommodated (Lambert, 1999). Clustering is represented by a converging AND/OR graph, which is associated with the disassembly AND/OR graph.

The *clustering problem* refers to the appropriate combination of different components and subassemblies separated in bins aimed at maximizing profit. In previously discussed models, the revenue from a component or a subassembly was externally determined. If materials recovery is aimed for, then a particular group of components that contain compatible materials composition is often blended. One reason is that the number of different streams (blends) has to be restricted for the operation to be cost-effective. In addition, the customer may require minimum quantities of secondary materials. Blending therefore is a tradeoff between these aspects and the required purity of the materials, which is dictated by the prices that customers pay based on qualitative and quantitative aspects of these recovered materials. Materials compatibility depends on the possibility of combined application, such as that of different kinds of plastics in lower grade applications, the ease of separation, or the seriousness of contamination. Evidently, a small quantity of hazardous material that is added to a large quantity of valuable materials will turn

**TABLE 7.5**
**Mass and Specific Revenue of the Components**

| Component | Mass (kg) | Specific Revenue ($/kg) |
|---|---|---|
| A | 2 | 2 |
| B | 5 | 4 |
| C | 1 | 1 |
| D | 1 | –3 |
| E | 2 | 2 |
| F | 1 | 2 |

the complete batch into hazardous waste, which causes a dramatic increase in the costs of final processing.

The following represents a hierarchy of revenue, based on clustering (Lambert, 1999):

1. Contaminated materials (considered hazardous waste)
2. Pure hazardous substances
3. Technologically incompatible materials, e.g., a mixture of glass and plastics
4. Mixed materials, e.g., "metals" or "synthetic materials."
5. Materials consisting of compatible substances, such as ferrous materials, or compatible types of plastic (see subsection 3.5.2)
6. Materials consisting of a homogeneous substance, e.g., a specific alloy
7. Components of the same kind, aimed at component reuse

### Example 13

Let us consider a batch of Bourjault's ballpoints (see Figure 5.1). Let the mass of the batch be 12 kg. The quantities mass share (in kg) and specific revenue (in $/kg) are assigned to the components according to Table 7.5. We calculated the mass here for a batch of 12 kg of ballpoints. A compatibility matrix is established and given in Table 7.6. This shows the specific revenue for mixtures of components. It is clear that,

**TABLE 7.6**
**Example of a Compatibility Matrix ($/kg)**

| | A | B | C | D | E | F |
|---|---|---|---|---|---|---|
| A | 2 | 2 | 1 | –3 | 2 | 2 |
| B | 0 | 4 | 1 | –3 | 2 | 2 |
| C | 0 | 0 | 1 | –3 | 1 | 1 |
| D | 0 | 0 | 0 | –3 | –3 | –3 |
| E | 0 | 0 | 0 | 0 | 2 | 2 |
| F | 0 | 0 | 0 | 0 | 0 | 2 |

if a hazardous substance is present, the specific revenue drops to that of the hazardous materials. The revenues in the diagonal elements are those from Table 7.5. Due to its symmetry, the lower left triangle of the table is not completed. Assume that no component reuse takes place. Further, a difference in revenue may exist between, for example, a batch of subassembly AB and a batch with a blend of the components A and B. Also, blends of more than two components may exist. However, we have neglected such cases for the sake of simplicity. In this example, a pessimistic value is selected. The revenue vector for the 15 subassemblies (viz., ABCDEF, ABCDE, ABCDF, ABCD, ABF, BCD, AB, AE, CD, A, B, C, D, E, F) is as follows:

$$R = (-36, -33, -30, -27, 16, -21, 14, 8, -6, 4, 20, 1, -3, 4, 2)$$

which is found, for subassemblies with one or two components, by multiplying the aggregate mass (in kg) and the appropriate specific revenue (in \$/kg).

Clustering is the blending of the different subassemblies that results from a selective disassembly process. Evidently, a subassembly that consists of two components with compatible materials needs no further disassembly. On the other hand, two components that consist of compatible materials can be merged. Feasibility conditions such as in the disassembly process do not play a role here. This means, for example, the mixture {A, C} can be created, although the subassembly AC is infeasible. With this in mind, a merging operation can be treated in the model similar to a reverted disassembly operation. We will assign a negative cost (bonus) to merging operations, as these reduce the number of bins. In disassembly operations, separation of two components with materials compatibility might still be required, for example, to deal with geometric constraints.

### Example 14

The clustering problem will be explained with an example that has the six different components of Bourjault's ballpoint (Figure 5.1) as an initial state. All possible mergers of two different components are potentially possible: {A, B}, {A, C}, etc., which results in $\frac{1}{2}N(N-1)=15$ possible mixtures. Part of the possible clustering operations is depicted in Figure 7.3.

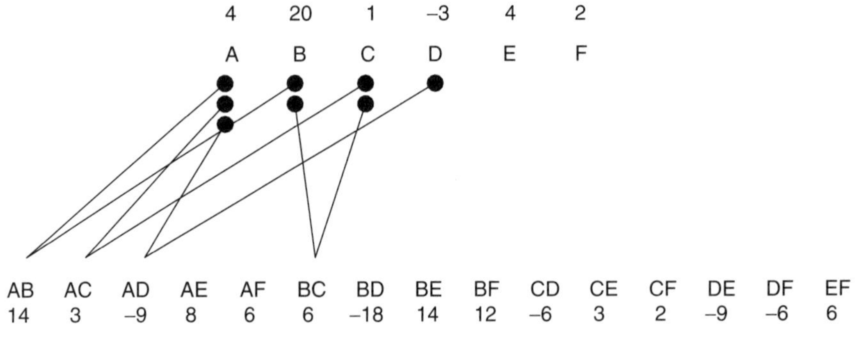


**FIGURE 7.3** Reduced clustering problem for the Bourjault's ballpoint example. Numbers indicate the revenue from the mixtures of components that can be derived from one batch.

The profit of clustering a combination of components, e.g., A and B, is given by

$$P_{A,B} = (m_A + m_B)R_{A,B} - (m_A \cdot R_A + m_B \cdot R_B + C_{A,B})$$

Here, the difference, compared to the disassembly problem, is that the clustering hyperarcs are directed *toward* the AND-relationship, which is connected to the mixture. Flow variables are connected to the hyperarcs. For instance, $x_{A,B}$ represents the clustering of components A and B. The flow variables are binary (0, 1) variables. The following constraints, therefore, have to be added to the model:

$$x_{h,i} = 0, \quad h \geq i \tag{7.6a}$$

$$\sum_h x_{h,i} \leq 1, \quad h < i \tag{7.6b}$$

The indices $h$ and $i$ represent components that have to be ordered. Alphabetical ordering is adapted here. Expression (7.6b) guarantees that every component appears at most once in the combined flow. Expression (7.6a) inhibits double counting. Specific revenues *R/m* are stored in a matrix (see Table 7.6), from which the strictly lower triangular matrix elements can be put equal to zero. The specific revenues of the single components are at the diagonal. Revenues are found by multiplying the specific revenues with the mass of the corresponding subassembly (see Table 7.5), thus $R_1 = 4$, $R_{1,3} = 3$, etc. The gain in revenue due to the clustering process is found by using the following expression:

$$R = \sum_{h,i} (R_{h,i} - R_h - R_i)x_{h,i} \tag{7.6c}$$

If there are three components, for example, A, B, and C, and the components A and B are merged, this means that $x_{A,B} = 1$. This agrees with the fact that {A,B} is now available, but after clustering the components A and B are no longer available. The cost can be found using the following expression:

$$C = \sum_{h,i} C_{h,i} \cdot x_{h,i} \tag{7.6d}$$

$C_{h,i}$ is the cost for the clustering operation, which is often a negative value that accounts for the decrease in the number of bins by 1. The objective function is the profit *P*, which has to be maximized

$$obj = P = R - C \tag{7.6e}$$

With the values of Figure 7.3, and $C_{h,i} = -5$ for all feasible pairs $h$ and $i$, the optimum value corresponds to a state in which {A, E}, {C, F} are clustered, and both B and D remain unclustered. The objective value equals 9. If clustering is considered extremely profitable, or there is a maximum number of three containers, the most profitable clustering scheme appears to be: {A, F}, {B, E}, and {C, D}.

This simple example was presented to demonstrate the clustering problem. It can be extended to include triple and higher-order combinations, etc. For instance, bins can be included that contain materials such as "metals" rather than a predefined combination of components. In this case, multiple components and subassemblies can enter the bin, which is related to the inverse problem of ternary and higher-order operations (see subsection 4.3.4.4). The clustering can easily be combined with the disassembly problem into one model.

### 7.2.6 Further Work on Mathematical Programming

Virtually no further literature on selecting the optimum disassembly sequence via mathematical programming is available so far. The article of Kanehara et al. (1993), which was a conference paper, remained relatively unknown, although it has been esteemed as the first paper to introduce the Petri net formalism for assembly systems (Tiwari et al., 2002). This is, however, not quite true. The first paper on this topic is due to Bourjault et al. (1987), but this is published in the French language, and therefore it has not been optimally disseminated. Many papers on disassembly Petri nets, which are not discussed here in detail, followed. Kanehara et al. were the first to publish on optimizing methods via mathematical programming. One additional paper on mathematical programming, also based on the paper by Kanehara et al. could be retrieved. This paper is due to Yee and Ventura (1999). This work will be discussed in the next section, which deals with sequence dependent costs.

## 7.3 SEQUENCE DEPENDENT COSTS, HEURISTICS, AND RESTRICTED EXACT METHODS

In practice, the costs of a given operation are not independent of the operations that have been carried out prior to this given operation. A typical task (such as a tool change or product reorientation or refixturing) depends on the state in which it is left by the previous operation. This is essential in determining the required disassembly costs and, consequently, in generating the optimal disassembly sequence. However, this significantly complicates the problem. In disassembly costing theory, this interdependency of costs has long been recognized (see chapter 3). To this end, the following three approaches are discussed here:

1. Heuristic approaches, which search for reasonable solutions, usually by making use of penalty functions
2. Exact approaches that are confined to sequential disassembly
3. Exact approaches that include parallel disassembly

### 7.3.1 Heuristic Approaches

A typical example of a heuristic approach of the problem of sequencing with sequence dependent costs is discussed by Gungor and Gupta (1997). We will follow their reasoning here. Two different penalty factors are introduced: The *direction factor* $\alpha$, which adds a penalty when the direction of disassembly is changed between two actions, and the *joint type change factor* $\beta$, which adds a penalty if a tool has to be changed. The disassembly cost is multiplied by the factor: $1 + \alpha + \beta$. This is incorporated in a heuristic, which is applied to the disassembly of a real world product, namely, a personal computer with 17 different components, for which the following data have been collected:

1. The components that have to be detached prior to the detachment of a specific component
2. The direction of detachment, which is along orthogonal axes $\pm x$, $\pm y$, $\pm z$
3. The type and number of fasteners
4. The cost of disassembly, which is expressed in the form of a difficulty rating

The 17 components of this example are enumerated as A through Q. The basic characteristics of the components are listed in Table 7.7. In this example product, two types of fasteners are used, viz., a screw (SC) and a snap-fit (SF), each of which is associated with a specific tool. The symbol 4SC means that four screws have to be released prior to the detachment of the component. Many tool changes might be required if the ordering of disassembly operations is not appropriate. Because the precedence relationships are given, a disassembly precedence graph can be composed (see Figure 7.4).

**TABLE 7.7**
**Properties of the Components for the PC Example**

| Component | Name | Preceding Components | Disassembly Direction | Fasteners | Average Difficulty |
|---|---|---|---|---|---|
| A | Top cover | — | −x | 4SC | 2.7 |
| B | FD cover | A | x | 2SF | 2.0 |
| C | FD drive | B and M | x | 1SF | 3.6 |
| D | HD cover | A | x | 2SF | 2.0 |
| E | HD drive | D and M | x | 1SF | 3.8 |
| F | Back panel | A | −x | 6SF | 3.9 |
| G | Cable bridge | A | y | 2SF | 1.0 |
| H | Network card | A | z | 1SC | 2.3 |
| I | Expansion card | G and H | y | 1SF | 2.0 |
| J | RAM chips | A | z | 4SF | 3.3 |
| K | Key lock | A | y | 2SC | 2.2 |
| L | Power Supply | F and G | y | 3SC+2SF | 2.9 |
| M | Flat cables | A | y | SF | 1.3 |
| N | Front panel | C and E | x | 7SF | 3.8 |
| O | Drive rack | N | y | 1SC+2SF | 2.2 |
| P | Motherboard | I and L and N | z | 7SC+2SF | 3.1 |
| Q | Power switch | L and N | x | 2SF | 2.5 |

*Source:* Gungor and Gupta, 1997

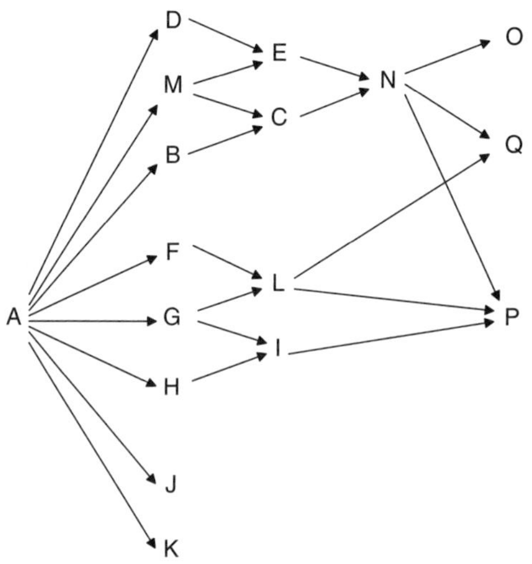


**FIGURE 7.4** Disassembly precedence graph for the PC example.

The *average* difficulty rating given in Table 7.7 has the disadvantage that the disassembly rating of the preceding component has no effect on the difficulty rating of the actual component. However, this is not so in practice. Therefore, *a posteriori* calculation takes place for evaluating the sequence that is generated this way. The quality of a sequence is expressed in one single figure, corresponding to the total disassembly time, which is considered proportional to the cost.

This method is restricted to *sequential* disassembly, which means that in every disassembly operation only one component or module is detached. In addition, complete disassembly is considered here. The optimum is found using a heuristic algorithm, which is outlined below:

The algorithm starts with the initialization of a sorted list, disassembly sequence (DS), which is initially empty, and an Unsorted List (UL) that contains the 17 different components.

The heuristics is based on the sorting of the components according to the fastener type. In this case, there are two types of fasteners ($m = 2$): SF and SC, out of which the disestablishment of a snap fit is considered easier than that of a screw.

Two lists are made: List P(1) represents a sorted list that contains the components with the easiest fastener, SF only, ordered according to the average difficulty, thus:

$$P(1) = \{G, M, B, D, I, Q, J, C, N\}$$

List P(2) represents a similar list, but with screws as fasteners. Those components with both screws and snap fit are also in P(2). The list is as follows:

$$P(2) = \{K, O, H, A, L, P, E, F\}$$

Next, the algorithm considers P(1) and checks whether preceding components constraints exist or not. If these are not present, the component is disassembled. It is then added to DS, and removed from UL and the P(1) list. It is also removed from the precedent components list. If no such component is present in P(1), a component of P(2) is considered. The appropriate component of P(2) is detached. This is repeated till so many components are removed from the precedent components list, that additional components of P(1) can be released.

As a result, the sequence A-G-M-B-D-J-C-K-H-I-E-N-O-F-L-Q-P is generated, which indeed corresponds to a rather good disassembly sequence.

Updating the average difficulty rates in the course of the calculation can further refine this method.

For the case of the AFI (Figure 5.7), the first operation may be the detachment of L, next J, next H, etc., where the motion is in the $+x$ direction. If G is finally detached, the disassembly of F, E, etc, can take place in the $-x$ direction. A sequence that consists of first detaching L, next F, next J, next E, etc., is also feasible. However, this will be more expensive, because the direction of motion changes multiple times, which takes more time and a more complex motion of the tools.

Another example, which can also be illustrated using the AFI, is the use of dedicated tools. This may, for example, require that first a wrench be used and all the bolts loosened, next a gripper be used, and so on.

If we use an AND/OR graph or a state diagram, distinction is made between similar actions (such as: detachment of component H) with respect to the subassembly on which this action is applied, e.g., GH, ABCDEFGHK, and so forth, but the cases that have been described here as an illustration, are not covered with standard methods.

### 7.3.2 Exact Approaches Confined to Sequential Disassembly

An exact approach that is confined to sequential disassembly has been presented by Yee and Ventura (1999). These authors closely follow the approach of Kanehara et al. (1993). In contrast to the sequence independent cost problem, this extended problem does not relax into a linear programming problem. The model such as the one described in subsection 7.2.3.1 is extended as follows.

Instead of the set of aggregate flow variables $x_j$, a set of partial flow variables $w_{j,k}$ is introduced. The indices $j$ and $k$ both refer to disassembly operations. These do not include an initial operation $j = 0$. These are (0,1) variables that are different from zero only if the operation $k$ is carried out immediately after performing operation $j$. In addition, the cost vector $C_j$ is extended to a cost matrix with elements $C_{j,k}$. Accordingly, the disassembly costs are represented by

$$C = \sum_{j,k} C_{j,k} \cdot w_{j,k} \tag{7.7}$$

Obviously, not every combination of $j$ and $k$ results in a feasible sequence. Therefore, only the feasible combinations are listed and enumerated. This is indicated

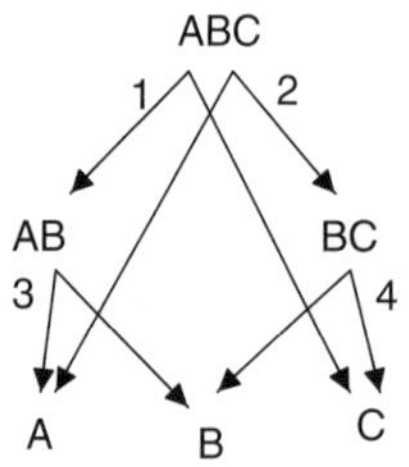


(a)

| $j$ | 1 | 2 | 3 | 4 | $M_F - M_I$ |
|---|---|---|---|---|---|
| ABC | −1 | −1 | 0 | 0 | −1 |
| AB | 1 | 0 | −1 | 0 | 0 |
| BC | 0 | 1 | 0 | −1 | 0 |
| A | 0 | 1 | 1 | 0 | 1 |
| B | 0 | 0 | 1 | 1 | 1 |
| C | 1 | 0 | 0 | 1 | 1 |

(b)

**FIGURE 7.5** (a) AND/OR graph; (b) Transition matrix and state vector in approach due to Kanehara et al. (1993).

with index $n$, which refers to all the feasible combinations of operations $j$ and $k$ that can be performed sequentially. Thus, the objective function is rewritten as

$$C = \sum_n C_n \cdot w_n \tag{7.8a}$$

The transition matrix $T_{i,j}$ in which the index $i$ refers to subassemblies, is also extended to $T_{i,n}$. Thus, the extended equivalent of the node balance constraint (7.2b) is

$$\sum_n T_{i,n} \cdot w_n = (M_F - M_I)_i \tag{7.8b}$$

which holds true for every node $i$.

The principle is illustrated in Figure 7.5 for a simple case. Note that only the final state is given here.

In Yee and Ventura's original approach, penalties are connected to those combinations of operations that require product reorientation, tool exchange, etc. The sum of these penalties should not exceed a predefined value. If, for example, the penalties that are related to tool exchange are represented by the vector $E_n$, and the maximum value is $E_{\max}$, the associated constraint is as follows:

$$\sum_n E_n \cdot w_n \le E_{\max} \tag{7.8c}$$

Obviously, this extended model has more variables, as the number of combinations exceeds that of single operations. In addition, the model is a full (0,1) integer-programming problem, which is NP complete. Even though efficient solvers exist for binary integer linear programming problems, they are only useful up to some degree of product complexity.

The method was applied by Yee and Ventura (1999) to sequential disassembly only aimed at assembly optimization. Therefore, only the final state is given, which corresponds to complete disassembly.

A more serious drawback of the method is that it is confined to sequential (dis-)assembly only. Extension to processes in which parallel disassembly is also permitted, needs rigorous reconsideration of the model. This will be outlined in section 7.4.

### Example 15

Consider the Bourjault's ballpoint example (Figure 5.1) and its disassembly AND/OR graph in Figure 5.5(b). If only sequential disassembly is considered, the operations 3, 6, and 8 are not permitted. Consequently, the subassemblies ABF, AB, and AE cannot be visited, and thus complete disassembly is inhibited. We are therefore only left with the operations 1, 2, 4, 5, 7, 9, and 11. If the initial action 0 (making the product available) is added, only the following subsequences are feasible:

0,1 0,2 1.4 2,5 4,7 5,7 7,9 9,11

This means that index $n$ in (7.8c) will run from 1 through 8. If parallel disassembly is permitted, the following 2-subsequences combinations are feasible:

0,1 0,2 1,3 1,4 2,5 2,6 3,9 3,12 4,7 4,8 5,7
5,8 6,10 6,11 7,9 8,11 8,13 9,11 9,12 10,11 10,13 11,10
11,12 11,13 12,9 12,11 13,11

If the direction of motion of the detached subassembly is considered as a driver for sequence dependent cost, it will be affected by the choice of the static component, i.e., the component that is in the fixture. If A is selected as the static component, and C is the static component in subassemblies that do not include A, we observe that the operations 2, 4, 6, 8, and 12 include motion of the detached component in the $+ x$ direction, and the operations 1, 3, 5, 7, 9, 10, and 13 include motion in the $- x$ direction. Operations 0 and 11 can be performed in both directions. Therefore, a penalty is assigned to the subsequence 1,4 and no penalty is assigned to the subsequence 5,7.

The set of feasible 2-subsequences can be graphically represented in a directed task precedence graph (see Figure 7.6).

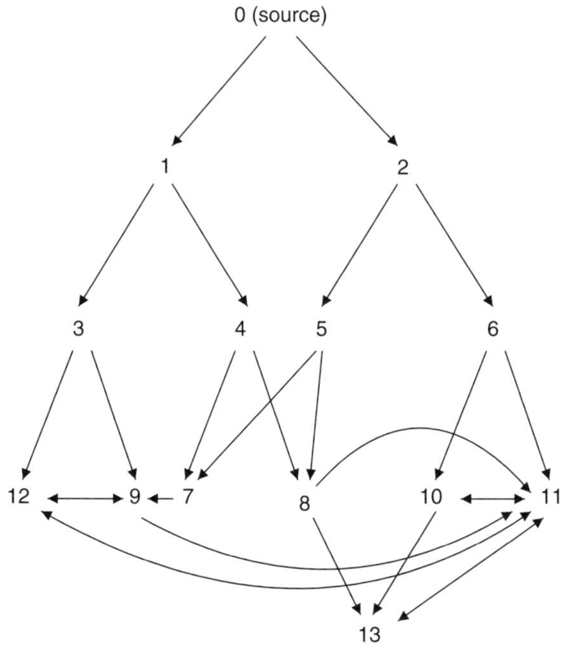


**FIGURE 7.6** Directed task precedence graph of Bourjault's ballpoint.

## 7.4 RIGOROUS EXACT METHODS

### 7.4.1 Introduction

The method discussed in subsection 7.3.2 was confined to sequential disassembly. Even then, the size of the problem increases exponentially with the number of components *N*. Therefore, several authors have chosen to solve the problem using heuristic methods (see, for example, Gungor and Gupta, 1997, 2001, and Rosell et al., 2003), or metaheuristic methods, including simulated annealing (Motavalli and Islam, 1997), and Lagrangian relaxation (Yee and Ventura, 1999). Developing the exact method, however, is useful, even though it cannot be applied to arbitrary complex products because of its inherent NP completeness. The exact method is useful for products of moderate complexity and for justifying the heuristic and metaheuristic methods in those cases. Therefore, attempts have been made to implement the exact method to its full extent to problems with parallel disassembly included. The first studies on this topic are due to Johnson and Wang (1998) for disassembly precedence graphs, and due to Kang et al. (2001) for AND/OR graphs. The problem turns out to be a constrained asymmetric TSP, which is a standard problem in mathematical programming. The symmetric TSP in its original formulation refers to the problem in which a person has to visit a number of cities in arbitrary order and finally has to return in his original place, with minimum distance traveled. The asymmetric TSP is similar to the symmetric one, but with different distances in different directions. In disassembly sequencing, the cities are replaced by disassembly operations and the distances by disassembly costs. As many subsequences of two operations cannot be performed due to structural constraints, they have to be added to the original asymmetric TSP. Therefore, a constrained TSP is needed.

### 7.4.2 Unconstrained Problem

#### 7.4.2.1 Rigorous Approach

The unconstrained asymmetric TSP can be formulated by introducing two virtual nodes; a source node and a sink node. We represent these nodes by the indices 0 and $s$, respectively. Both the source and the sink are connected to every node and to each other. However, the arcs between the source and the other nodes are directed *from* the source. The arcs between the other nodes and the sink are directed *toward* the sink. A model can be formulated as follows.

Let $J$ be the set of nodes including the virtual nodes. Let $M$ be the subset of $J$ with source and sink not included. The indices $j$ and $k$ refer to elements in $J$. The indices $m$, $n$, and $p$ refer to elements in $M$. Costs are represented by the elements $C_{j,k}$ of the cost matrix that reflect the costs of performing operations $k$ immediately following operation $j$. Flow variables $w_{j,k}$ are (0,1) variables. These are 1 if and only if operation $k$ is performed immediately after operation $j$.

Obviously, for the diagonal elements

$$w_{j,j} = 0 \tag{7.9a}$$

The objective function is given by

$$obj = \sum_{j,k} C_{j,k} \cdot w_{j,k} \tag{7.9b}$$

For every node in $M$, the following node equilibrium holds:

$$\sum_{j} w_{j,m} = \sum_{j} w_{m,j} \tag{7.9c}$$

This guarantees that a trajectory cannot end in a real node. It enables the possibility, however, that visiting every node is not obligatory.

The arcs pointing toward the source and starting from the sink are inhibited by

$$\sum_{j} w_{j,0} = \sum_{j} w_{s,j} = 0 \tag{7.9d}$$

The following guarantees that a trajectory starts at the source node

$$\sum_{j} w_{0,j} = 1 \tag{7.9e}$$

Unfortunately, this model does not inhibit subtours, which are cyclic trajectories that visit a subset of $M$ and include neither the source nor the sink node. Explicitly inhibiting every cycle is possible. However, this requires an unmanageable number of constraints. For instance, the following constraints inhibit 2-cycles and 3-cycles, respectively:

$$w_{m,n} + w_{n,m} \leq 1 \tag{7.9f}$$

and

$$w_{m,n} + w_{n,p} + w_{p,m} \leq 2 \tag{7.9g}$$

#### 7.4.2.2 Iterative Approach

In practice, not all the possible cycles have to be explicitly inhibited. Instead it is possible to use an iterative procedure as follows:

The procedure starts without any cycle inhibitors in the model. A solution is returned that may include one or more cyclic solutions. There will always be a linear solution from the source toward the sink. If cycles appear, the linear solution is not optimum, because the cycles contribute to the objective.

The smallest cycle that appears in the linear solution is inhibited next by adding a constraint similar to (Expression 7.9f) or (Expression 7.9g). The procedure is repeated again and again till a solution without any cycle is obtained, at which point the iterative procedure stops.

**EXAMPLE 16**

We will apply the iterative approach on an unconstrained problem with 13 operations. The cost matrix $C_{j,m}$ has been randomly generated and is as follows:

$$
\begin{bmatrix}
70 & 61 & 53 & 6 & 29 & 14 & 63 & 52 & 21 & 81 & 23 & 63 & 55 \\
70 & 81 & 70 & 11 & 81 & 56 & 94 & 51 & 65 & 91 & 48 & 36 & 52 \\
45 & 55 & 79 & 9 & 26 & 45 & 69 & 47 & 79 & 3 & 16 & 57 & 33 \\
43 & 46 & 25 & 14 & 14 & 5 & 51 & 53 & 7 & 72 & 48 & 25 & 75 \\
84 & 31 & 54 & 26 & 56 & 97 & 43 & 1 & 17 & 84 & 73 & 88 & 53 \\
86 & 36 & 20 & 21 & 53 & 7 & 86 & 27 & 16 & 29 & 27 & 34 & 69 \\
4 & 21 & 5 & 24 & 16 & 8 & 73 & 80 & 22 & 9 & 5 & 96 & 4 \\
96 & 23 & 55 & 12 & 19 & 5 & 81 & 49 & 9 & 58 & 30 & 61 & 3 \\
1 & 88 & 70 & 98 & 16 & 99 & 32 & 3 & 54 & 96 & 54 & 3 & 41 \\
72 & 83 & 19 & 70 & 84 & 17 & 84 & 18 & 72 & 53 & 70 & 17 & 54 \\
60 & 97 & 73 & 91 & 54 & 28 & 10 & 48 & 62 & 67 & 10 & 3 & 82 \\
38 & 19 & 32 & 93 & 68 & 17 & 60 & 50 & 80 & 98 & 98 & 41 & 73 \\
40 & 76 & 66 & 83 & 73 & 38 & 82 & 87 & 88 & 16 & 84 & 31 & 62 \\
57 & 79 & 4 & 46 & 68 & 5 & 98 & 41 & 48 & 69 & 20 & 21 & 5
\end{bmatrix}
$$

It should be stressed that the row numbers of the cost matrix run from 0 through 13 and the column numbers run from 1 through 13. The row that corresponds to $C_{s,m}$ has all elements equaling zero, and is not included here. Note that $C_{0,m}$ is the cost of operation $m$, if it has no precursor.

For guaranteeing that a nontrivial solution is returned, we will maximize the costs. The *first iteration* returns the quasi-optimum solution with costs 1135, consisting of a linear trajectory, a 4-cycle, and a 2-cycle. These are as follows:

0-10-2-3-13-7-1-5-*s*

4-6-12-8-4

9-11-9

The costs can be easily found from the matrix elements. For instance, the costs that are connected with the linear trajectory are $c_{0,10} = 81$, $c_{10,2} = 97$, $c_{2,3} = 79$, etc.

In the *second iteration*, the 2-cycle is inhibited by adding the following constraint:

$$w_{9,11} + w_{11,9} \leq 1$$

With this included, the following solution is found:

0-10-2-9-5-*s*
1-3-13-7-1
6-12-8-6
4-11-4

The corresponding objective function value is 1128.

In the *third iteration*, the newly found 2-cycle is inhibited, by adding the following constraint:

$$w_{4,11} + w_{11,4} \leq 1$$

This results in the following quasi-solution:

0-8-4-6-12-11-9-5-*s*
2-3-13-7-1-10-2

The corresponding objective function value is 1127.

In the *fourth iteration*, the 6-cycle is inhibited, via the following constraint:

$$w_{2,3} + w_{3,13} + w_{13,7} + w_{7,1} + w_{1,10} + w_{10,2} \leq 5$$

This results in the following optimal solution:

0-8-4-6-12-9-11-10-2-3-13-7-1-5-*s*

The corresponding objective function value is 1125.

This example illustrates that the number of iterations (although strictly speaking, it should depend on the type of the problem) is relatively low. This seems strange at first glance, because a multitude of cycles are possible. Two reasons why this happens should be mentioned.

First, the maximum number of possible linear sequences $A_{max}$ is much larger than that of possible cyclic sequences $C_{max}$. If $M$ is the number of nonvirtual nodes, we have

$$A_{max}(M) = \sum_{j=0}^{M} \frac{M!}{j!} = \sum_{j=0}^{M-1} \frac{M!}{j!} + 1 \quad (7.10a)$$

$$C_{max}(M) = \sum_{j=0}^{M-1} \frac{M!}{j!(M-j)} - M \quad (7.10b)$$

*Explanation of (Expression 7.10a)*

First the number of noncyclic solutions that visit all the nodes is determined. There are $M$ possible trajectories from the source to a node. Once a node is visited, it cannot be visited again, thus for the second operation in the sequence, only $M-1$ nodes are left. This proceeds till all the nodes are visited. Consequently, there are $M!/0!$ acyclic solutions that visit all the nodes. Next, the number of acyclic solutions is determined that visit all nodes but one. Following an analogous reasoning, it is clear that there are $M!/1!$ such acyclic solutions. Similarly, there are $M!/2!$ acyclic solutions that visit all the nodes but 2, and so on. Finally, there are $M!/(M-1)!$acyclic solutions that visit only one node, and there is 1 acyclic solution that does not visit any node, which corresponds to the direct connection from source to sink.

*Explanation of* (*Expression 7.10b*)

The *explanation of* (*Expression 7.10b*) proceeds in a similar way, except that the number of detected *k*-cycles must be divided by *k*, to avoid multiple counts. Note, for example, that the 3-cycle 1-2-3-1 is equivalent to both 2-3-1-2 and 3-1-2-3.

The number $A_{max}$ $(M)$ equals the total number of arrangements of a set with $M + 1$ elements; the number $C_{max}$ $(M)$ is called the *logarithmic number* (Sloane, 2003). These are decreased by $M$, because the diagonal elements are not counted.

Although both numbers grow exponentially with $M$, the ratio $A_{max}/C_{max}$ tends to $M$ with increasing $M$. In addition, it can be concluded from considering the terms of (7.10a) separately, that the majority of the acyclic solutions incorporates a large number of operations. Some values of $A_{max}$ and $C_{max}$ are presented in Table 7.8.

Second, only those sequences that correspond to a relatively high contribution to the objective will appear with some preference in both the quasi solutions and the physically relevant solutions. It can be seen, for example, that the sequence 10-2-3-13-7-1 appears repeatedly, in both cyclic solutions and linear solutions. In terms of genetic algorithms, these sequences (or parts of it) can be considered "fittest" genes (Caccia and Pozzetti, 2000).

**Example 17**

The 16 linear sequences with $M = 3$ are

| | | | | | |
|---|---|---|---|---|---|
| 0-*s* | | | | | |
| 0-1-*s* | 0-2-*s* | 0-3-*s* | | | |
| 0-1-2-*s* | 0-1-3-*s* | 0-2-1-*s* | 0-2-3-*s* | 0-3-1-*s* | 0-3-2-*s* |
| 0-1-2-3-*s* | 0-1-3-2-*s* | 0-2-1-3-*s* | 0-2-3-1-*s* | 0-3-1-2-*s* | 0-3-2-1-*s* |

The 5 cyclic sequences with $M = 3$ are

1-2-1 1-3-1 2-3-1 1-2-3-1 1-3-2-1

**TABLE 7.8**
**Maximum Number of Linear Solutions and Cycles in an Unconstrained Network**

| M | 1 | 2 | 3 | 4 | 5 | 6 | 7 | 8 | 9 | 10 |
|---|---|---|---|---|---|---|---|---|---|---|
| $A_{max}(M)$ | 2 | 5 | 16 | 65 | 326 | 1,956 | 13,700 | 109,601 | 986,410 | 9,864,101 |
| $C_{max}(M)$ | 0 | 1 | 5 | 20 | 84 | 409 | 2,365 | 16,064 | 125,664 | 1,112,073 |
| $A_{max}/C_{max}$ | ∞ | 5 | 3.2 | 3.25 | 3.88 | 4.78 | 5.79 | 6.82 | 7.85 | 8.87 |

**TABLE 7.9**
**Iterations for an Unconstrained Problem with 20 Nodes**

| Trajectory | Cycle | Objective | Next Disable | Remark |
|---|---|---|---|---|
| 0-16-11-1-13-15-4-14-*s* | 6-19-9-6 and 17-8-20-3-7-18-5-10-2-12-17 | 180 | 6-19-9-6 | infeasible |
| 0-4-14-*s* | 2-12-17-8-20-10-2 and 1-13-7-18-5-3-15-6-19-9-16-11-1 | 180 | 2-12-17-8-20-10-2 | infeasible |
| 0-4-14-9-16-11-1-13-7-18-5-10-2-8-20-3-15-6-19-*s* | 12-17-12 | 180 | 12-17-12 | infeasible |
| 0-4-14-9-16-11-1-13-7-18-5-10-2-12-17-8-20-3-15-6-19-*s* | | 180 | | optimum |

We repeated the experiment for various cost matrices with values of $M$ up to 20. The number of iterations in all instances remained reasonable (see Table 7.9, for example, for a problem with $M = 20$). Here again, we see a sequence 16-11-1-13 reappearing in both cyclic and linear solutions. The required CPU time, however, increased here due to the increasing number of (0,1)-variables. In addition, it appears that the CPU time increases with an increasing number of constraints. However, for this kind of relatively small problem, it never exceeded two seconds.

### 7.4.3 Constrained Problem in Disassembly AND/OR Graph, Rigorous Approach

#### 7.4.3.1 Introduction

Kang et al. (2001) applied a modified TSP approach on a full disassembly sequencing problem based on an AND/OR graph. The authors transformed the AND/OR graph into a task precedence graph. Unfortunately, such a graph has a complicated structure, even for a rather simple product such as the Bourjault's ballpoint (see Figure 5.5(b)). It has 13 operations. Consequently, $M = 13$. The number of (0,1)-variables is 27 + 14, the latter accounting for flows toward the sink.

Two kinds of constraints have to be dealt with: cycles and exclusive OR relationships. Cycles may occur if parallel disassembly is included. For example, we see that operation 11 may be followed by operation 12, and vice versa. This means that cycle 11-12-11 is not *a priori* excluded. The exclusive OR relationships are dealt with in a natural way, because the sum of the incoming arcs of a node cannot exceed 1 and, consequently, the sum of the departing nodes also cannot exceed 1, which means that only one choice is selected. If an arc toward the sink is selected, the sequence stops. However, precedence relationships can still be violated. For instance, the set of feasible subsequences still enables sequences such as (see Figure 7.6)

$$0\text{-}1\text{-}4\text{-}8\text{-}11\text{-}10\text{-}13\text{-}s$$

It can be concluded from the AND/OR graph of Figure 5.5(b) that this sequence, although not a cycle, is infeasible because it violates the precedence relationship $6 \rightarrow 10$. Therefore, precedence relationships have to be explicitly accounted for. In the unconstrained TSP, precedence relationships did not occur.

Kang et al. (2001) applied a *two-commodity network flow problem*, which was initially proposed by Johnson and Wang (1998) and taken from machine scheduling problems. In such a problem, two integer variables are assigned to every node. These act as two counters, viz., an ascending and a descending counter. The counters are zero for a node that is not visited. The ascending counter increases by 1 for every subsequent node along the trajectory, in the forward direction. The descending counter starts with a value that equals at least the maximum number of nodes that can be visited, and it decreases by 1 for every subsequent node that is visited along the trajectory. If a precedence relationship $j \rightarrow k$ between two operations $j$ and $k$ exists, this implies that the descending counter that corresponds to operation $j$ should exceed that of operation $k$. This also accounts for the case in which operation $j$ is performed, but operation $k$ is not performed at all. In this case, the counter that corresponds to $k$ equals zero.

This method was applied by Johnson and Wang (1998) to a fairly simple divergent disassembly precedence graph such as the one depicted in Figure 5.15. The method was extended by Kang et al. (2001) to include general disassembly AND/OR graphs. Therefore, the AND/OR graph was transformed into the directed task precedence graph (see Figure 7.6) which exhibits an increased complexity.

We will outline here a modified approach that only has a descending counter and that deals with the features of the AND/OR graph. There are multiple formulations. If a quasi flow from the sink to the source is added, such as in the simple graph of Figure 7.7, node balance equations are equivalent for real and virtual nodes.

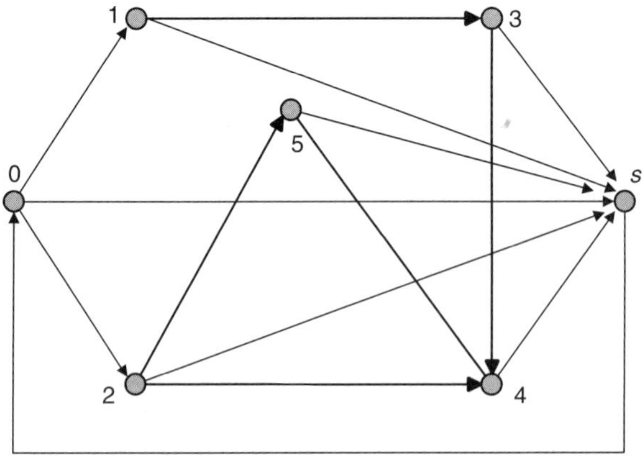


**FIGURE 7.7** Constrained process graph in which the linear solution is made cyclic.

#### 7.4.3.2 Model Formulation

The mathematical model is structured as follows:

Let $I$ be the set of feasible subassemblies including single components that appear in the AND/OR graph. Let $J$ be the set of operations including the virtual operations (source and sink). Let $M$ be the subset of $J$ consisting of all the operations, except the virtual ones. The indices $i$ refer to $I$. The indices $j$ and $k$ refer to $J$. The indices $m$ and $n$ refer to $M$. Revenues are represented by the components $R_i$ of the revenue vector that represent the revenue of subassembly $i$. Costs are represented by the elements $C_{j,k}$ of the cost matrix that reflect the costs of performing operation $k$ immediately following operation $j$. The transition matrix $T$ with elements $T_{i,j}$ is as defined in subsection 7.2.3. The succession matrix $S$ has elements $S_{j,k}$ which equal 1 only if the ordered pair of operations $j$ and $k$ is a feasible 2-sequence. Other elements are zero. Partial flow variables $w_{j,k}$ are (0,1) variables, which are different from zero if operation $k$ is performed next to operation $j$. Aggregate flow variables $x_k$ are (0,1) variables which are different from zero if operation $k$ is performed. Partial (descending) counters $p_{j,k}$ are variables that are zero if the subsequence $j-k$ is not part of the trajectory, and have an integer value if the subsequence is in the trajectory. Aggregate (descending) counters $a_j$ are integer values that are different from zero only if operation $j$ is performed. These start from the value $N - 1$, which is the maximum number of partitions that can be made in a set of $N$ components, which are in the original product. The aggregate counter is decreased by 1 for every subsequent operation on the trajectory.

The model is formulated as follows:

Relationship between partial and aggregate flow variables

$$x_j = \sum_k w_{k,j} \tag{7.11a}$$

Diagonal flow variables are set equal to zero

$$w_{j,j} = 0 \tag{7.11b}$$

Flow initialization

$$w_{s,0} = 1 \tag{7.11c}$$

No partial flows point to the source

$$\sum_m w_{m,0} = 0 \tag{7.11d}$$

No partial flows point from the sink

$$\sum_m w_{s,m} = 0 \tag{7.11e}$$

Node balance equation

$$\sum_{k} w_{k,j} = \sum_{k} w_{j,k} \tag{7.11f}$$

Partial counter initialization

$$\sum_{j} p_{0,j} = N - 1 \tag{7.11g}$$

Relationship between partial and aggregate counters

$$a_m = \sum_{j} p_{j,m} \tag{7.11h}$$

Counter decrement

$$\sum_{j} p_{m,j} = a_m - x_m \tag{7.11i}$$

Connection between partial counters and partial flows

$$p_{j,k} \leq Q \cdot w_{j,k} \tag{7.11j}$$

where $Q$ is a large number, exceeding the number of nodes.
Profit function

$$obj = \sum_{i,j} T_{i,j} \cdot R_i \cdot x_j - \sum_{j,k} C_{j,k} \cdot w_{j,k} \tag{7.11k}$$

Succession

$$S_{j,k} - w_{j,k} \geq 0 \tag{7.11l}$$

AND/OR graph structure, which is equivalent to Expression (7.4a)

$$\sum_{j} T_{i,j} \cdot x_j \geq 0 \tag{7.11m}$$

Additional information on precedence

Let the indices $q$ and $r$ refer to operations that belong to $M$. The arcs that correspond to $q$, point to a node in the AND/OR graph and the operations that correspond to $r$, point from the same node. One operation $q$ has to be performed prior to performing an operation $r$. This is reflected by

$$\sum_{q} a_q - \sum_{r} a_r \geq \sum_{r} x_r \tag{7.11n}$$

#### 7.4.3.3 Alternative Formulations

Alternative formulation 1:

The sink and the source can be merged. In this case, maximum symmetry is obtained. The corresponding mathematical model is equivalent to system (Expression 7.11a–Expression 7.11n), except that the relationships (Expression 7.11b–Expression 7.11e) are replaced with

$$x_0 = 1 \tag{7.12}$$

Alternative formulation 2:

The sink can also be completely removed. In this case, the set (Expression 7.11a through Expression 7.11n) is modified as follows:

The flow initialization, Expression 7.11d, is replaced with Expression 7.12; Expression 7.11e is removed; and the node balance equation, Expression 7.11f, is replaced with:

$$\sum_k w_{k,j} \geq \sum_k w_{j,k} \tag{7.13}$$

This enables a trajectory to stop at any node, because it permits that the sum of the incoming flows is equal to 1, with the sum of the outgoing flows being equal to zero.

#### 7.4.3.4 Discussion

The introduction of counters inhibits cyclic solutions, at the cost of increased CPU time. This is due to the introduction of integer variables $p_{j,m}$. Although these have not to be explicitly defined as such, because the model Expression 7.11g and Expression 7.11i guarantee that they can only have integer values, this does not result in significantly reduced CPU time, which still is prohibitive if complex models are investigated. The CPU time further increases with an increasing number of constraints of the type given in Expression 7.11n. In general, it is not necessary to include the complete set. We can start with no such equation, investigate the resulting solution on violations of precedence relationships, adding the corresponding relationship of the set (Expression 7.11n) and repeat the calculations. The number of iterations is restricted, because the number of relevant nodes in the AND/OR graph is restricted.

Expression 7.11n needs some additional discussion. There are four possibilities:

- If none of the operations $q$ and $r$ is performed, the relationship is satisfied because both sides equal zero.
- If one of the operations $q$ is performed and none of the operations $r$ is performed, the left hand side of Expression 7.11n is positive and the right hand is zero, thus the relationship is satisfied.
- If one of the operations $q$ and one of the operations $r$ are performed, the right hand side of Expression 7.11n is 1 and the left hand is only $\geq 1$ if

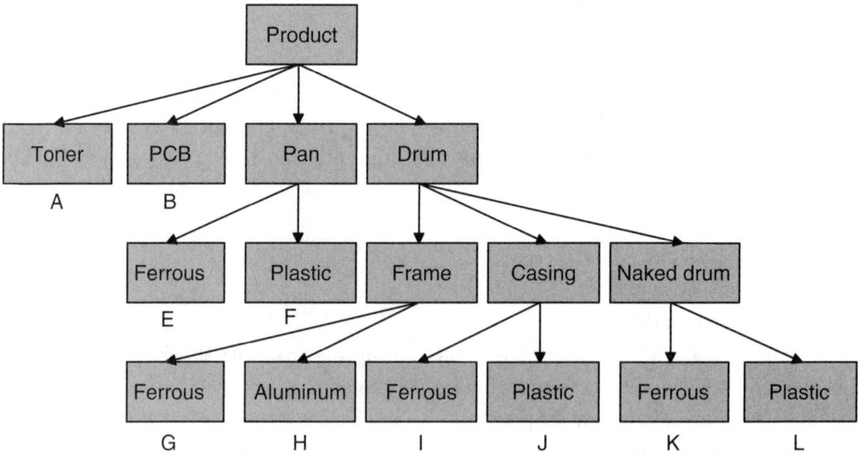


**FIGURE 7.8** Bill of materials of a module in a photocopying machine (Kang et al., 2001).

$q$ is performed prior to $r$. Therefore, the relationship inhibits performing $r$ prior to $q$.

- If operation $q$ is not performed and operation $r$ is performed, the left hand side of (7.11n) is negative and the right hand side is 1, which means that the relationship is violated.

We will illustrate this method with the help of an example that is inspired from a case due to Kang et al. (2001), which represents a module of a photocopier. The hierarchical tree structure of the photocopier is given in Figure 7.8 and corresponding AND/OR graph in Figure 7.9.

The model equations guarantee that the returned solution is always a connected subgraph of the original AND/OR graph. Table 7.10 shows the needed precedence constraints that have to be added if a specific instance of the cost and revenue elements is considered. The calculations have been done on a fairly primitive computer (200 MHz Pentium) and the XA mixed integer programming solver, which is integrated in the AIMMS interface (Bisschop and Entriken, 1993).

### 7.4.4 Constrained Problem in Disassembly AND/OR Graph, Iterative Approach

The iterative approach takes place with the reduced model, with no counters involved. This considerably moderates the needed CPU time, but various types of erroneous solutions, such as cycles, are not inhibited. The reduced model is as follows:

$$x_j = \sum_j w_{j,k} \tag{7.14a}$$

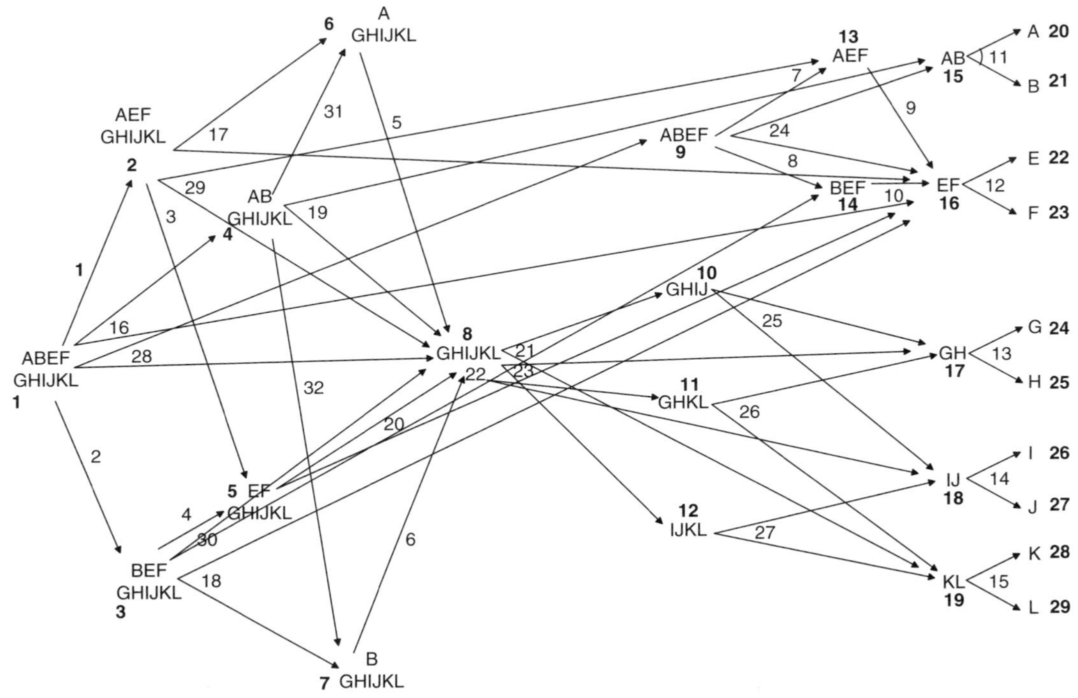


**FIGURE 7.9** AND/OR graph of the module in Figure 7.8. Note that the operations are enumerated using normal numbers and the subassemblies are enumerated using boldface numbers.

**TABLE 7.10**
**Iterations for Optimum Disassembly Sequence Generation of the Product in Figure 7.8**

| Trajectory | Violated Relation | Objective | Precedence Constraint | CPU Time |
|---|---|---|---|---|
| 0-28-8-13-27-15-14-12-23-10-*s* | 23→13 | 698 | $a_{23} + a_{25} + a_{26} - a_{13} \geq x_{13}$ | 7 |
| 0-28-8-15-14-12-23-13-27-10-*s* | 27→15 | 689 | $a_{21} + a_{26} + a_{27} - a_{15} \geq x_{15}$ | 51 |
| 0-28-8-15-14-12-23-13-27-15-*s* | 27→14 | 698 | $a_{22} + a_{25} + a_{27} - a_{14} \geq x_{14}$ | 44 |
| 0-28-8-23-10-13-27-14-15-12-*s* | optimum | 683 | | 186 |

Diagonal flow variables are set equal to zero

$$w_{j,j} = 0 \tag{7.14b}$$

Flow initialization

$$w_{s,0} = 0 \tag{7.14c}$$

No partial flows point to the source

$$\sum_m w_{m,0} = 0 \tag{7.14d}$$

No partial flows point from the sink

$$\sum_m w_{s,m} = 0 \tag{7.14e}$$

Node balance equation

$$\sum_j w_{j,k} = \sum_j w_{k,j} \tag{7.14f}$$

Profit function

$$obj = P = \sum_{i,j} T_{i,j} \cdot R_i \cdot x_j - \sum_{j,k} C_{j,k} \cdot w_{j,k} \tag{7.14g}$$

Succession

$$S_{j,k} - w_{j,k} \geq 0 \tag{7.14h}$$

AND/OR graph structure, which is equivalent to Expression (7.4a)

$$\sum_j T_{i,j} \cdot x_j \geq 0 \tag{7.14i}$$

**EXAMPLE 18**

The above mentioned method is applied to an instance of the case of Figure 7.8. The results of the calculations are listed in Table 7.11.

This table of results illustrates the basic features of the iterative method. With all the operations that are returned, a connected subgraph of the original AND/OR graph is found. However, both cycles and erroneous subsequences may appear. Also a cycle itself, or part of it, can be considered an erroneous subsequence. The idea is not to disable a specific cycle if it appears, but instead disable the shortest possible erroneous subcycle, which is likely to appear in many solutions. Since erroneous 2-subsequences are wiped out via the *S*-matrix, the 3-subsequences have to be evaluated first. If there are none, 4-subsequences have to be considered, and so on.

It can be seen from Table 7.11 that in the first iteration, subsequence 29-21-12 is erroneous. In iteration 14, we even need to consider a 6-subsequence, which is inhibited via adding the following constraint:

$$w_{14,8} + w_{8,10} + w_{10,12} + w_{12,15} + w_{15,25} \leq 4$$

In iteration 4 it is demonstrated that a 2-subcycle is inhibited by considering it an erroneous 3-subsequence, in which a specific operation appears twice.

In iteration 8, iteration10, and iteration 20 no cycles appear in the solutions. Nevertheless, these linear sequences do not represent feasible solutions, because erroneous subsequences are still present.

It can also be observed that specific erroneous subsequences, such as 10-13-14-8, appear multiple times (here in iteration 2 and iteration 6) before they are selected as a dominant erroneous subsequence, which occurs in iteration 15.

Finally, we observe that the required CPU time gradually increases, although not dramatically, with growing number of constraints.

### 7.4.5 CONSTRAINED PROBLEMS WITH DISASSEMBLY PRECEDENCE GRAPHS

Using the methods that have been outlined in the preceding subsections, one can also approach the problem with assembly precedence graphs in a similar way, via both the rigorous and the iterative methods. The mathematical model used here is identical to the one that was used for AND/OR graph representation, except for the definition of the AND/OR graph structure, which is reflected in Expression 7.11m. In fact, we will demonstrate in the example below that the precedence relationships here can be expressed via a set of relationships that are much simpler than Expression 7.11n. Usually, only sequential operations are included in a disassembly precedence graph. A disassembly precedence graph has $N$ nodes, which is far less than the number of nodes in a corresponding AND/OR graph. However, a simple disassembly precedence graph represents a large number of feasible disassembly sequences. This is reflected in the *S*-matrix, which typically has a larger share of elements equaling 1 compared to that of an AND/OR graph with equal number of nodes.

**TABLE 7.11**
**Iterations for Optimum Disassembly Sequence Generation of the Product in Figure 7.8**

| Iteration | Linear Trajectories | Cycles | Objective | Erroneous Sequence | CPU Time(s) |
|---|---|---|---|---|---|
| 1 | 0-1-29-21-12-15-25-*s* | 9-13-14-9 | 602 | 29-21-12 | 1.92 |
| 2 | 0-28-21-12-15-25-*s* | 8-10-13-14-8 | 592 | 28-21-12 | 2.97 |
| 3 | 0-28-21-15-25-*s* | 8-10-12-13-14-8 | 592 | 12-13-14-8 | 1.81 |
| 4 | 0-28-21-15-25-*s* | 8-10-14-8<br>12-13-12 | 592 | 12-13-12 | 3.63 |
| 5 | 0-28-21-15-25-*s* | 8-10-13-12-14-8 | 592 | 12-14-8 | 1.87 |
| 6 | 0-28-22-12-15-*s* | 8-26-10-13-14-8 | 580 | 28-22-12 | 3.84 |
| 7 | 0-1-29-23-12-15-*s* | 9-27-13-14-9 | 578 | 29-23-12 | 2.20 |
| 8 | 0-1-29-23-13-14-9-27-12-15-*s* | none | 578 | 23-13-14 | 3.35 |
| 9 | 0-1-29-23-13-15-*s* | 9-27-12-14-9 | 578 | 23-13-15 | 4.23 |
| 10 | 0-1-29-23-13-12-14-9-27-15-*s* | none | 578 | 12-14-9 | 4.83 |
| 11 | 0-1-29-23-13-12-15-*s* | 9-27-14-9 | 578 | 14-9-27 | 3.40 |
| 12 | 0-28-21-*s* | 8-10-12-14-8<br>13-15-25-13 | 577 | 12-14-8 | 2.20 |
| 13 | 0-28-21-*s* | 8-10-12-13-15-25-14-8 | 577 | 13-15-25 | 2.19 |
| 14 | 0-28-21-*s* | 8-10-12-15-25-13-14-8 | 577 | 14-8-10-12-15-25 | 3.46 |
| 15 | 0-28-21-*s* | 8-10-13-14-8<br>12-15-25-12 | 577 | 10-13-14-8 | 5.17 |
| 16 | 0-28-21-*s* | 8-10-12-13-14-8<br>15-25-15 | 577 | 15-25-15 | 3.57 |
| 17 | 0-28-21-*s* | 8-10-15-25-12-13-14-8 | 577 | 12-13-14-8 | 3.46 |
| 18 | 0-28-21-*s* | 8-10-13-12-15-25-14-8 | 577 | 13-12-15-25 | 3.30 |
| 19 | 0-1-29-*s* | 9-23-13-12-27-15-14-9 | 576 | 14-9-23 | 2.53 |
| 20 | 0-1-29-22-12-15-13-14-9-26-*s* | none | 572 | 29-22-12 | 4.45 |
| 21 | 0-28-23-12-15-*s* | 8-10-27-13-14-8 | 564 | 28-23-12 | 6.82 |
| 22 | 0-28-23-13-12-15-*s* | 8-10-27-14-8 | 564 | 28-23-13-12 | 6.59 |
| 23 | 0-28-21-11-15-25-*s* | 12-13-14-24-12 | 560 | 28-21-11 | 4.72 |
| 24 | 0-28-23-*s* | 8-10-27-14-8<br>12-15-13-12 | 558 | 10-27-14-8 | 3.46 |
| 25 | 0-28-23-*s* | 8-10-27-13-14-8<br>12-15-12 | 558 | 12-15-12 | 6.37 |
| 26 | 0-28-23-*s* | 8-10-27-12-15-13-14-8 | 558 | 14-8-10-27 | 6.86 |
| 27 | 0-28-21-10-12-15-25-*s* | 8-13-14-8 | 553 | 28-21-10 | 8.08 |
| 28 | 0-28-21-24-11-15-25-12-13-14-*s* | none | 552 | optimum | 7.79 |

**EXAMPLE 19**

The disassembly precedence graph in Figure 7.4 shows 17 components A through Q, and a precedence structure that is more general than the simple divergent disassembly graph studied by Johnson and Wang (1998), because AND relationships are also permitted, such as

$$(\text{I and L and N}) \rightarrow \text{P}$$

Some cases are considered where the product is completely disassembled. Therefore, component revenues are not accounted for, as these are the same for every complete sequence. We used cost maximization for guaranteeing complete sequences. Alternatively, this can always be guaranteed by putting all $x_j$ equal to 1 beforehand. In addition, one of the operations from the list, J, K, O, P, Q, has to be the final operation, which can also be introduced as a constraint.

If no counters are included, we have to rely on the iterative method. Because the precedence relationships are not explicitly included in the model, cycles can occur in the solution. An example of such cycle is CGC. It turns out that the optimum solution is found in about 40 to 50 iterations, depending on the case considered. Each iteration took about 1 to 3 sec of CPU time.

If counters are introduced, cycles are inhibited and complete sequences are returned. If no additional constraints are included, we have to deal with erroneous subsequences only. There are still many iterations to account for, and the required CPU time is about 20 sec for a single iteration, gradually increasing with the number of additional constraints inhibiting erroneous subsequences. With the counters present, a rigorous calculation can be carried through, because explicit inclusion of precedence relationships is possible. These include:

$$\begin{gathered} a_H > a_I \quad a_G > a_I \quad a_G > a_L \quad a_F > a_L \quad a_B > a_C \quad a_M > a_C \\ a_M > a_E \quad a_D > a_E \quad a_E > a_N \quad a_C > a_N \quad a_N > a_O \quad a_N > a_Q \\ a_L > a_Q \quad a_N > a_P \quad a_L > a_P \quad a_I > a_P \end{gathered} \tag{7.15}$$

Introduction of these relationships returns the optimum solution in one straightforward calculation, using a mixed integer-programming solver. However, substantial CPU time is required, here (about 3000 sec on a 200 MHz PC).

## 7.4.6 CONCLUSIONS

We discussed methods for finding the optimum disassembly sequence of a product when represented with either an AND/OR graph or a disassembly precedence graph. We observed that the problem is a constrained traveling salesperson problem (TSP), which suffers from NP-completeness. We confined ourselves to solutions obtained via standard solvers based on branch and bound algorithms. It appeared that the required CPU time for constrained TSPs still exceeded that of unconstrained problems, although the search space is considerably restricted. We saw that additional algorithms, such as the iterative approach, which can proceed automatically, can considerably decrease the calculation effort. Although viable methods are presented, research on more efficient algorithms for this specific case is warranted, particularly for increasingly complex products. The inefficiency of the mixed integer-programming solver in these types of problems appears if one expands the AND/OR graph via a state diagram into a Bourjault's tree (see subsection 5.3.3). In this case, the problem relaxes to a linear programming problem and the optimum solution for problems such as that in Figure 7.8 can be found almost instantaneously. Unfortunately, the modeling effort is quite substantial, because this method of graphically modeling fails compactness. This is, however, useful in benchmarking the results obtained using equivalent heuristic and metaheuristic methods.

## 7.5 DEMAND-DEPENDENT PROBLEMS

### 7.5.1 INTRODUCTION

In the preceding theory, the focus has been on the planning aspects of assembly, repair, and end-of-life disassembly with an emphasis on materials recycling. In chapters 4 through 6, and in the preceding subsections of this chapter, the mechanical structure of a product has been the principal starting point. Thus we have been taking a *mechanical approach*. In chapter 3, many products were described as a *hierarchical tree structure* (see Figure 3.8 and Figure 3.9). In such a structure, not every detail is incorporated, for it is based on a distinct modular hierarchy of the product. This kind of representation is fruitful if the product has such a structure, and if disassembly is aimed at module recovery. A tree consists of *roots*, which are the branching points of the tree, and *leaves*, which represent the modules that can be "harvested" if the roots are disassembled. This can be elucidated with the monitor as an example (see Figure 3.9). If one wants the gun, the following sequence of operations has to be performed:

- Root 1: Making the monitor available
- Root 2: Unscrewing and detachment of the cover
- Root 3: Unscrewing the CRT unit and disconnecting cables; detachment of CRT unit
- Root 4: Detachment of anode loop, unplugging, cutting wires; detachment of electronics module
- Root 5: Unscrewing and removal of clamps; detachment of neck rings
- Root 6: Breaking of the neck glass

In the hierarchical tree structure, there is only one possible disassembly sequence for the detachment of the desired component or module. This simplification clearly relaxes the planning problem and makes it suitable for scheduling purposes. For the case of component reuse, there is usually a demand on specific components that can be present in multiple product types (*commonality*) and in each of these product types at different places and in different quantities (*multiplicity*). The problem is to determine the mix of products that have to be partially dismantled, and the corresponding disassembly depth in such a way that demand on components is met at minimum costs. Costs are linked to the disassembly of roots. This kind of problem is called *lot-size-balancing* (Veerakamolmal and Gupta, 1998c).

The hierarchical trees, such as the one in Figure 3.9, are actually task precedence diagrams, as the disassembly of a root often corresponds to a number of tasks of preparatory nature, such as the loosening of fasteners or the detachment of constraining components that are irrelevant with respect to component reuse. Therefore only those components are considered leaves that are relevant from a component demand point of view. This relates the tree to an uncomplicated disassembly precedence graph, which is completely divergent (see Figure 4.13(c) and Figure 5.15). Obviously, the unit operations may be different from the detachment of a component only. Another link is with reverse *materials requirements planning* (MRP), for the pattern of leaves in the hierarchical tree structure is related to the *bill of materials* (BOM). Many authors apply the hierarchical tree structure for product representation. Research on this topic has been published in papers due to Gupta and Taleb (1994), Taleb and Gupta (1997), and Taleb et al. (1997). Authors such as Spengler et al. (1997) and Krikke et al. (1998, 1999) also discuss applications of the hierarchical tree approach. Spengler et al. focus on the appropriate disassembly depth in the demolition of houses with regard to a demand-constrained problem. Krikke et al. introduce uncertainty and use optimization models for online adaptation to test results for modules that influence the expected value of submodules thus governing the appropriate disassembly depth. This kind of problem requires the on-line adaptiveness of the disassembly schedule.

Veerakamolmal and Gupta (1998abc, 1999) present mathematical models for solving the lot-size balancing problem including multiplicity and commonality. This is known as the component-disassembly optimization model. The authors apply the model to several configurations from practice, which include a set of three different PC types, consisting of 27 relevant components in all. Various components are subjected to multiplicity and commonality, which means that they can occur multiple times in a specific product type and can be present in multiple product types as well. The original model that was proposed is based on a combination of linear programming and several nonlinear operations, including the ceil operation (the search for the smallest integer larger than a specific expression). This model was later modified to avoid the ceil operation (Lambert and Gupta, 2002). This resulted in an approach that could be solved using standard mixed integer programming software. The method is based on a modification of the previously discussed optimization model (see subsection 7.2.3 and Expressions 7.4ab). This model is applied to Veerakamolmal and Gupta's case, which is presented in a graphical representation in Figure 7.10. The roots, which are arbitrary sets of tasks here rather than disassembly operations, are represented by directed arcs. If these tasks are accomplished, the leaves can be harvested, if desired, with no additional costs. Three indices are used here: the product type index $i$, the root number $r$, and the component type index $c$. It can be observed, for example, that component 5 can be obtained via disassembling root 1 and 2 in product type 1 or, alternatively, via disassembling root 1 and 2 in product type 2 or, alternatively, via disassembling root 1 and 2 in product type 3. The method for detecting the optimum solution will be explained for the case of a single product in example 20 first and for the case of multiproduct in example 21.

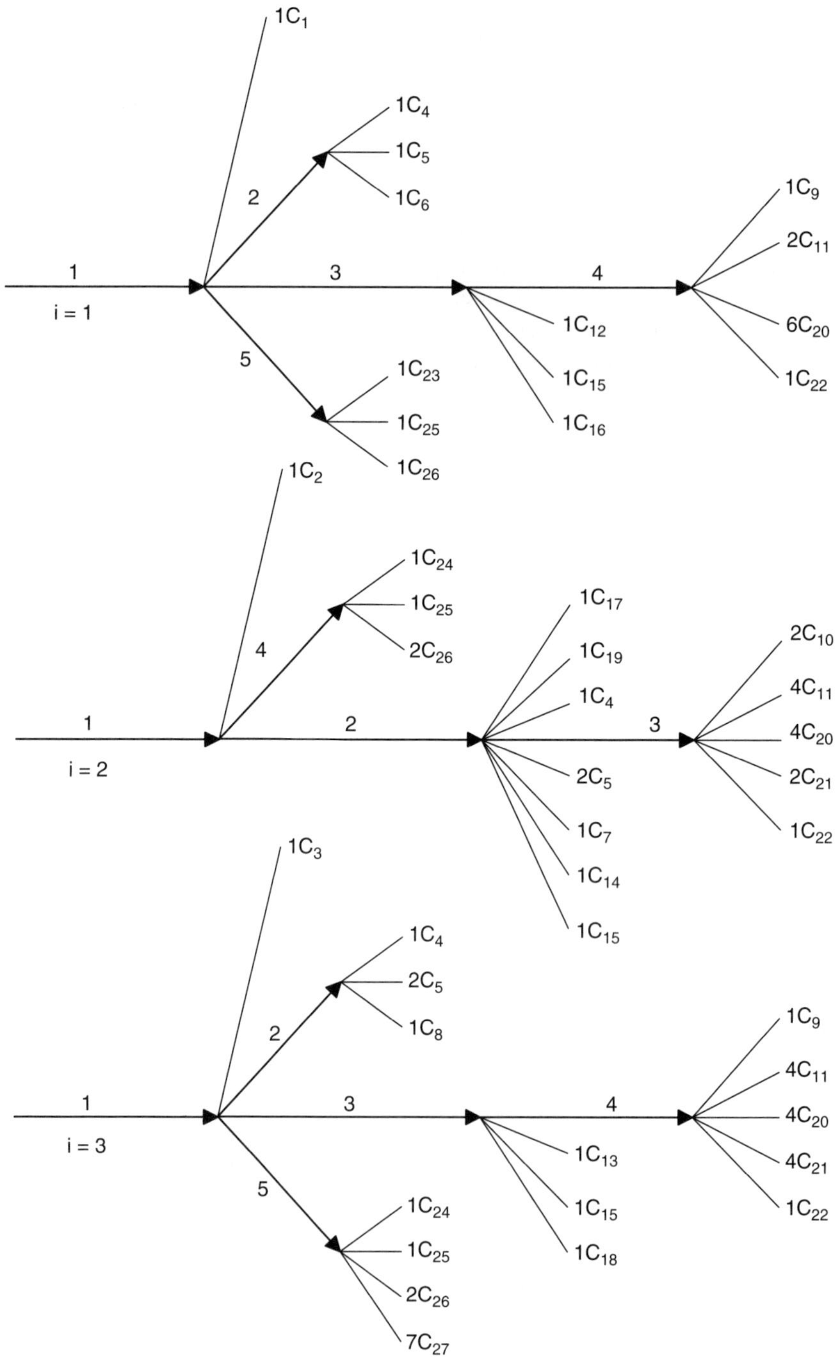


**FIGURE 7.10** The hierarchical tree structures for the three product types.

## 7.5.2 Single-Product Case

### Example 20

For reasons of clarity, a single-product system will be studied first. Consider the product type $i = 1$ in Figure 7.10. First, a flow vector with components $x_r$ is defined. These components are flow variables that are connected to the roots $r$. These are integer variables, which are nonzero if root $r$ is disassembled, and 0 if not. Similar to the AND/OR graph, constraints are defined as follows:

$$x_2 + x_3 + x_5 \leq x_1$$

$$x_4 \leq x_3$$

This set of constraints can be condensed into a single expression via the introduction of the tree structure matrix $S$ with (0,1)-elements $S_{r,r'}$. The element $S_{r,r'}$ equals 1 if the disassembly of root $r'$ can immediately succeed that of root $r$. The element $S_{1,1}$ is always put equal to 1 for the sake of convenience. Consequently, for product $i = 1$:

$$S = \begin{bmatrix} 1 & 1 & 1 & 0 & 1 \\ 0 & 0 & 0 & 0 & 0 \\ 0 & 0 & 0 & 1 & 0 \\ 0 & 0 & 0 & 0 & 0 \\ 0 & 0 & 0 & 0 & 0 \end{bmatrix}$$

Thus, the set of constraints reduces to

$$x_{r'} \leq \sum_{r} S_{r,r'} \cdot x_r, \forall r' \tag{7.16a}$$

Putting $S_{1,1}$ equal to 1 thus enables $x_1$ to be different from zero.

The yield matrix $Y_{r,c}$ refers to the number of components $c$ that are leaves of root number $r$. Its elements are integers. For example, here $Y_{3,12} = 1$, $Y_{4,20} = 6$, $Y_{1,3} = 0$.
The complete yield matrix for the product with $i = 1$ is as follows:

$$Y = \begin{bmatrix} 1 & 0 & 0 & 0 & 0 & 0 & 0 & 0 & 0 & 0 & 0 & 0 & 0 & 0 & 0 & 0 & 0 & 0 & 0 & 0 & 0 & 0 & 0 & 0 & 0 & 0 & 0 \\ 0 & 0 & 0 & 1 & 1 & 1 & 0 & 0 & 0 & 0 & 0 & 0 & 0 & 0 & 0 & 0 & 0 & 0 & 0 & 0 & 0 & 0 & 0 & 0 & 0 & 0 & 0 \\ 0 & 0 & 0 & 0 & 0 & 0 & 0 & 0 & 0 & 0 & 0 & 1 & 0 & 0 & 1 & 1 & 0 & 0 & 0 & 0 & 0 & 0 & 0 & 0 & 0 & 0 & 0 \\ 0 & 0 & 0 & 0 & 0 & 0 & 0 & 0 & 1 & 0 & 2 & 0 & 0 & 0 & 0 & 0 & 0 & 0 & 0 & 6 & 0 & 1 & 0 & 0 & 0 & 0 & 0 \\ 0 & 0 & 0 & 0 & 0 & 0 & 0 & 0 & 0 & 0 & 0 & 0 & 0 & 0 & 0 & 0 & 0 & 0 & 0 & 0 & 0 & 0 & 1 & 0 & 1 & 1 & 0 \end{bmatrix}$$

Next, the yield vector $y_c$ is introduced. Its components are integer variables, each corresponding to a specific component $c$. The following constraint holds

$$y_c \leq \sum_{r} Y_{r,c} \cdot x_r, \forall c \tag{7.16b}$$

This constraint guarantees that the yield of a component $c$ cannot exceed the maximum yield that can be obtained if the roots are disassembled according to the particular disassembly plan.

The demand vector $D_c$ is a given parameter that reflects the existing demand on components $c$. It thus consists of integers. Meeting the demand is accomplished if

$$y_c = D_c, \forall c \tag{7.16c}$$

Meeting the demand can usually be achieved in different ways. For arriving at the optimum way, costs are assigned to the disassembly of the roots, which are reflected in the cost vector $C_r$. Minimizing the objective function

$$obj = \sum_r C_r \cdot x_x \tag{7.16d}$$

and solving the problem using an integer linear programming software, results in the optimum disassembly plan. The value of $x_1$ reflects the number of product items that has to be disassembled for meeting the demand. If costs $C$ are assigned for making the products available, these can be included in the objective function without violating the linear character of the problem.

### 7.5.3 Multiproduct Case

The multiproduct case proceeds along the same lines as the single-product case. The only extension is the introduction of an extra index $i$ to the matrices. Because the number of roots is not equal for all the product types, the maximum number of roots has to be chosen. Redundant elements of matrices are thus set equal to zero.

The model is expressed by constraints that are similar to Expression 7.16a through Expression 7.16d as follows:

$$x_{i,r'} \leq \sum_r S_{i,r,r'} \cdot x_{i,r}, \forall i, r' \tag{7.17a}$$

$$y_c \leq \sum_{i,r} Y_{i,r,c} \cdot x_{i,r}, \forall c \tag{7.17b}$$

$$y_c = D_c, \forall c \tag{7.17c}$$

Minimize

$$obj = \sum_i \left( C_i x_{i,1} + \sum_r C_{i,r} \cdot x_{i,r} \right) \tag{7.17d}$$

The parameter $C_i$ reflects the costs of making the product type $i$ available.

**TABLE 7.12a**

| Root *r* | 1 | 2 | 3 | 4 | 5 |
|---|---|---|---|---|---|
| $i = 1$ | 237 | 0 | 237 | 237 | 237 |
| $i = 2$ | 200 | 200 | 199 | 180 | — |
| $i = 3$ | 163 | 75 | 163 | 163 | 350 |

## Example 21

For the case of the three products that are given in Figure 7.10, the following parameter values are assumed (Veerakamolmal and Gupta, 1998abc; Lambert and Gupta, 2002):

$$C_i = [140 \quad 120 \quad 135]$$

$$C_{i,r} = \begin{bmatrix} .73 & .50 & .38 & .42 & .34 \\ .90 & .62 & .46 & .46 & 0 \\ .74 & .52 & .40 & .42 & .34 \end{bmatrix}$$

$$D_c = [0 \quad 0 \quad 0 \quad 0 \quad 550 \quad 0 \quad 0 \quad 0 \quad 400 \quad 120 \quad 1,150 \quad 0 \quad 0 \quad 0 \quad 0 \quad 0 \quad 0 \quad 0 \quad 200 \quad 1,250 \quad 1,050 \quad 0 \quad 0 \quad 0 \quad 580 \quad 450 \quad 350]$$

$$C_c = [1.25 \quad 2.25 \quad 1.5 \quad .5 \quad 2.5 \quad .75 \quad .75 \quad 1 \quad 1.25 \quad 1.25 \quad .5 \quad .25 \quad .25 \quad .25 \quad .25 \quad .75 \quad .75 \quad .75 \quad .5 \quad .25 \quad .25 \quad 1.25 \quad 1.5 \quad `1.5 \quad 1.25 \quad 1.75]$$

$$R_c = [0 \quad 0 \quad 0 \quad 0 \quad 2 \quad 0 \quad 0 \quad 0 \quad 15 \quad 18 \quad 18 \quad 0 \quad 0 \quad 0 \quad 0 \quad 0 \quad 0 \quad 0 \quad 14 \quad 15 \quad 25 \quad 0 \quad 0 \quad 0 \quad 6 \quad 15 \quad 15]$$

The values $C_c$ represent the disposal costs of the redundant components. The revenues $R_c$ of the components can also be included. The aggregate revenue of the components, however, does not depend on the specific solution, because it is entirely determined by the demand. In addition, capacity constraints can be introduced, and the maximum value can be assumed for the supply of products of type *i*.

For the three product case, with the parameter values listed above, the optimum solution shown in Table 7.12a and Table 7.12b is achieved.

In the tables, the symbol "–" means that the corresponding root or component is not present in that particular product. This is different from zero, which refers to an

**TABLE 7.12b**

| Component *c* | 5 | 9 | 10 | 11 | 19 | 20 | 21 | 25 | 26 | 27 |
|---|---|---|---|---|---|---|---|---|---|---|
| $i = 1$ | 0 | 237 | — | 474 | — | 1250 | — | 237 | 237 | — |
| $i = 2$ | 400 | — | 120 | 24 | 200 | 0 | 398 | 180 | 213 | — |
| $i = 3$ | 150 | 163 | — | 652 | — | 0 | 652 | 163 | 0 | 350 |

object (root or component) that is present but that is not disassembled because the demand is met by similar components from another product or root.

It can be observed that the column of Table 7.12a that corresponds to $r = 1$ reveals the number of products of the different types that have to be partially disassembled. Usually, the potential yield is larger than the demand, which results in incomplete disassembly of the available products.

## 7.6 CONCLUSION

It is demonstrated in this chapter that exact methods can flawlessly be applied in many domains of disassembly sequencing and planning problems. Mathematical programming is combined here with network analysis, which together provide a good insight into the problem structure. This chapter focused on exact methods, for these are promising both for direct application in practice and for supporting heuristic methods evaluation. Obviously, this is beneficial to assembly, maintenance and end-of-life processing. We have made an attempt in presenting an overview of the different formalisms used in this field, which should provide a starting point for further research. Such an overview has not been presented so far.

We learned that methods in one domain of disassembly theory could be extended to other domains. This refers to disassembly and clustering, to sequencing with sequence independent and sequence dependent costs, and to disassembly planning. The product structure was a central issue in the approach of this chapter. More aggregate scheduling issues, which do not account for the product's structure, are beyond the scope of this book. Part of the scheduling domain is involved with optimization over a fixed period of time, a planning horizon, which requires dynamic optimization or simulation. Although this book has been partly written with reverse logistics as a point of departure, we did not include specific logistic issues such as storage and allocation. However, we acknowledge that the product structure and, consequently, the design of the disassembly process, is at the very core of reverse logistics as well. Without pretending to present a complete overview of this domain, we mention here the studies of Hoshino et al. (1995), Clegg et al. (1995), Spengler et al. (1997), Guide et al. (1997ab; 1999), and Uzsoy and Venkatachalam (1998), which provide a good introduction to this field. For an overview of this field, see Dekker et al., 2004.

## REFERENCES

Baldwin, D.F., Abell, T.E., Lui, M.M., De Fazio, T.L., and Whitney, D.E., 1991, An integrated computer aid for generating and evaluating assembly sequences for mechanical products. *IEEE Transactions on Robotics and Automation*, **7**, 78–94.

Bisschop, J. and Entriken, R., 1993, *AIMMS, the Modeling System*. Paragon Decision Technology Haarlem, The Netherlands.

Bourjault, A., Chappe, D., and Henrioud, J.M., 1987, Elaboration automatique des gammes d'assemblage à l'aide de résaux de Petri (Automatic determination of assembly sequence with Petri nets). *RAIRO(APII)*, **21**, 323–342 (in French; Abstract in English).

Caccia, C. and Pozzetti, A., 2000, A genetic algorithm for disassembly strategy definition. *Proceedings of SPIE Conference on Environmentally Conscious Manufacturing*, 68–77.

Clegg, A.J., Williams, D.J., and Uzsoy, R., 1995, Production planning for companies with remanufacturing capability. *Proceedings of IEEE International Symposium on Electronics and the Environment*, 186–191.

Dekker, R., Fleischmann, M., Inderfurth, K., and Van Wassenhove, L.N., 2004, *Reverse Logistics, Quantitative Models for Closed-Loop Supply Chains*. Springer-Verlag Berlin (D).

Guide, V.D.R., Srivastava, R., and Spencer M.S., 1997a, An evaluation of capacity planning techniques in a remanufacturing environment. *International Journal of Production Research*, **35**(1), 67–82.

Guide, V.D.R., Srivastava, R., and Kraus M.E., 1997b, Product structure complexity and scheduling of operations in recoverable manufacturing. *International Journal of Production Research*, **35**(11), 3179–3199.

Guide, V.D.R., Jayaraman, V., and Srivastava, R., 1999, The effect of lead time variation on the performance of disassembly release mechanisms. *Computers and Industrial Engineering*, **36**, 759–779.

Gungor, A. and Gupta, S.M., 1997, An evaluation methodology for disassembly processes. *Computers and Industrial Engineering*, **33**(1–2), 329–332.

Gungor, A. and Gupta, S.M., 2001, Disassembly sequence plan generation using a branch-and-bound algorithm. *International Journal of Production Research*, **39**(3), 481–509.

Gupta, S.M. and Taleb, K.N., 1994, Scheduling disassembly. *International Journal of Production Research*, **32**(8), 1857–1866.

Homem De Mello, L.S. and Sanderson, A.C., 1990, AND/OR graph representation of assembly plans. *IEEE Transactions on Robotics and Automation*, **6**(2), 188–189.

Homem De Mello, L.S. and Sanderson, A.C., 1991, A correct and complete algorithm for the generation of mechanical assembly sequences. *IEEE Transactions on Robotics and Automation*, **7**(2), 228–240.

Hoshino, T., Yura, K., and Hitomi, K., Optimization analysis for recycle-oriented manufacturing systems. *International Journal of Production Research*, **33**, 2069–2078.

Johnson, M.R. and Wang, M.H., 1998, Economical evaluation of disassembly operations for recycling, remanufacturing and reuse. *International Journal of Production Research*, **36**(12), 3227–3252.

Kanehara, T., Suzuki, T., Inaba, A., and Okuma, S., 1993, On algebraic and graph structural properties of assembly Petri net. *Proceedings. of the 1993 IEEE/RSJ International Conference on Intelligent Robots and Systems, Yokohama*, July 26–30, 2286–2293.

Kang, J.G., Lee, D.H., Xirouchakis, P., and Persson, J.G., 2001, Parallel disassembly sequencing with sequence-dependent operation times. *Annals of the CIRP*, **50**(1), 343–346.

Krikke, H.R., Van Harten, A., and Schuur, P.C., 1998, On a medium term product recovery and disposal strategy for durable assembly products. *International Journal of Production Research*, **36**(1), 111–139.

Krikke, H.R., Van Harten, A., and Schuur, P.C., 1999, Business case Roteb: recovery strategies for monitors. *Computers and Industrial Engineering*, **36**(4), 739–757.

Lambert, A.J.D., 1994, Optimal disassembly of complex products. *Proceedings of Northeast Decision Sciences Institute*, Portsmouth, NH, 74–80.

Lambert, A.J.D., 1997, Optimal disassembly of complex products. *International Journal of Production Research*, **35**(9), 2509–2523.

Lambert, A.J.D., 1999, Linear programming in disassembly/clustering sequence generation. *Computers and Industrial Engineering*, **36**, 723–738.

Lambert, A.J.D., 2002, Determining optimum disassembly sequences in electronic equipment. *Computers and Industrial Engineering*, **43**, 553–575.

Lambert, A.J.D. and Gupta, S.M., 2002, Demand-driven disassembly optimisation for electronic consumer goods. *Journal of Electronics Manufacturing*, **11**(2), 121–135.

Lambert, A.J.D., 2003, Optimum disassembly sequence with sequence-dependent disassembly costs. *Proceedings of IEEE International Symposium on Assembly and Task Planning*, 151–156.

Motavalli, S. and Islam, A.U., 1997, Multi-criteria assembly sequencing. *Computers and Industrial Engineering*, 32(4), 743–751.

Moore, K.E. and Gupta, S.M., 1996, Petri net models of flexible and automated manufacturing systems: a survey. *International Journal of Production Research*, **34**, 3001–3035.

Navin-Chandra, D., 1994, The recovery problem in product design. *Journal of Engineering Design*, **5**(1), 65–86.

Penev, K.D. and De Ron, A.J., 1996, Determination of a disassembly strategy. *International Journal of Production Research*, **34**(2), 495–506.

Rosell, J., Muñoz, N., and Gambin, A., 2003, Robot task sequence planning using Petri nets. *Proceedings of IEEE International Symposium on Assembly and Task Planning*, 24–29.

Shapiro, J.F., 1979, *Mathematical programming: structures and algorithms*, John Wiley and Sons, New York.

Sloane, N.J.A., 2003, *The On-line Encyclopedia of Integer Sequences*. AT&T Research. Available on Internet via: http://www.research.att.com/~njas/sequences

Spengler, Th., Püchert, H., Penkuhn, T., and Rentz, O., 1997, Environmental integrated production and recycling management. *European Journal of Operational Research*, **97**, 308–326.

Taleb, K.N. and Gupta, S.M., 1997, Disassembly of multiple product structures. *Computers and Industrial Engineering*, **32**(4), 949–961.

Taleb, K.N., Gupta, S.M., and Brennan, L., 1997, Disassembly of complex product structures with parts and materials commonality. *Production Planning and Control*, **8**(3), 255–269.

Tiwari, M.K., Sinha, N., Kumar, S., Rai, R., and Mukhopadhyay, S.K., 2002, A Petri net based approach to determine the disassembly strategy of a product. *International Journal of Production Research*, **40**(5), 1113–1129.

Uzsoy, R. and Venkatachalam, G., 1998, Supply chain management for companies with product recovery and remanufacturing capability. *International Journal of Environmentally Conscious Design and Manufacturing*, **7**(1), 59–72.

Veerakamolmal, P. and Gupta, S.M., 1998a, Design of integrated component recovery system. *Proceedings of IEEE Symposium on Electronics and the Environment*, 264–269.

Veerakamolmal, P. and Gupta, S.M., 1998b, High-mix/low-volume batch of electronic equipment disassembly. *Computers and Industrial Engineering*, **35**(1–2), 65–68.

Veerakamolmal, P. and Gupta, S.M., 1998c, Optimal analysis of lot-size balancing for multiproducts selective disassembly. *International Journal of Flexible Automation and Integrated Manufacturing*, **6**(3–4), 245–269.

Veerakamolmal, P. and Gupta, S.M., 1999, Analysis of design efficiency for the disassembly of modular electronic products. *Journal of Electronics Manufacturing*, **9**(1), 79–95.

Yee, S.T. and Ventura, J.A., 1999, A Petri net model to determine optimal assembly sequences with assembly operation constraints. *Journal of Manufacturing Systems*, **18**(3), 203–213.

# Part III

## *Disassembly Planning*

# 8 Disassembly to Order Problems: Multicriteria Methods

## 8.1 INTRODUCTION

The two categories of problems covered in this book are disassembly sequencing and disassembly planning. Disassembly sequencing addresses the question, "how to disassemble?" while disassembly planning delineates "how much to disassemble?" The majority of chapters 4 through 7 were devoted to disassembly sequencing. For a thorough survey of disassembly sequencing problems reported in the literature, see Lambert (2003). A form of a disassembly planning problem was mentioned in the introduction of last chapter, the disassembly-to-order (DTO) problem. DTO problems typically deal with situations where a variety of end-of-life (EOL) products are taken back from the last users or collectors and brought into the disassembly facility. At the disassembly facility, the EOL products are sorted and prepared for disassembly operation. As we know, disassembly is the process of systematic removal of the desired items (components or subassemblies) or materials from the EOL product so that the items or materials are obtained in the desired form. Complex components such as motherboards are frequently considered as items to be disassembled. Disassembly can be *selective* (only a selection of items in the product are disassembled) or *complete* (all items in the product are disassembled). In addition, disassembly can be *nondestructive* (focusing on items rather than materials) or *destructive* (focusing on materials rather than items). After nondestructive disassembly, the items are sold, reused, recycled, stored for future use, or disposed of. Similarly, after destructive disassembly, the materials are either recycled or disposed of.

Section 7.5 dealt with the demand-dependent problems (or lot-size balancing problems), which are essentially the DTO problems (see Veerakamolmal and Gupta, 1998abc, 1999; Lambert and Gupta, 2002). These models are either solved using a combination of linear programming and several nonlinear operations or a mixed integer program. While such methods provide the problem's so-called "optimal solution," Herbert Simon, the Nobel Prize winning economist, challenged the whole concept of optimal decision, which he claimed is only for an imaginary simplified world! He advocated that, in most cases, the decisions need to be only "good enough." To this end, he coined the word *satisfice*, which is the combination of *satisfy* and *suffice*. He used this word to describe a kind of decision-making strategy by which an individual chooses the first choice that is both satisfactory and sufficient instead of choosing the "best" strategy.

The whole area of multicriteria methods is encompassed by the concept of satisfycing. Multicriteria methods contain and aim at simultaneously improving multiple objective functions. However, they usually consist of conflicting goals so that optimal solutions in the conventional sense do not exist. Instead, one aims at something like Pareto optimality, where a state $P$ is said to be Pareto optimal, if there is no other state $Q$ dominating the state $P$ with respect to a set of objective functions. The state $P$ dominates the state $Q$, if $P$ is better than $Q$ in at least one objective function and not worse with respect to all other objective functions.

In this chapter, we will highlight two multicriteria methods, goal programming and linear physical programming, to address the DTO problem.

## 8.2 MULTICRITERIA METHODOLOGIES

Most real world decision-making and design problems are inherently multiobjective. Yet, many mathematical programming models, such as linear programming (LP) models, require that the decision maker (DM) concoct one aggregate objective function that is usually subjected to constraints. Additionally, most real world objectives exist in an imprecise environment. And yet, mathematical programming frameworks often require precise statements of the objectives at the problem formulation stage. These observations bring to light serious impediments to the practical application of mathematical programming by both novices and experts. Thus, the practitioner would benefit immensely from a general framework for decision-making that allows the problem formulation to retain its multiobjective character (i.e., the DM should not be forced to form a weighted sum of several criteria for the sole purpose of conforming to the prevailing mathematical paradigm). Both goal programming (GP) and linear physical programming (LPP) possess this feature. In addition, LPP allows for the possibility of deliberate imprecision in the statement of the preferences.

## 8.3 GOAL PROGRAMMING

### 8.3.1 INTRODUCTION

Goal programming, generally applied to linear and integer problems, deals with the achievement of prescribed goals or targets. First reported by Charnes et al. (1955) and Charnes and Cooper (1961), it was extended in the 1960s and 1970s by Ijiri (1965), Lee (1972), and Ignizio (1976). Since then, there has been an explosion of areas where goal programming has been applied (Schniederjans, 1995). Both academicians and practitioners have embraced this technique.

The basic purpose of goal programming is to simultaneously satisfy several goals relevant to the decision-making situation. To this end, a set of attributes to be considered in the problem situation is established. Then for each attribute, a target value (or appraisal level) is determined. Next, the deviation variables are introduced. These deviation variables may be negative or positive (represented by $\eta_k$ and $\rho_k$, respectively). The negative deviation variable $\eta_k$ represents the quantification of the under-achievement of the $k$th goal. Similarly, $\rho_k$ represents the quantification of the over-achievement of the $k$th goal (Romero, 1991). Finally for each attribute, the desire to overachieve (minimize $\eta_k$) or underachieve (minimize $\rho_k$), or satisfy the target value exactly (minimize $\eta_k + \rho_k$) is articulated (Ignizio, 1982).

### 8.3.2 Goal Programming Model for a DTO System

The objective of the particular DTO system presented here is to determine the best combination of the number of each product type to be taken back and disassembled to meet the selective demand for items and materials obtained from the products under a variety of physical, financial, and environmental constraints so as to achieve the preemptive goals of maximum total profit, maximum sales from materials, minimum number of disposed items, minimum number of stored items, minimum cost of disposal, and minimum cost of preparation, in that order. Hence, no weights are assigned to any of the goals but the achievement order is characterized, defining a preemptive goal programming problem. When solved, the model provides the number of reused, recycled, stored, and disposed items as well as the values of a host of other performance measures (Kongar and Gupta, 2002a).

#### 8.3.2.1 The Goals

The first goal is to maximize the overall profit from DTO activities. In goal programming terms, our desire for the total profit variable (*TPR*) would be to aspire for a total profit of at least $TPR^*$ and exceed it as much as possible. Mathematically, this can be achieved by forcing the negative deviation ($\eta_1$) from the predetermined value $TPR^*$ to secure a value equal to zero. Note also that by putting no restrictions on the positive deviation ($\rho_1$), the model would put no ceiling on the total profit value to exceed $TPR^*$. In other words, this goal guarantees a best profit level for the DTO process while articulating a lower aspiration perimeter. This goal can be formulated as follows:

$$\begin{aligned} &\min \eta_1 \\ &s.t. \quad TPR + \eta_1 - \rho_1 = TPR^* \end{aligned} \tag{8.1}$$

The second goal is related to the revenue from the sales of demanded materials (*RMS*). Here also, the negative deviation ($\eta_2$) from the predetermined value $RMS^*$ is to be minimized. This goal can be expressed as follows:

$$\begin{aligned} &\min \eta_2 \\ &s.t. \quad RMS + \eta_2 - \rho_2 = RMS^* \end{aligned} \tag{8.2}$$

The third goal, which has an "environmentally benign" character rather than a financial bias, addresses our desire to minimize the number disposed items (*NDIS*) and to limit the disposed items to a predefined quantity $NDIS^*$. Mathematically, this can be achieved by forcing the positive deviation ($\rho_3$) from $NDIS^*$ to achieve a value equal to zero. This goal can be formulated as follows:

$$\begin{aligned} &\min \rho_3 \\ &s.t. \quad NDIS + \eta_3 - \rho_3 = NDIS^* \end{aligned} \tag{8.3}$$

The fourth goal is to minimize the number of stored items (*NSTR*). This goal addresses the desire to minimize the inventory level, reducing both the financial burden and eliminating the occupation of extra space. Hence, the positive deviation ($\rho_4$) from the upper storage limit $NSTR^*$ is minimized. This can be expressed mathematically as follows:

$$\begin{aligned} &\min \rho_4 \\ &s.t. \quad NSTR + \eta_4 - \rho_4 = NSTR^* \end{aligned} \tag{8.4}$$

The fifth goal minimizes and limits the cost of disposal (*CDI*) to an upper bound $CDI^*$. Hence, the positive deviation ($\rho_5$) from the upper bound is minimized and can be formulated as follows:

$$\begin{aligned} &\min \rho_5 \\ &s.t. \quad CDI + \eta_5 - \rho_5 = CDI^* \end{aligned} \tag{8.5}$$

The sixth and last goal minimizes and limits the positive deviation ($\rho_6$) of the cost of preparation (including sorting, refurbishing, and cleaning) of EOL products *CAC* to a predefined level $CAC^*$. This goal guarantees that the cost to prepare the EOL products will not exceed a certain level and thus help search for EOL products that are cheaper to prepare while meeting the demand of materials and components. This goal can be formulated as follows:

$$\begin{aligned} &\min \rho_6 \\ &s.t. \quad CAC + \eta_6 - \rho_6 = CAC^* \end{aligned} \tag{8.6}$$

#### 8.3.2.2 Total Profit Value and Related Terms

The *TPR* is the difference between all the revenues and all the costs considered in the model. There are two sources of revenues considered in the DTO system, viz., *RMS* and the revenue from the sales of demanded items (*RPS*). The various costs considered in the model are: the take back cost (*TB*), transportation cost from collectors to the facility (*CTRCF*), transportation cost from facility to outside recycling plant (*CTRFR*), transportation cost from facility to disposal site (*CTRFD*), transportation cost from facility to storage location (*CTRFS*), the cost of preparation of EOL products *CAC*, the cost of destructive disassembly (*CDD*), the cost of nondestructive disassembly (*CND*), recycling cost (*CRE*), storage cost (*CST*), and *CDI*. Therefore, *TPR* can be written as follows:

$$\begin{aligned} TPR = RMS + RPS - TB - CTRCF - CTRFR - CTRFD - CTRFS \\ - CAC - CDD - CND - CRE - CST - CDI \end{aligned} \quad (8.7)$$

Each term that contributes to the *TPR* function is elaborated below.

*RMS* is a function of the amount of materials sold and the market value of material obtained from each item type $j$ ($PM_j$). The amount of materials sold is a function of the number of item $j$ (obtained from every EOL product $i$) recycled ($\Sigma_i IR_{ij}$), the weight of item $j$ ($W_j$), and the percentage of marketable material obtained from item $j$ ($PRC_j$).

Therefore, by summing the revenue over all items, *RMS* can be obtained as follows:

$$RMS = \sum_j \left( \sum_i IR_{ij} \right) \cdot W_j \cdot PRC_j \cdot PM_j \quad (8.8)$$

*RPS* is a function of the number of item type $j$ (obtained from every EOL product $i$) reused ($\Sigma_i X_{ij}$) and the unit sale price for item type $j$ ($PRM_j$). Therefore, *RPS* can be mathematically expressed as follows:

$$RPS = \sum_j \left( \sum_i X_{ij} \right) \cdot PRM_j \quad (8.9)$$

TB is a function of the number of EOL products ordered ($Y_i$) and the cost of each product ($UTB_i$). Therefore

$$TB = \sum_i Y_i \cdot UTB_i \quad (8.10)$$

*CTRCF* is a function of the number of EOL products ordered ($Y_i$) and the transportation cost per unit from collectors to the facility ($UCTRCF_i$). Therefore

$$CTRCF = \sum_i Y_i \cdot UCTRCF_i \quad (8.11)$$

*CTRFR* is a function of the number of items (obtained from every EOL product $i$) sent to the outside recycler ($\Sigma_i OR_{ij}$) and the transportation cost per item of type $j$ from the facility to the outside recycler ($UCTRFR_j$).

Therefore

$$CTRFR = \sum_j \left( \sum_i OR_{ij} \cdot UCTRFR_j \right) \tag{8.12}$$

*CTRFD* is a function of the *NDIS* and the transportation costs per unit from facility to the disposal site ($UCTRFD_j$). The number of items sent to disposal include the nondemanded items $j$ (obtained from every EOL product $i$) ($L_{ij}$), functionally defective items ($\alpha_{ij} \cdot X_{ij}$, where $\alpha_{ij}$ is the functionally defective rate for item $j$ in product $i$), the items damaged during disassembly ($\beta_{ij} \cdot X_{ij,}$ where $\beta_{ij}$ is the rate at which item $j$ in product $i$ gets damaged during disassembly), and items replaced by the customer with other types that are neither demanded as parts nor contain the demanded type of materials recycled ($\gamma_{ij} \cdot (X_{ij} + R_{ij})$ where $\gamma_{ij}$ is the rate at which customers replace item $j$ in product $i$). Therefore

$$NDIS = \sum_j \sum_i (L_{ij} + (\alpha_{ij} + \beta_{ij}) \cdot X_{ij} + \gamma_{ij} \cdot (X_{ij} + R_{ij})) \tag{8.13}$$

and

$$CTRFD = \sum_j \left( \left( \sum_i (L_{ij} + (\alpha_{ij} + \beta_{ij}) \cdot X_{ij} + \gamma_{ij} \cdot (X_{ij} + R_{ij})) \right) \cdot UCTRFD_j \right) \tag{8.14}$$

*CTRFS* is a function of the *NSTR* and the transportation cost per unit from the facility to the storage location ($UCTRFS_j$). Therefore

$$NSTR = \sum_j \sum_i V_{ij} \tag{8.15}$$

and

$$CTRFS = \sum_j \left( \sum_i V_{ij} \right) \cdot UCTRFS_j \tag{8.16}$$

*CAC* is a function of the number of EOL products $i$ ordered ($Y_i$) and the cost of preparing each product ($UCAC_i$). Therefore

$$CAC = \sum_i Y_i \cdot UCAC_i \tag{8.17}$$

*CDD* is the cost of destructive disassembly (considered for the items that are recycled for their material content or the items that are sent to landfills for proper disposal) and is a function of number of items type $j$ (obtained from every EOL product $i$) to be recycled and disposed ($\Sigma_i (R_{ij} + L_{ij})$), the cost per hour ($cd$) and the time of disassembling each item ($ddt_j$). Therefore

$$CDD = \sum_j \left( \sum_i (R_{ij} + L_{ij}) \right) \cdot cd \cdot ddt_j \tag{8.18}$$

*CND* is the cost of nondestructive disassembly (considered for the items that are reused or the items that are sent to storage) and is a function of number of items type $j$ (obtained from every EOL product $i$) to be reused and stored $(\Sigma_i (X_{ij} + V_{ij}))$, the cost per hour ($cnd$) and the time of disassembling each item ($dt_j$). Therefore

$$CND = \sum_j \left( \sum_i \left( X_{ij} + V_{ij} \right) \right) \cdot cnd \cdot dt_j \tag{8.19}$$

*CRE* is a function of the number of items type $j$ (obtained from every EOL product $i$) recycled in plant $(\Sigma_i IR_{ij})$ and the corresponding unit recycling cost ($UCRE_j$). Therefore

$$CRE = \sum_j \left( \sum_i IR_{ij} \right) \cdot UCRE_j \tag{8.20}$$

*CST* is a function of the number of stored items type $j$ (obtained from every EOL product $i$) $(\Sigma_i V_{ij})$ and the corresponding unit holding cost ($h_j$). Therefore

$$CST = \sum_j \left( \sum_i V_{ij} \right) \cdot h_j \tag{8.21}$$

*CDI* is a function of the number of disposed items type $j$ (obtained from every EOL product $i$) $\Sigma_i (L_{ij} + (\alpha_{ij} + \beta_{ij}) \cdot X_{ij} + \gamma_{ij} \cdot (X_{ij} + R_{ij}))$ and the corresponding unit disposal cost ($UCDI_j$). Therefore

$$CDI = \sum_j \left( \left( \sum_i (L_{ij} + (\alpha_{ij} + \beta_{ij}) \cdot X_{ij} + \gamma_{ij} \cdot (X_{ij} + R_{ij})) \right) \cdot UCDI_j \right) \tag{8.22}$$

#### 8.3.2.3 The Constraints

The number of items type $j$ (obtained from every EOL product $i$) recycled $(\Sigma_i R_{ij})$ has to be equal to the sum of items recycled both in plant $(\Sigma_i IR_{ij})$ and by outside contractor $(\Sigma_i OR_{ij})$. Therefore

$$\sum_i R_{ij} = \left( \delta_j \cdot \sum_i IR_{ij} \right) + \left( (1 - \delta_j) \cdot \sum_i OR_{ij} \right), \forall j \tag{8.23}$$

where $\delta_j$ is a binary variable, which takes a value of 0 when an item $j$ is sent to outside recycler and a value of 1 when an item is recycled in-house.

Note that, the total number of items recycled in plant (*NIRE*) is

$$NIRE = \sum_i \sum_j IR_{ij} \tag{8.24}$$

and the total number of items that are sent to outside contractor (*NORE*) is

$$NORE = \sum_i \sum_j OR_{ij} \tag{8.25}$$

The number of items type $j$ (obtained from every EOL product $i$) recycled ($\Sigma_i R_{ij}$) also has to be equal to the corresponding demand of items for recycling. Therefore

$$DR_j = \sum_i R_{ij}, \forall j \tag{8.26}$$

The total number of items recycled (*NRC*) can be expressed as follows:

$$NRC = \sum_j DR_j \tag{8.27}$$

The total volume (*TS*) occupied by the stored items have to be less than or equal to the total available volume in storage (*AS*). *TS* is a function of the number of stored item $j$ (obtained from every EOL product $i$) ($\Sigma_i V_{ij}$) and its corresponding volume ($v_j$). Therefore

$$TS = \sum_j (v_j \cdot \sum_i V_{ij}) \tag{8.28}$$

and

$$TS \leq AS \tag{8.29}$$

The number of items type $j$ retrieved from all EOL products $i$ ordered (each with multiplicity $Q_{ij}$) ($\Sigma_i Y_i \cdot Q_{ij}$) has to be greater than or equal to the number of demanded items ($D_j$). Therefore

$$\sum_i Y_i \cdot Q_{ij} \geq D_j, \forall j \tag{8.30}$$

where

$$D_j \leq \sum_i X_{ij}, \forall j \tag{8.31}$$

Note that the total number of reused items (*NRU*) is

$$NRU = \sum_i \sum_j X_{ij} \tag{8.32}$$

The number of items recycled in plant ($NIRE$) should be less than or equal to the plant recycling capacity ($RCAP$):

$$NIRE \leq RCAP \tag{8.33}$$

The total number of disassembled items type $j$ retrieved from all EOL products $i$ ordered (each with multiplicity $Q_{ij}$) ($\Sigma_i Y_i \cdot Q_{ij}$) should be equal to the items type $j$ that are reused ($\Sigma_i X_{ij}$), recycled ($\Sigma_i R_{ij}$), stored ($\Sigma_i V_{ij}$), and disposed ($\Sigma_i L_{ij}$).

$$\sum_i Y_i \cdot Q_{ij} = \sum_i (X_{ij} + R_{ij} + V_{ij} + L_{ij}), \forall j \tag{8.34}$$

All the variables must be nonnegative integers. Thus

$$\{Y_i\}, \{X_{ij}\}, \{R_{ij}\}, \{V_{ij}\}, \{L_{ij}\} \geq 0 \text{ and integer; for all } i \text{ and } j. \tag{8.35}$$

#### 8.3.2.4 The GP Model

The GP model can now be written as follows:

Find ($Y_i$, $X_{ij}$, $R_{ij}$, $V_{ij}$, $L_{ij}$) so as to

$$\text{Lexicographically minimize } u = \{(\eta_1), (\eta_2), (\rho_3), (\rho_4), (\rho_5), (\rho_6)\} \tag{8.36}$$

where $\{\eta_k, \rho_k\}$ are defined in Equation 8.1 through Equation 8.6 (the terms of which are explained in Equation 8.7 through Equation 8.22),

subject to

constraints defined in Equation (8.23) through Equation (8.35) and $\{\eta_k, \rho_k\} \geq 0$ $\forall\ k = 1, \ldots, 6$.

#### 8.3.2.5 Procedure to Solve the GP Model

The following steps are used to solve the GP model:

*Step 1.* Read in all the relevant data. Set the first goal as current goal.

*Step 2.* Obtain a linear programming (LP) solution with the current goal as the objective function.

*Step 3.* If the current goal is the last goal, set it equal to the LP objective function value found in Step 2. STOP. Otherwise, go to Step 4.

*Step 4.* If the current goal is achieved or overachieved, set it equal to its aspiration level and add this to the constraint set. Go to Step 2. Otherwise, if the value of the current goal is underachieved, set the aspiration level of the current goal equal to the LP objective function value found in Step 2 and add this to the constraint set. Go to Step 5.

*Step 5.* Set the next goal of importance as the current goal. Go to Step 2.

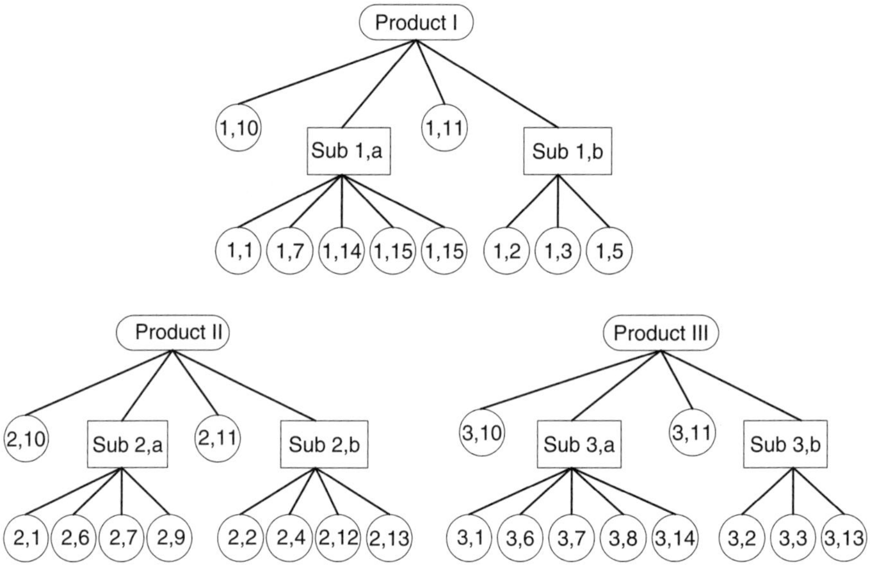


**FIGURE 8.1** Product structures for the case example.

### 8.3.3 A Case Example

We employ a case example to demonstrate the application of the GP model. Consider three different products (Products I, II, and III with $i = 1$, 2, and 3, respectively), which are disassembled for their items and materials (see Figure 8.1 for their product structures). The objective of the model is to determine the best combination of the number of each product type to be taken back and disassembled to meet the selective demand for items and materials obtained from the products under a variety of physical, financial, and environmental constraints so as to achieve the preemptive goals of maximum total profit, maximum sales from materials, minimum number of disposed items, minimum number of stored items, minimum cost of disposal, and minimum cost of preparation, in that order. When solved, the model provides the number of reused, recycled, stored and disposed items as well as the values of all performance measures: *TPR, RMS, NDIS, NSTR, CDI CAC, RPS, TB, CST, CTRCF. CTRFR, CTRFD, CTRFS, CDD, CND, NRU, NIRE, NORE,* and *NRC.*

#### 8.3.3.1 Input Data for the Case Example

Table 8.1 and Table 8.2 exhibit the data needed for the case example. Note that the three different products are made up of various combinations of 15 different items. Table 8.1 lists the various items in each product structure and their component commonality. For each item *j*, Table 8.2 shows its resale item value, resale material value, recyclable percentage, volume, weight, item demand as part, item demand as material, disposal cost, recycling cost, non-destructive disassembly time, and destructive disassembly time. Additional data include: $\alpha_{ij} = 0.05$, $\beta_{ij} = 0.01$, and

**TABLE 8.1**
**Product Structure Data for the Case Example**

| Description | Product i | Item j | $Q_{ij}$ |
|---|---|---|---|
| Full hgt 4 drv ext SCSI | 1 | 1 | 1 |
| Internal Case | 1 | 2 | 1 |
| 56 K win modem | 1 | 3 | 1 |
| Networking PCI card | 1 | 5 | 1 |
| 1.44 MB –3.5" | 1 | 7 | 1 |
| DC Power Supply | 1 | 10 | 1 |
| Pentium II Proc. | 1 | 11 | 1 |
| CD-R-8X4 | 1 | 14 | 1 |
| 40 GB[2] hard drive | 1 | 15 | 2 |
| Full hgt 4 drv ext SCSI | 2 | 1 | 1 |
| Internal Case | 2 | 2 | 1 |
| 64 v PCI Soundcard | 2 | 4 | 1 |
| 80 GB[2] ultrahard drive | 2 | 6 | 1 |
| 1.44 MB –3.5" | 2 | 7 | 1 |
| CD-RW-8X8X32 | 2 | 9 | 1 |
| DC Power Supply | 2 | 10 | 1 |
| Pentium II Proc. | 2 | 11 | 1 |
| 28 K win modem | 2 | 12 | 1 |
| Video Card | 2 | 13 | 1 |
| Full hgt 4 drv ext SCSI | 3 | 1 | 1 |
| Internal Case | 3 | 2 | 1 |
| 56 K win modem | 3 | 3 | 1 |
| 80 GB[2] ultrahard drive | 3 | 6 | 1 |
| 1.44 MB –3.5" | 3 | 7 | 1 |
| 250 MB Built-in Zip Dr | 3 | 8 | 1 |
| DC Power Supply | 3 | 10 | 1 |
| Pentium II Proc. | 3 | 11 | 1 |
| Video Card | 3 | 13 | 1 |
| CD-R-8X4 | 3 | 14 | 1 |

$\gamma_{ij} = 0.01$ for all $i$ and $j$, $UCTRFR_j$ = \$20/item, $UCTRFS_j$ = \$19/item, and $UCTRFD_j$ = \$10/item for all $j$, $UTB_i$ = {\$35, \$48, \$37}/product for $i$ = 1, 2, and 3, $UCTRCF_i$ = {\$10, \$20, \$10}/product for $i$ = 1, 2, and 3, $UCAC_i$ = {\$5, \$4, \$5}/product for $i$ = 1, 2, and 3, $cnd$ = \$14.69/h, $cd$ = \$12.50/h, $h_j$ = \$0.03/unit for all $j$, $RCAP$ = 45,000 units, $AS$ = 81,940,347 cubic cm. The various aspiration levels are as follows: $TPR^*$ = \$100,000, $RMS^*$ = \$240,000, $NDIS^*$ = 71,600 units, $NSTR^*$ = 2,600 units, $CDI^*$ = \$443,700, $CAC^*$ = \$60,624.

### 8.3.3.2 Results of the Case Example

The above data were used to solve the GP Model using LINGO (version 6). The results obtained are summarized in Table 8.3, Table 8.4, and Table 8.5. The optimum numbers of items to be reused, recycled, stored, and disposed are given in Table 8.3

**TABLE 8.2**
**Input Data for the Case Example**

| $j$ | $PRM_j$ (\$/unit) | $PM_j$ (\$/unit.kg) | $PRC_j$ (%) | $V_j$ (cu. cm/unit) | $W_j$ (kg/unit) | $D_j$ (units) | $DR_j$ (units) | $UCDI_j$ (\$/unit) | $UCRE_j$ (\$/unit) | $dt_j$ (h) | $ddt_j$ (h) |
|---|---|---|---|---|---|---|---|---|---|---|---|
| 1 | 95 | 0.0250 | 85 | 21798.7000 | 1.8120 | 1800 | 2500 | 8 | 7 | 0.005 | 0.003 |
| 2 | 85 | 0.0552 | 95 | 19668.0000 | 2.2650 | 2000 | 3000 | 7 | 9 | 0.006 | 0.005 |
| 3 | 98 | 0.1104 | 95 | 0.0049 | 0.0680 | 1800 | 2000 | 6 | 7 | 0.002 | 0.001 |
| 4 | 82 | 0.1104 | 95 | 0.0049 | 0.0906 | 1800 | 1800 | 6 | 8 | 0.001 | 0.001 |
| 5 | 48 | 0.1104 | 95 | 0.0049 | 0.0680 | 2000 | 2000 | 7 | 8 | 0.004 | 0.003 |
| 6 | 39 | 0.1435 | 55 | 57.3650 | 0.4530 | 2000 | 2800 | 4 | 9 | 0.003 | 0.001 |
| 7 | 49 | 0.1214 | 65 | 57.3650 | 0.4530 | 1500 | 1900 | 7 | 8 | 0.004 | 0.002 |
| 8 | 95 | 0.1104 | 85 | 40.9750 | 0.0453 | 2000 | 2400 | 2 | 5 | 0.005 | 0.004 |
| 9 | 188 | 0.1104 | 85 | 106.5350 | 0.0453 | 1900 | 1800 | 8 | 10 | 0.002 | 0.001 |
| 10 | 68 | 1.7660 | 45 | 1639.0000 | 0.5481 | 1900 | 2200 | 5 | 12 | 0.006 | 0.004 |
| 11 | 105 | 1.1038 | 25 | 409.7500 | 0.4530 | 2000 | 3000 | 4 | 9 | 0.004 | 0.003 |
| 12 | 56 | 0.1104 | 95 | 0.0049 | 0.0680 | — | 2000 | 6 | 5 | 0.002 | 0.001 |
| 13 | 94 | 0.1104 | 85 | 0.0049 | 0.0453 | — | 2500 | 7 | 5 | 0.003 | 0.002 |
| 14 | 112 | 0.1104 | 85 | 106.5350 | 0.0453 | — | 1800 | 8 | 8 | 0.004 | 0.003 |
| 15 | 29 | 0.1435 | 55 | 57.3650 | 0.4530 | — | 2800 | 3 | 4 | 0.003 | 0.001 |

**TABLE 8.3**
**Overall Numbers of Items to be Reused, Recycled, Stored, and Disposed**

| Item No. ($j$) | Reused Items (*units*) ($\Sigma_i X_{ij}$) | Recycled Items (*units*) In Plant ($\Sigma_i IR_{ij}$) | Recycled Items (*units*) Contractor ($\Sigma_i OR_{ij}$) | Stored Items (*units*) ($\Sigma_i V_{ij}$) | Disposed Items (*units*) ($\Sigma_i L_{ij}$) |
|---|---|---|---|---|---|
| 1 | 1926 | 2500 | — | 101 | 8384 |
| 2 | 2140 | 3000 | — | 100 | 7671 |
| 3 | 1926 | — | 2000 | 100 | 4954 |
| 4 | 1926 | 1800 | — | 200 | 5 |
| 5 | 2140 | — | 2000 | 100 | 0 |
| 6 | 2140 | 2800 | — | 112 | 3619 |
| 7 | 1605 | 1900 | — | 200 | 9206 |
| 8 | 2140 | 2400 | — | 200 | 0 |
| 9 | 2033 | 1800 | — | 51 | 47 |
| 10 | 2033 | 2200 | — | 100 | 8578 |
| 11 | 1498 | — | 1900 | 200 | 9313 |
| 12 | 0 | 2000 | — | 100 | 1831 |
| 13 | 0 | 2500 | — | 400 | 5771 |
| 14 | 0 | 1800 | — | 338 | 6842 |
| 15 | 0 | 2800 | — | 301 | 5379 |

and Table 8.4. This requires 4240, 3931, and 4740 units of EOL Products I, II and III, respectively to be acquired and disassembled. We note from Table 8.5 that the DTO system is profitable and all the aspiration levels are met.

## 8.4 LINEAR PHYSICAL PROGRAMMING

### 8.4.1 INTRODUCTION

LPP is a recently developed multicriteria technique, which operates in the environment of multiple criteria and uses a utility function to represent the DM's preference. The main difficulty associated with the formulation of a utility function is in determining the correct weights. The key distinguishing feature of LPP is that the DM is entirely removed from the process of choosing weights (Messac, Gupta and Akbulut, 1996).

The original approach to goal programming by Charnes et al. (1955) and Charnes and Cooper (1961), more popularly known as the Archimedean approach, is non-preemptive in that goals are comparable and differ only by numerical weights and can thus be represented by a utility function. In fact, this is the ideal way to solve a multiple objective problem (Steuer, 1989). Traditionally, it has been considered nearly impossible to obtain a mathematical representation of the DM's preference (utility function) (Cohon, 1978; Stadler, 1988). Due to this difficulty, utility theory has not had a strong practical impact in the field of multicriteria optimization. The main difficulty associated with the formulation of Archimedean goal programming

**TABLE 8.4**
**Detailed Numbers of Items to be Reused, Recycled, Stored, and Disposed of for Each Product**

| | Product I $Y_1 = 4240$ | | | | Product II $Y_2 = 3931$ | | | | Product III $Y_3 = 4740$ | | | |
|---|---|---|---|---|---|---|---|---|---|---|---|---|
| *j* | Reused | Recycled | Stored | Disposed | Reused | Recycled | Stored | Disposed | Reused | Recycled | Stored | Disposed |
| 1 | 0 | 2500 | 0 | 1740 | 1926 | 0 | 101 | 1904 | 0 | 0 | 0 | 4740 |
| 2 | 1140 | 3000 | 100 | 0 | 0 | 0 | 0 | 3931 | 1000 | 0 | 0 | 3740 |
| 3 | 1926 | 2000 | 100 | 214 | — | — | — | — | 0 | 0 | 0 | 4740 |
| 4 | — | — | — | — | 1926 | 1800 | 200 | 5 | — | — | — | — |
| 5 | 2140 | 2000 | 100 | 0 | — | — | — | — | — | — | — | — |
| 6 | — | — | — | — | 0 | 2800 | 0 | 1131 | 2140 | 0 | 112 | 2488 |
| 7 | 0 | 0 | 200 | 4040 | 0 | 0 | 0 | 3931 | 1605 | 1900 | 0 | 1235 |
| 8 | — | — | — | — | — | — | — | — | 2140 | 2400 | 200 | 0 |
| 9 | — | — | — | — | 2033 | 1800 | 51 | 47 | — | — | — | — |
| 10 | 0 | 0 | 0 | 4240 | 0 | 2200 | 100 | 1631 | 2033 | 0 | 0 | 2707 |
| 11 | 1498 | 1900 | 200 | 642 | 0 | 0 | 0 | 3931 | 0 | 0 | 0 | 4740 |
| 12 | — | — | — | — | 0 | 2000 | 100 | 1831 | — | — | — | — |
| 13 | — | — | — | — | 0 | 2500 | 0 | 1431 | 0 | 0 | 400 | 4340 |
| 14 | 0 | 0 | 338 | 3902 | — | — | — | — | 0 | 1800 | 0 | 2940 |
| 15 | 0 | 2800 | 301 | 5379 | — | — | — | — | — | — | — | — |

**TABLE 8.5**
**Values of Various Variables at Different Steps of the Procedure to Solve the GP Model**

| | Description | Variable | Aspiration Level | Current Goals and Steps of the Procedure | | | | | | |
|---|---|---|---|---|---|---|---|---|---|---|
| | | | | Goal 1 Step 2 | Goal 2 Step 4 | Goal 2 Step 4 | Goal 3 Step 4 | Goal 4 Step 4 | Goal 5 Step 4 | Goal 6 Step 3 |
| Goals | Total Revenue | *TPR* | 100000 | 111296 | 100000 | 100000 | 100000 | 100000 | 100000 | 100000 |
| | Revenue from Material Sales | *RMS* | 240000 | 219466 | 246902 | 240000 | 240000 | 240000 | 240000 | 240000 |
| | Number of Disposed Items | *NDIS* | 71600 | 71703 | 71703 | 71450 | 71600 | 71600 | 71600 | 71600 |
| | Number of Stored Items | *NSTR* | 2600 | 2450 | 2450 | 2703 | 2553 | 2600 | 2600 | 2600 |
| | Cost of Disposal | *CDI* | 443700 | 444570 | 444570 | 442546 | 443849 | 443636 | 443700 | 443700 |
| | Cost of Acquisition | *CAC* | 60624 | 60604 | 60604 | 60604 | 60604 | 60629 | 60629 | 60624 |
| Other Results | Revenue from Product Sales | *RPS* | — | 1662500 | 1662500 | 1662500 | 1662500 | 1662500 | 1662500 | 1662500 |
| | Take-back Cost | *TB* | — | 512228 | 512228 | 512228 | 512228 | 512403 | 512403 | 512468 |
| | Cost of Storage | *CST* | — | 8166 | 8166 | 8216 | 8166 | 8752 | 8184 | 8235 |
| | Transportation Cost from Customer to Facility | *CTRCF* | — | 168320 | 168320 | 168320 | 168320 | 168370 | 168370 | 168420 |
| | Transportation Cost from Facility to Recycling | *CTRFR* | — | 118000 | 118000 | 118009 | 119606 | 118030 | 118534 | 118373 |
| | Transportation Cost from Facility to Disposal | *CTRFD* | — | 201000 | 201000 | 201000 | 201000 | 201000 | 201000 | 201000 |
| | Transportation Cost from Facility to Storage | *CTRFS* | — | 46550 | 46550 | 51357 | 48507 | 49457 | 49457 | 49457 |
| | Cost of Destructive Disassembly | *CDD* | — | 6398 | 6398 | 6398 | 6398 | 6401 | 6401 | 6401 |
| | Cost of Nondestructive Disassembly | *CND* | — | 1032 | 1032 | 1032 | 1032 | 1032 | 1032 | 1032 |
| | Number of Reused Items | *NRU* | — | 21507 | 21507 | 21507 | 21507 | 21507 | 21507 | 21507 |
| | Number of Recycled Items (In-plant) | *NIRE* | — | 27500 | 27500 | 27500 | 27500 | 27500 | 27500 | 27500 |
| | Number of Recycled Items (Contractor) | *NORE* | — | 5900 | 5900 | 5900 | 5980 | 5901 | 5926 | 27500 |
| | Number of Recycled Items (Total) | *NRC* | — | 33400 | 33400 | 33400 | 33400 | 33400 | 33400 | 33400 |

is in determining correct weights and correct goals. The more recent lexicographic (preemptive) approach (as discussed in section 8.3) ranks goals in order of importance. The key advantage of the lexicographic approach is the disappearance of the weight between different-priority goals. Unfortunately, this advantage comes at a cost in that a higher-priority goal is assumed to be infinitely more important than a lower priority goal.

The major advantage of the utility function approach is that if the utility function is correctly assessed and used, it will ensure the most satisfactory solution to the DM. The solution will be a point at which the nondominated solution-curve and the indifference-curve (i.e., contours of equal utility) are tangent to each other. Thus the solution will have the highest utility to the DM, and will also be nondominated. Unfortunately, no general methodical approach exists for forming utility functions in a multiobjective setting. To fully understand the associated difficulties, we observe that the DM is required to articulate the preferential judgment in an information maze. Unlike the single objective utility function case, the multiobjective utility function must properly account for the intersensitivity (or interdependence) among its individual terms/criteria (Hwang and Masud, 1979). The key distinguishing feature of LPP is that the DM is entirely removed from the process of choosing weights.

### 8.4.2 Procedural Overview Of Linear Physical Programming

Within the linear physical programming procedure, the DM expresses his/her preferences with respect to each criterion using four different *classes*. The decision variable vector is denoted as $x$, and the generic criterion $u$ as $g_u(x)$. The value of the criterion under consideration, $g_u$, is on the horizontal axis, and the function that will be minimized for that criterion, $z_u$, hereby called the *class function*, is on the vertical axis. A lower value of the class function is better than (more valuable than) a higher value thereof. The ideal value of the class function is zero. Each *class* comprises two cases, *hard* and *soft*, referring to the sharpness of the preference. All *soft* class functions become constituent components of the aggregate objective function that is minimized. See Figure 8.2 for both the qualitative and quantitative depiction of each *soft class* function.

The linear physical programming application procedure entails four concise and distinct steps:

1. For each criterion, the DM determines which one of the four *soft* and *hard* classes apply.
2. For each criterion, the DM defines the limits of the ranges of differing degrees of desirability: the *target*-values. (For classes 1S through 4S, there are respectively five, five, nine, and ten such values. For classes 1H through 4H, these values are respectively $t_{p,\max}$ ; $t_{p,\min}$ ; $t_{p,\text{val}}$ ; and $t_{p,\min}$ and $t_{p,\max}$.)
3. Given the DM's input in the form of range boundaries (or targets) for each criterion (or goal), the following LPP weight algorithm is used to generate the weights (Messac, Gupta and Akbulut, 1996).

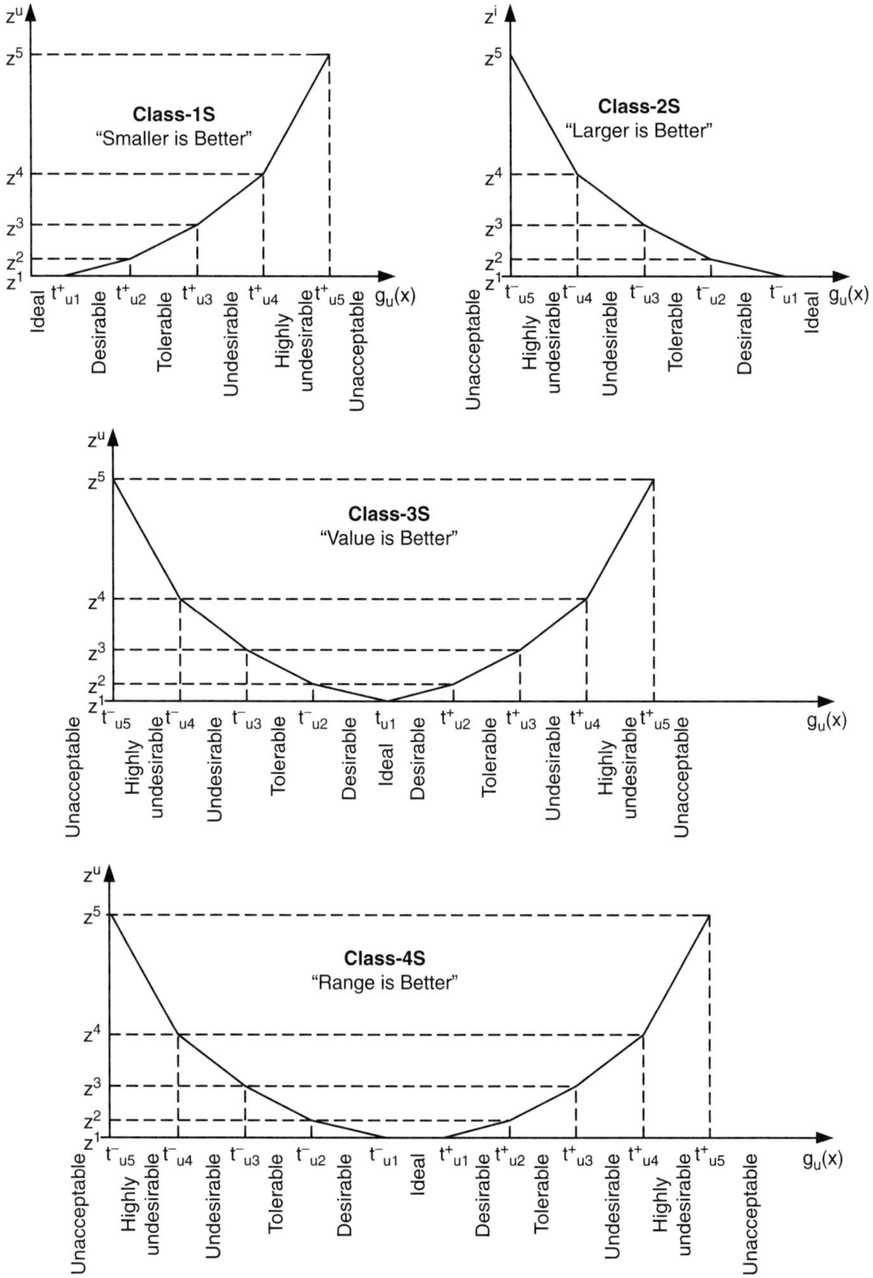


**FIGURE 8.2** Soft class functions.

STEP # Action ('$q$' values correspond to *soft* criteria)

I. Initialize:

$$\beta = 1.1; w_{q1}^{+} = 0, w_{q1}^{-} = 0; \tilde{z}^{2} = \text{small positive number (say, 0.1)}$$

$$q = 0;\ s = 1,\ n_{sc} = \text{number of } soft \text{ criteria.}$$

II. Set $q = q + 1$

III. Set $s = s + 1$

Evaluate, in sequence,

$$\tilde{z}^{s}, \tilde{t}_{qs}^{+}, \tilde{t}_{qs}^{-}, w_{qs}^{+}, w_{qs}^{-}, \tilde{w}_{qs}^{+}, \tilde{w}_{qs}^{-}, \tilde{w}_{\min}$$

If $\tilde{w}_{\min}$ is less than some chosen small positive number (say, 0.01), then increase $\beta$, and go to step II.

IV. If $s \neq 5$, go to step III.

V. If $i \neq n_{sc}$, go to step II.

4. The following LP problem is then solved:

$$\min_{d_{qs}^{-}, d_{qs}^{+}, x} J = \sum_{q=1}^{n_{sc}} \sum_{s=2}^{5} \left( \tilde{w}_{qs}^{-} d_{qs}^{-} + \tilde{w}_{qs}^{+} d_{qs}^{+} \right) \tag{8.37}$$

subject to

$$g_q - d_{qs}^{+} \leq t_{q(s-1)}^{+}; d_{qs}^{+} \geq 0; g_q \leq t_{q5}^{+},\ \forall\ q \text{ in 1S, 3S, 4S,}$$
$$q = 1,2,\ldots,n_{sc},\ s = 2,\ldots,5 \tag{8.38}$$

$$g_q + d_{qs}^{-} \geq t_{q(s-1)}^{-}; d_{qs}^{-} \geq 0; g_q \geq t_{q5}^{-},\ \forall\ q \text{ in 2S, 3S, 4S,}$$
$$q = 1,2,\ldots,n_{sc},\ s = 2,\ldots,5 \tag{8.39}$$

and

$$g_p \leq t_{p,\max} \quad \forall\ p \text{ in classes 1H, } p = 1,2,\ldots,n_{hc} \tag{8.40}$$

$$g_p \geq t_{p,\min} \quad \forall\ p \text{ in classes 2H, } p = 1,2,\ldots,n_{hc} \tag{8.41}$$

$$g_p = t_{p,\text{value}} \quad \forall\ p \text{ in classes 3H, } p = 1,2,\ldots,n_{hc} \tag{8.42}$$

$$t_{p,\min} \leq g_p \leq t_{p,\max} \quad \forall\ p \text{ in classes 4H, } p = 1,2,\ldots,n_{hc} \tag{8.43}$$

$$x_{\min} \leq x \leq x_{\max} \tag{8.44}$$

where $\beta$ is a convexity parameter; $q$ is the index for the soft criteria; $p$ is the index for the hard criteria; $s$ is the index for the range; $w_{qs}^{+}$ and $w_{qs}^{-}$ are positive and negative weights, respectively, for criteria $q$ in the $s$th range; $z_q$ is the class function for criteria $q$; $\tilde{z}^s$ is the change in $z_q$ that takes place as one travels across the $s$th range; $\tilde{t}^+_{qs}$ and $\tilde{t}^-_{qs}$ are the lengths of the $s$th ranges on the positive and negative sides of the $q$th criteria; $\tilde{w}^+_{qs}$ and $\tilde{w}^-_{qs}$ are positive and negative normalized weights, respectively, for criteria $q$ in the $s$th range; $\tilde{w}_{\min}$ is the minimum of $\tilde{w}^+_{qs}$ and $\tilde{w}^-_{qs}$; $d^+_{qs}$ and $d^-_{qs}$ are positive and negative deviational variables, respectively; $J$ is the objective function that has to be minimized; and $n_{sc}$ and $n_{hc}$ denote the number of soft and hard criteria, respectively.

## 8.4.3 Linear Physical Programming Model for a DTO System

In this section, we present a DTO system to determine the number of EOL products to process to fulfill a certain demand for products, parts and/or materials under a variety of objectives and constraints using an LPP model. Besides addressing problems involving multiple objectives and constraints, the LPP model allows the decision maker to express his or her value-system in a realistic manner for each objective of interest. The model also provides the number of items to be disassembled for remanufacturing, recycling, storage, and disposal. We illustrate the impementation of the model using a case example (Kongar and Gupta, 2002b).

### 8.4.3.1 A Case Example

Consider four computer workstations (hereby called Product I, Product II, Product III, and Product IV, which are designated by index $i$ = 1, 2, 3, and 4, respectively) that are available as EOL products (see Figure 8.3 for their product structures). Table 8.6 exhibits the data related to the case example. Note that the four products are made up of various combinations of 17 different items. Table 8.6 displays the component commonality data ($Q_{ij}$) for each item $j$ in product $i$. For each item $j$, Table 8.6 also shows its resale item value, volume, item demand as part, disposal cost, recycling cost, nondestructive disassembly time, and destructive disassembly time, respectively. Additional data is given as follows: Demand for recycling ($DRE_j$) is 100 units/item for all $j$. Recycling revenue ($PM_j$) is \$1/unit for each item $j$. Take back cost for each product $UTB_i$ = \${35, 48, 37, 40}/product, respectively. Unit transportation cost for each product from collectors to facility $UCTRCF_i$ = \${10, 20, 10, 15}/product, respectively. Unit transportation cost for each item from facility to outside recycler $UCTRFR_j$ = \$20/item for all $j$. Unit transportation cost for each item from facility to disposal site $UCTRFD_j$ = \$15/item for all $j$. Unit transportation cost for each item from facility to storage $UCTRFS_j$ = \$19/item for all $j$. Unit preparation cost for each product $UCAC$ = \${5, 4, 5, 4}/product respectively. $cdd$ = \$12.5/h and $cd$ = \$14.59/h. Available space ($AS$) is given as 500,000 cubic cm.

We consider four intangible measures as follows:

*Degree of Environmental Damage* (a function of hazardous material content in each item ($ued_{ij}$), represented on a 10-point scale with a low value representing a lower degree of environmental damage and a high value representing a higher degree of environmental damage).

*Degree of Environmental Benefit* (a function of environmental benefit from recycling of each item ($ueb_{ij}$), represented on a 10-point scale with a low

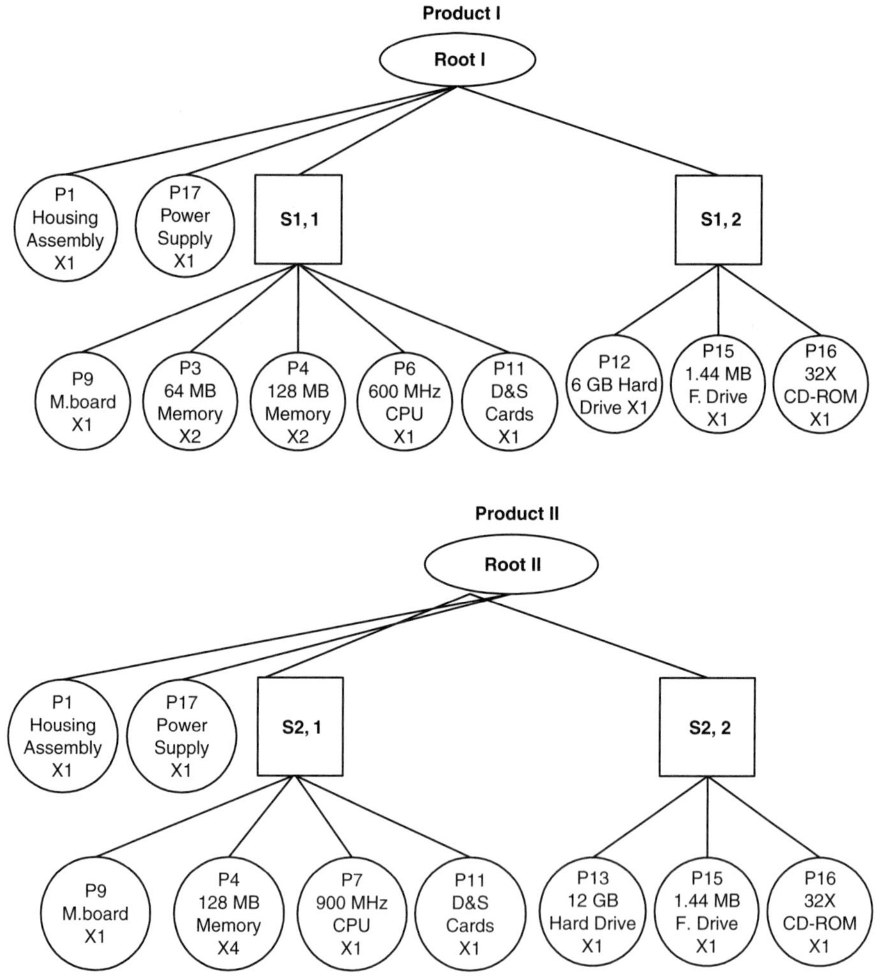


**FIGURE 8.3** EOL product structures.

value representing a lower degree of environmental benefit and a high value representing a higher degree of environmental benefit).

*Degree of Customer Satisfaction* (a function of the technological sophistication and other physical characteristics of each item ($ucs_{ij}$), represented on a 10-point scale with a low value representing a lower degree of customer satisfaction and a high value representing a higher degree of customer satisfaction).

*Degree of Quality Achievement* (a function of reusability of each item over time ($uqa_{ij}$), represented on a 10-point scale with a low value representing a lower quality item and a high value representing a higher quality item).

Table 8.7 exhibits the intangible measures of each item $j$ regardless of the product type $i$.

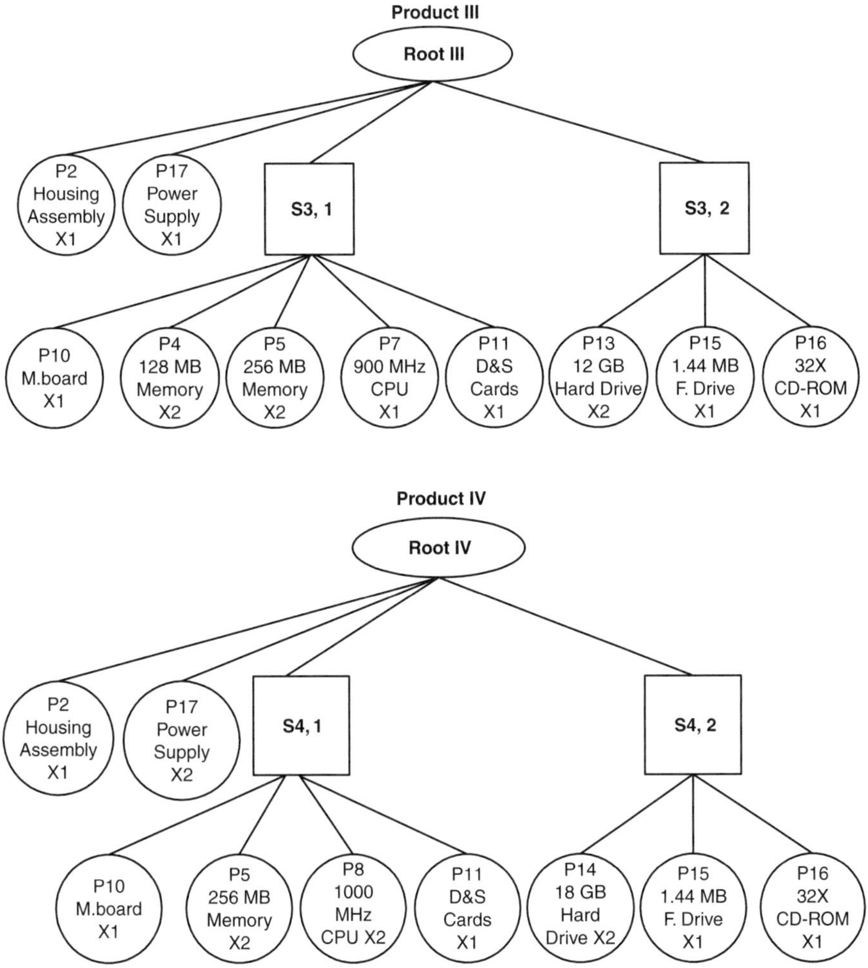


**FIGURE 8.3** (Continued)

#### 8.4.3.2 The Goals

We define goals 1, 2, 3, 4, 5, and 8 as Class-2S type (i.e., "Larger is Better"). Hence,

$$g_1 - d^+_{1s} \le t^+_{1(s-1)}; d^+_{1s} \ge 0; g_1 \le t^+_{15},\ s = 2, \ldots, 5 \tag{8.45}$$

$$g_2 - d^-_{2s} \ge t^-_{2(s-1)}; d^-_{2s} \ge 0; g_2 \ge t^-_{25},\ s = 2, \ldots, 5 \tag{8.46}$$

$$g_3 - d^-_{3s} \ge t^-_{3(s-1)}; d^-_{3s} \ge 0; g_3 \ge t^-_{35},\ s = 2, \ldots, 5 \tag{8.47}$$

**TABLE 8.6[a]**
**Economics and Production Data**

| j | $Q_{1j}$ (units) | $Q_{2j}$ (units) | $Q_{3j}$ (units) | $Q_{4j}$ (units) | $PRM_j$ ($/unit) | $V_j$ (cu. cm/unit) | $D_j$ (units) | $UCDI_j$ ($/unit) | $UCRE_j$ ($/unit) | $dt_j$ (hrs) | $ddt_j$ (hrs) |
|---|---|---|---|---|---|---|---|---|---|---|---|
| 1 | 1 | 1 | — | — | 50 | 100 | — | 5 | 5 | .005 | .004 |
| 2 | — | — | 1 | 1 | 50 | 120 | — | 5 | 5 | .006 | .005 |
| 3 | 2 | — | — | — | 100 | 100 | — | 4 | 4 | .002 | .001 |
| 4 | 2 | 4 | 2 | — | 250 | 150 | 390 | 4 | 4 | .001 | .001 |
| 5 | — | — | 2 | 4 | 300 | 100 | 390 | 5 | 4 | .004 | .004 |
| 6 | 1 | — | — | — | 100 | 150 | — | 5 | 5 | .003 | .002 |
| 7 | — | 1 | 1 | — | 250 | 100 | 95 | 6 | 5 | .004 | .003 |
| 8 | — | — | — | 2 | 320 | 150 | 150 | 5 | 6 | .005 | .004 |
| 9 | 1 | 1 | — | — | 250 | 150 | 95 | 5 | 3 | .002 | .001 |
| 10 | — | — | 1 | 1 | 280 | 100 | 100 | 4 | 5 | .006 | .005 |
| 11 | 1 | 1 | 1 | 1 | 190 | 100 | — | 5 | 5 | .004 | .003 |
| 12 | 1 | — | — | — | 155 | 150 | — | 5 | 5 | .003 | .002 |
| 13 | — | 1 | 2 | — | 180 | 150 | 140 | 4 | 6 | .003 | .002 |
| 14 | — | — | — | 2 | 200 | 100 | 100 | 5 | 6 | .004 | .003 |
| 15 | 1 | 1 | 1 | 1 | 100 | 100 | 195 | 4 | 5 | .003 | .002 |
| 16 | 1 | 1 | 1 | 1 | 250 | 150 | — | 5 | 5 | .003 | .002 |
| 17 | 1 | 1 | 1 | 2 | 180 | 120 | — | 5 | 5 | .003 | .002 |

[a] *Adapted from* (Veerakamolmal et al., 2001).

**TABLE 8.7**
**Initial EOL Data for Products**

| j | $ued_{ij}$ | $ueb_{ij}$ | $ucs_{ij}$ | $uqa_{ij}$ |
|---|---|---|---|---|
| 1 | 5 | 5 | 5 | 5 |
| 2 | 5 | 5 | 10 | 10 |
| 3 | 5 | 5 | 1 | 10 |
| 4 | 5 | 5 | 5 | 10 |
| 5 | 5 | 5 | 10 | 10 |
| 6 | 5 | 5 | 5 | 5 |
| 7 | 5 | 5 | 5 | 10 |
| 8 | 5 | 5 | 5 | 10 |
| 9 | 10 | 5 | 5 | 5 |
| 10 | 10 | 5 | 5 | 5 |
| 11 | 5 | 5 | 5 | 1 |
| 12 | 5 | 1 | 1 | 10 |
| 13 | 5 | 1 | 5 | 10 |
| 14 | 5 | 1 | 10 | 10 |
| 15 | 5 | 1 | 1 | 10 |
| 16 | 5 | 1 | 5 | 10 |
| 17 | N/A | 1 | 5 | 5 |

*Note:* N/A = Not Allowed

$$g_4 - d^-_{4s} t^-_{4(s-1)}; d^-_{4s} \geq 0; g_4 \geq t^-_{45}\text{, s} = 2,\ldots,5 \tag{8.48}$$

$$g_5 - d^-_{5s} \geq t^-_{5(s-1)}; d^-_{5s} \geq 0; g_5 \geq t^-_{55}\text{, s} = 2,\ldots,5 \tag{8.49}$$

$$g_8 - d^-_{8s} \geq t^-_{8(s-1)}; d^-_{8s} \geq 0; g_8 \geq t^-_{85}\text{, s} = 2,\ldots,5 \tag{8.50}$$

Goals 6 and 7 are defined as Class-4S type (i.e., "Range is Better"). Hence

$$g_6 - d^+_{6s} \leq t^+_{6(s\text{-}1)}; d^+_{6s} \geq 0; g_6 \leq t^+_{65}\text{, s} = 2,\ldots,5 \tag{8.51}$$

$$g_6 - d^-_{6s} \geq t^-_{6(s-1)}; d^-_{6s} \geq 0; g_6 \geq t^-_{65}\text{, s} = 2,\ldots,5 \tag{8.52}$$

$$g_7 - d^+_{7s} \leq t^+_{7(s-1)}; d^+_{7s} \geq 0; g_7 \leq t^+_{75}\text{, s} = 2,\ldots,5 \tag{8.53}$$

$$g_7 - d^-_{7s} \geq t^-_{7(s-1)}; d^-_{7s} \geq 0; g_7 \geq t^-_{75}\text{, s} = 2,\ldots,5 \tag{8.54}$$

Goal 9 is defined as Class-1S, (i.e., "Smaller is Better"). Hence

$$g_9 - d^+_{9s} \leq t^+_{9(s-1)}; d^+_{9s} \geq 0; g_9 \leq t^+_{95}\text{, s} = 2,\ldots,5 \tag{8.55}$$

The first goal ($g_1$) is for average customer satisfaction (*CS*), which is a ratio of total customer satisfaction (*TCS*) and the number of reused items (*NRES*) plus recycled items (*NRC*). Therefore, we can write

$$CS = TCS/(NRES + NRC) \tag{8.56}$$

where *TCS* is the sum of customer satisfaction levels for each item reused and recycled $(\Sigma_{ij} ucs_{ij})$. Therefore

$$TCS = \sum_{ij} ucs_{ij} \tag{8.57}$$

$$NRES = \sum_{j}\left(\sum_{i} X_{ij}\right) \tag{8.58}$$

$$NRC = \sum_{j}\left(\sum_{i} R_{ij}\right) \tag{8.59}$$

where $X_{ij}$ and $R_{ij}$ are the quantities of disassembled items ($j$) for resale and recycling from products ($i$), respectively.

The second goal ($g_2$) is for average quality achievement (*QA*), which is a ratio of total quality achievement (*TQA*) and the number of reused items. Therefore, we can write

$$QA = TQA/(NRES) \tag{8.60}$$

where *TQA* is the sum of quality achievement levels for each item reused $(\Sigma_{ij} uqa_{ij})$. Therefore

$$TQA = \sum_{ij} uqa_{ij} \tag{8.61}$$

The third goal ($g_3$) is for resale revenue (*RPS*), which is a function of the number of item type $j$ reused $(\Sigma_i X_{ij})$ and the unit sale price for item type $j$ ($PRM_j$). Therefore

$$RPS = \sum_{j}\left(\sum_{i} X_{ij}\right)\cdot PRM_j \tag{8.62}$$

The fourth goal ($g_4$) is for the recycling revenue (*RMS*), which is a function of the number of each item $j$ sold for recycling $(\Sigma_i R_{ij})$ and the selling price of each item $j$ ($PM_j$). Therefore, by summing up the revenue over all items, *RMS* can be obtained as follows:

$$RMS = \sum_{j}\left(\sum_{i} R_{ij}\right)\cdot PM_j \tag{8.63}$$

The fifth goal ($g_5$) is for total profit function (*TPR*), which is the difference between all the revenues and all the costs considered in the model. Therefore, *TPR* can be written as follows:

$$TPR = RMS + RPS - TB - CTRCF - CTRFR - CTRFD - CTRFS \\ - CAC - CDD - CND - CRE - CST - CDI \tag{8.64}$$

where *TB* is a function of the number of EOL products ordered ($Y_i$) and the cost of each product ($UTB_i$). Therefore

$$TB = \sum_i Y_i \cdot UTB_i \tag{8.65}$$

*CTRCF* is a function of the number of EOL products ordered ($Y_i$) and the transportation cost per unit from collectors to the facility ($UCTRCF_i$). Therefore

$$CTRCF = \sum_i Y_i \cdot UCTRCF_i \tag{8.66}$$

*CTRFR* is a function of the number of items sent to the recycling facility $(\Sigma_i R_{ij})$ and the transportation cost per unit from the facility to the recycling facility ($UCTRFR_j$). Therefore

$$CTRFR = \sum_j \left( \sum_i R_{ij} \cdot UCTRFR_j \right) \tag{8.67}$$

*CTRFD* is a function of the number of items sent to disposal (*NDIS*) and the transportation costs per unit from facility to the disposal site ($UCTRFD_j$). The number of items sent to disposal includes the nondemanded items ($L_{ij}$). Therefore

$$NDIS = \sum_j \sum_i (L_{ij}) \tag{8.68}$$

and

$$CTRFD = \sum_j \left( \sum_i L_{ij} \cdot UCTRFD_j \right) \tag{8.69}$$

*CTRFS* is a function of the number of items sent to storage (*NSTR*) and the transportation cost per unit from the facility to the storage location ($UCTRFS_j$). Therefore

$$NSTR = \sum_j \sum_i V_{ij} \tag{8.70}$$

and

$$CTRFS = \sum_j \left( \sum_i V_{ij} \right) \cdot UCTRFS_j \tag{8.71}$$

*CAC* is a function of the number of EOL products ordered ($Y_i$) and the cost of preparing each product ($UCAC_i$). Therefore

$$CAC = \sum_i Y_i \cdot UCAC_i \tag{8.72}$$

*CDD* is the cost of destructive disassembly (considered for the items that are recycled for their material content or the items that are sent to landfills for proper disposal) and is a function of number of items to be recycled and disposed of $(\Sigma_i (R_{ij} + L_{ij}))$, the cost per hour (*cd*) and the time of disassembling each item ($ddt_j$). Therefore

$$CDD = \sum_j \left( \sum_i (R_{ij} + L_{ij}) \right) \cdot cd \cdot ddt_j \tag{8.73}$$

*CND* is the cost of nondestructive disassembly (considered for the items that are reused or the items that are sent to storage) and is a function of number of items to be reused and stored $(\Sigma_i (X_{ij} + V_{ij}))$, the cost per hour (*cnd*) and the time of disassembling each item ($dt_j$). Therefore

$$CND = \sum_j \left( \sum_i (X_{ij} + V_{ij}) \right) \cdot cnd \cdot dt_j \tag{8.74}$$

*CRE* is a function of the number of items recycled in plant $(\Sigma_i R_{ij})$ and the corresponding unit recycling cost ($UCRE_j$). Therefore

$$CRE = \sum_j \left( \sum_i R_{ij} \right) \cdot UCRE_j \tag{8.75}$$

*CST* is a function of the number of stored items $(\Sigma_i V_{ij})$ and the corresponding unit holding cost ($h_j$). Therefore

$$CST = \sum_j \left( \sum_i V_{ij} \right) \cdot h_j \tag{8.76}$$

*CDI* is a function of the number of disposed items $(\Sigma_i L_{ij})$ and the unit disposal cost ($UCDI_j$). Therefore

$$CDI = \sum_j \left( \sum_i L_{ij} \right) \cdot UCDI_j \tag{8.77}$$

The sixth goal ($g_6$) is for the average environmental damage (*ED*), which is a ratio of total environmental damage (*TED*) and the number of disposed items (*NDIS*). Therefore

$$ED = TED/NDIS \tag{8.78}$$

where *TED* is the sum of all environmental damage levels $\Sigma_{ij} ued_{ij}$ for all disposed items. Therefore

$$TED = \sum_{ij} ued_{ij} \tag{8.79}$$

$$NDIS = \sum_{j}\sum_{i} L_{ij} \tag{8.80}$$

The seventh goal ($g_7$) is for the average environmental benefit (*EB*), which is a ratio of total environmental benefit (*TEB*) and the number of recycled items (*NRC*). Therefore

$$EB = TEB/NRC \tag{8.81}$$

where *TEB* is the sum of all environmental benefit levels ($\Sigma_{ij} ueb_{ij}$) for all recycled items. Therefore

$$TEB = \sum_{ij} ueb_{ij} \tag{8.82}$$

$$NRC = \sum_{j}\left(\sum_{i} R_{ij}\right) \tag{8.83}$$

The eighth and ninth goals ($g_8$, $g_9$) are for the number of recycled items (*NRC*) and the number of disposed items (*NDIS*), respectively.

### 8.4.3.3 The Constraints

The model also involves some constraints as follows:

The total space (*TS*) occupied by the stored items has to be less than or equal to the total available space in the warehouse (*AS*). *TS* is a function of the number of stored item *j* ($\Sigma_i V_{ij}$) and its corresponding volume ($v_j$). Therefore

$$TS = \sum_{j}(v_j \cdot \sum_{i} V_{ij}) \tag{8.84}$$

and

$$TS \le AS \tag{8.85}$$

There is also a limit on the number of items that can be disassembled in the facility. This constraint can be expressed mathematically as

$$\sum_i \sum_j (X_{ij} + R_{ij} + I_{ij} + W_{ij}) \leq DCA \tag{8.86}$$

Where $X_{ij}$, $R_{ij}$, and $W_{ij}$ are the quantities of disassembled items (*j*) for resale, recycling, and disposal from products (*i*), respectively. *DCA* denotes the disassembly capacity.

The number of items retrieved from EOL products ordered $(\Sigma_i Y_i . Q_{ij})$ has to be greater than or equal to the number of demanded items ($D_j$). Therefore

$$\sum_i Y_i \cdot Q_{ij} \geq D_j, \forall j \tag{8.87}$$

where

$$D_j \leq \sum_i X_{ij}, \forall j \tag{8.88}$$

The total number of disassembled items $(\Sigma_i Y_i \cdot Q_{ij})$ should be equal to the items that are reused $(\Sigma_i X_{ij})$, recycled $(\Sigma_i R_{ij})$, stored $(\Sigma_i V_{ij})$, and disposed of $(\Sigma_i L_{ij})$, where $Q_{ij}$ is the multiplicity factor. Therefore

$$\sum_i Y_i \cdot Q_{ij} = \sum_i (X_{ij} + R_{ij} + V_{ij} + L_{ij}), \forall j \tag{8.89}$$

All the variables must be nonnegative integers. Thus

$$\{Y_i\}, \{X_{ij}\}, \{R_{ij}\}, \{V_{ij}\}, \{L_{ij}\} \geq 0 \text{ and integer; for all } i \text{ and } j. \tag{8.90}$$

The tolerance limits for each goal are given in Table 8.8.

#### 8.4.3.4 Results of the Case Example

After the LPP weight algorithm is used, the weights shown in Table 8.9 are obtained for each objective.

Using these weights the linear optimization model is built. We solved the LPP mathematical model using LINGO (version 6). The model includes 263 constraints and 368 variables. A LP model with an objective to maximize *TPR* was also run for comparison purposes. The results of both LP and LPP models are provided in Table 8.10.

It is clear from Table 8.10 that the DTO system is highly sensitive to the introduction of the targets. Of the nine goals, only two of them remained unchanged in the LPP model. Quality achievement (*QA*) and revenue from product sales (*RPS*) remained unchanged. *RMS,* however, increased about 60% even though the LPP provides a lower overall profit (*TPR*). This is because the related cost values also increase while trying to achieve higher recycling material sale revenue (such as transportation, disassembly cost and so on).

**TABLE 8.8**
**Targets and LPP Classes for the DTO System**

| | | | | | | |
|---|---|---|---|---|---|---|
| $t^+_{91}$ | 10 | | | | | Class-1S |
| $t^+_{92}$ | 20 | | | | | |
| $t^+_{93}$ | 100 | | | | | |
| $t^+_{94}$ | 250 | | | | | |
| $t^+_{95}$ | 350 | | | | | |
| $t^-_{11}$ | 9 | $t^-_{21}$ | 8 | $t^-_{31}$ | 300,000 | Class-2S |
| $t^-_{12}$ | 8 | $t^-_{22}$ | 7 | $t^-_{32}$ | 250,000 | |
| $t^-_{13}$ | 7.5 | $t^-_{23}$ | 5 | $t^-_{33}$ | 200,000 | |
| $t^-_{14}$ | 7 | $t^-_{24}$ | 4 | $t^-_{34}$ | 100,000 | |
| $t^-_{15}$ | 6.5 | $t^-_{25}$ | 3 | $t^-_{35}$ | 50,000 | |
| $t^-_{41}$ | 6,000 | $t^-_{51}$ | 250,000 | $t^-_{81}$ | 400 | |
| $t^-_{42}$ | 5,000 | $t^-_{52}$ | 230,000 | $t^-_{82}$ | 300 | |
| $t^-_{43}$ | 3,000 | $t^-_{53}$ | 180,000 | $t^-_{83}$ | 200 | |
| $t^-_{44}$ | 2,000 | $t^-_{54}$ | 80,000 | $t^-_{84}$ | 150 | |
| $t^-_{45}$ | 1,000 | $t^-_{55}$ | 40,000 | $t^-_{85}$ | 100 | |
| $t^-_{61}$ | 4 | $t^+_{61}$ | 5 | | | Class-4S |
| $t^-_{62}$ | 3 | $t^+_{62}$ | 6 | | | |
| $t^-_{63}$ | 2 | $t^+_{63}$ | 7 | | | |
| $t^-_{64}$ | 1 | $t^+_{64}$ | 8 | | | |
| $t^-_{65}$ | 0 | $t^+_{65}$ | 10 | | | |
| $t^-_{71}$ | 4 | $t^+_{71}$ | 5 | | | |
| $t^-_{72}$ | 3 | $t^+_{72}$ | 6 | | | |
| $t^-_{73}$ | 2 | $t^+_{73}$ | 7 | | | |
| $t^-_{74}$ | 1 | $t^+_{74}$ | 8 | | | |
| $t^-_{75}$ | 0 | $t^+_{75}$ | 10 | | | |

**TABLE 8.9**
**LPP Weights for DTO System Obectives**

| | | | |
|---|---|---|---|
| $\tilde{w}^-_{12} = 0.698$ | $\tilde{w}^-_{13} = 0.258$ | $\tilde{w}^-_{14} = 0.038$ | $\tilde{w}^-_{15} = 0.006$ |
| $\tilde{w}^-_{22} = 0.925$ | $\tilde{w}^-_{23} = 0.053$ | $\tilde{w}^-_{24} = 0.019$ | $\tilde{w}^-_{25} = 0.003$ |
| $\tilde{w}^-_{32} = 0.849$ | $\tilde{w}^-_{33} = 0.140$ | $\tilde{w}^-_{34} = 0.008$ | $\tilde{w}^-_{35} = 0.003$ |
| $\tilde{w}^-_{42} = 0.925$ | $\tilde{w}^-_{43} = 0.053$ | $\tilde{w}^-_{44} = 0.019$ | $\tilde{w}^-_{45} = 0.003$ |
| $\tilde{w}^-_{52} = 0.940$ | $\tilde{w}^-_{53} = 0.057$ | $\tilde{w}^-_{54} = 0.002$ | $\tilde{w}^-_{55} = 0.001$ |
| $\tilde{w}^-_{62} = 0.376$ | $\tilde{w}^-_{63} = 0.038$ | $\tilde{w}^-_{64} = 0.041$ | $\tilde{w}^-_{65} = 0.045$ |
| $\tilde{w}^+_{62} = 0.351$ | $\tilde{w}^+_{63} = 0.042$ | $\tilde{w}^+_{64} = 0.049$ | $\tilde{w}^+_{65} = 0.058$ |
| $\tilde{w}^-_{72} = 0.376$ | $\tilde{w}^-_{73} = 0.038$ | $\tilde{w}^-_{74} = 0.041$ | $\tilde{w}^-_{75} = 0.045$ |
| $\tilde{w}^+_{72} = 0.351$ | $\tilde{w}^+_{73} = 0.042$ | $\tilde{w}^+_{74} = 0.049$ | $\tilde{w}^+_{75} = 0.058$ |
| $\tilde{w}^-_{82} = 0.849$ | $\tilde{w}^-_{83} = 0.106$ | $\tilde{w}^-_{84} = 0.039$ | $\tilde{w}^-_{85} = 0.006$ |
| $\tilde{w}^+_{92} = 0.001$ | $\tilde{w}^+_{93} = 0.003$ | $\tilde{w}^+_{94} = 0.036$ | $\tilde{w}^+_{95} = 0.960$ |

**TABLE 8.10**
**LP and LPP Model Results for the Goals**

| # | Goal | LP | LPP |
|---|---|---|---|
| 1 | CS | 5.122 | 9.852 |
| 2 | *QA* | 9.411 | 9.411 |
| 3 | *RPS* | 402,700 | 402,700 |
| 4 | *RMS* | 8,520 | 13,408 |
| 5 | *TPR* | 366,382 | 250,000 |
| 6 | *ED* | 6.75 | 4.75 |
| 7 | *EB* | 2.191 | 6 |
| 8 | *NRC* | 1,405 | 3,356 |
| 9 | *NDIS* | 270 | 10 |

Customer satisfaction (*CS*) has increased to 9.852 (high) from 5.122 (medium). Intangible variables reflect the effect of the LPP programming more clearly than the tangible ones. Environmental Damage (*ED*) resulted in 4.75 points on a ten point scale and is in the ideal range for a Class –4S function. Environmental Benefit (*EB*) has also been raised up to 6 from 2.191 points, reaching the ideal range.

The number of EOL products to be disassembled also varies depending on the model. As for the LPP model, 124, 0, 195, and 412 units of EOL products I, II, III, and IV should be taken-back, respectively. The corresponding values are 90, 5, 145, and 50 units, respectively for the LP model. As expected, the total profit also varies depending on the solution methodology. The LPP results in a \$250,000 profit while LP provides a higher profit of \$366,382. The revenue from the material sales (*RMS*) are \$13,408 and \$8,520 for LPP and LP, respectively.

## 8.5 CONCLUSION

In this chapter, we introduced a disassembly-to-order system that sought environmentally benign solutions under a variety of constraints and goals. The decision for selecting the most appropriate goals is based on the decision-maker's preference. In an environmentally conscious manufacturing environment, it is no longer realistic to use a single objective (such as maximization of profit) because of the introduction of restrictive regulations and the desire to protect the environment. This makes the decision procedure more complicated and introduces a motive to fulfill multiple objectives. The models presented in this chapter were an attempt to achieve a desired level of profit while also satisfying additional goals simultaneously.

## REFERENCES

Charnes, A., Cooper, W.W., and Ferguson, R.O., 1955, Optimal estimation of executive compensation by linear programming. *Management Science*, **1**(2), 138–151.

Charnes, A. and Cooper, W.W., 1961, *Management Models and Industrial Applications of Linear Programming*. **I**, John Wiley and Sons, New York.

Cohon, J. L., 1978, *Multiobjective Programming and Planning*. Mathematics in Science and Engineering, **140**, Academic Press, 163–179.

Hwang, C.L. and Masud, A.S.M., 1979, *Multiple Objective Decision Making: Methods and Applications*. Springer-Verlag, Berlin.

Ignizio, J.P., 1976, *Goal Programming and Extensions*. Lexington Books, D. C. Heath and Company, Massachusetts, Toronto, London.

Ignizio, J.P., 1982, *Linear Programming in Single and Multiple Objective Systems*. Englewood Cliffs, NJ: Prentice Hall.

Ijiri, Y., 1965, *Management Goals and Accounting for Control*. Rand McNally, Chicago.

Kongar, E. and Gupta, S.M., 2002a, A multi-criteria decision making approach for disassembly-to-order systems. *Journal of Electronics Manufacturing*, **11**(2), 171–183.

Kongar, E. and Gupta, S.M., 2002b, Disassembly-to-order system using linear physical programming. *Proceedings of IEEE Symposium on Electronics and the Environment*, 312–317.

Lambert, A.J.D. and Gupta, S.M., 2002, Demand-driven disassembly optimization for electronic consumer goods. *Journal of Electronics Manufacturing*, **11**(2), 121–135.

Lambert, A.J.D., 2003, Disassembly Sequencing: a survey. *International Journal of Production Research*, **41**(16), 3721–3759.

Lee, S.M., 1972, *Goal Programming for Decision Analysis*. Auerbach Publishers, Philadelphia.

Messac, A., Gupta, S.M., and Akbulut, B., 1996, Linear physical programming: effective optimization for complex linear systems. *Transactions on Operational Research*, **8**(2), 39–59.

Romero, C., 1991, *Handbook of Critical Issues in Goal Programming*. Oxford, Pergamon Press.

Schniederjans, M.J., 1995, *Goal Programming: Methodology and Applications*. Boston, MA: Kluwer.

Stadler, S., 1988, Multicriteria optimization in engineering and in the sciences. *Mathematical Concepts and Methods in Science and Engineering Series*, **37**, Plenum Press, 120–121.

Steuer, R.E., 1989, *Multiple Criteria Optimization: Theory, Computation, and Application*. Reprint Edition, Krieger Publishing Company, Malabar, FL.

Veerakamolmal, P. and Gupta, S.M., 1998a, Design of integrated component recovery system. *Proceedings of IEEE Symposium on Electronics and the Environment*, 264–269.

Veerakamolmal, P. and Gupta, S.M., 1998b, High-mix/low-volume batch of electronic equipment disassembly. *Computers and Industrial Engineering*, **35**(1–2), 65–68.

Veerakamolmal, P. and Gupta, S.M., 1998c, Optimal analysis of lot-size balancing for multiproducts selective disassembly. *International Journal of Flexible Automation and Integrated Manufacturing*, **6**(3–4), 245–269.

Veerakamolmal, P. and Gupta, S.M., 1999, Analysis of design efficiency for the disassembly of modular electronic products. *Journal of Electronics Manufacturing*, **9**(1), 79–95.

Veerakamolmal, P., Lee, Y.-J., Fasano, J.P., Hale, R., and Jacques, M., 2001, Cost-benefit study of consumer product take back programs using IBM's WIT reverse logistics optimization tool. *Proceedings of the SPIE International Conference on Environmentally Conscious Manufacturing* II, 13–22.

# 9 Disassembly Line Balancing Problems

## 9.1 INTRODUCTION

The preceding chapter articulated the difference between disassembly sequencing (how to disassemble) and disassembly planning (how much to disassemble). While chapters 4 through 7 mostly dealt with disassembly sequencing, chapter 8 dealt with disassembly planning. This chapter discusses the disassembly line-balancing problem, which falls within the domains of both disassembly sequencing and disassembly planning.

A disassembly line setting provides the most efficient way to perform disassembly. It is the best choice for automated disassembly, a feature that is essential for disassembling products in large quantities (Gungor and Gupta, 2002). A disassembly line represents an ordered sequence of workstations where a specific set of predetermined disassembly tasks are performed at each workstation such that the precedence relationships between the tasks are satisfied and some measures of effectiveness are optimized.

## 9.2 AN ASSEMBLY LINE VS. A DISASSEMBLY LINE

### 9.2.1 Introduction

A disassembly line is much more complex than a traditional assembly line and encounters many challenges. For example, a disassembly line has serious inventory problems due to the disparity between the demands for certain components or subassemblies and their yield from the disassembly process. The flow process is also peculiar. As opposed to the normal "convergent" flow in regular assembly environment, in disassembly, the flow process is "divergent" (a single product is broken down into many subassemblies and components) (Brennan et al., 1994). There is also a high degree of uncertainty in the structure and the quality of the returned products. The condition of the products received are usually unknown and the reliability of the components is a suspect. In addition, some components of the product may cause pollution or may even be hazardous. These components tend to have a higher chance of being damaged and hence may require special handling, which can also influence the utilization of the disassembly workstations. Various types of demand sources may also lead to difficulties in managing a disassembly line. For example, unlike in the traditional assembly line where the external demand normally occurs only at the last station, the demand in the disassembly case usually occurs at any of the intermittent stations. The reason for this is that as the product moves on the disassembly line, various components are disassembled at every station and accumulated at that station. The demands for those components are fulfilled from that station. Thus, there could be as many demand sources as there are number of stations. This process creates unique, intriguing, and unpredictable conditions.

### 9.2.2 Challenges

It is important that a disassembly line be balanced so that it can work as efficiently as possible. Balancing a disassembly line involves assigning disassembly tasks required to disassemble a product to various workstations so that the total times required to accomplish the work to be done at every workstation are as close to each other as possible. Let us take a closer look at various kinds of challenges associated with a disassembly line.

#### 9.2.2.1 Challenges Associated with the Product

Changing characteristics of products complicate the operations on a disassembly line. Balancing the disassembly line used in such cases can be very complex. Such a line may be balanced for a group of products yet may become unbalanced when a new type of product is received.

#### 9.2.2.2 Challenges Associated with the Disassembly Line

Various line configurations may be possible. They are used to cope with the irregularities and product variability in the disassembly system. One important consideration is the line speed. It can be dynamically modified to minimize the effects of varying demands for subassemblies and/or components on the disassembly line.

#### 9.2.2.3 Challenges Associated with the Components

1. *Quality of incoming products:* There is a high level of uncertainty in the quality of the products received and their constituent components. They may be either physically defective or functionally defective or both.
2. *Quantity of components in incoming products:* Due to upgrading or downgrading of the product during its use, the actual number of components in it may be more or less than expected when the product is received.

#### 9.2.2.4 Operational Challenges

1. *Variability of disassembly task times*: The disassembly task times may vary depending on several factors that are related to the condition of the product and the state of the disassembly workstation (or worker). Dynamic learning is possible, which allows systematic reduction in disassembly times.
2. *Early leaving work-pieces*: If one or more (not all) tasks of a work-piece, which have been assigned to the current workstation, cannot be completed due to some defect (that might be related to one or more of the tasks),

the work-piece might leave the workstation early. We term this phenomenon as the *early-leaving work-piece* (*EP*). Due to EP, the workstation experiences an unscheduled idle time for the duration of the tasks that causes the work-piece to leave early.

3. *Self-skipping work-pieces*: If all tasks of a work-piece, which have been assigned to the current workstation, are disabled due to some defect of their own and/or precedence relationships, the work-piece leaves the workstation early without being worked on. We term this phenomenon as *self-skipping work-piece* (*SSWP*).
4. *Skipping work-pieces*: At workstation $m$, if one or more defective tasks of a work-piece directly or indirectly precede all the tasks of workstation $m + 1$ (i.e., the workstation immediately succeeding workstation $m$), the work-piece "skips" workstation $m + 1$ and moves on to workstation $m + 2$. We term this phenomenon as *skipping work-piece* (*SWP*). In addition to unscheduled idle time, both SSWP and SWP experience added complexities in material handling and the status of the downstream workstation.
5. *Disappearing work-pieces*: If a defective task disables the completion of all the remaining tasks on a work-piece, the work-piece may simply be taken off the disassembly line before it reaches any downstream workstation. In other words, the work-piece "disappears"! Therefore, we term this phenomenon as the *disappearing work-piece* (*DWP*). DWP may result in starvation of subsequent workstations leading to a higher overall idle time.
6. *Revisiting work-pieces*: Work-piece currently at workstation $w$, may *revisit* a preceding workstation ($w$-$\alpha$) where ($w$-$\alpha$) $\geq 1$ and $\alpha \geq 1$ and integer, to perform task $f$ if the completion of current task $i$ enables one to work on task $f$ which was originally assigned to workstation ($w$-$\alpha$), and was, however, disabled due to the failure of another preceding task. We term this as *revisiting work-pieces* (*RWP*). An RWP results in overloading one of the previous workstations.
7. *Exploding work-pieces*: A work-piece may split into two or more work-pieces (subassemblies) as it moves on the disassembly line due to the disassembly of certain components that hold the work-piece together. Each of these subassemblies acts as an individual work-piece on the disassembly line. We term this phenomenon as the *exploding work-pieces* (*EWP*). The EWP complicates the flow mechanism of the disassembly line.

#### 9.2.2.5 Demand Challenges

In disassembly, the following demand scenarios are possible: Demand for one component only (single component disassembly—a special case of selective disassembly); demand for multiple components (selective disassembly); and demand for all components (complete disassembly). Possible physical and functional defects in the demanded components or the components preceding the demanded components may complicate the situation further.

#### 9.2.2.6 Assignment Challenges

Certain tasks must be grouped and assigned to a specific workstation for reasons such as requirement of similar operating conditions for them and availability of special machining and tooling at certain workstations.

#### 9.2.2.7 Other Challenges

There are additional uncertainty factors associated with the reliability of the disassembly workstations. For example, hazardous components may require special handling, which can also influence the utilization of the workstations. Some of the assembly line balancing factors, which are presented by Ghosh and Gagnon (1989) in their comprehensive literature survey, can also be important in the disassembly line balancing case.

A comparison of the assembly and the disassembly lines both from technical and operational points of view is summarized in Table 9.1. It clearly demonstrates that there are a lot of differences between the two lines and that disassembly lines are much more complex.

## 9.3 BALANCING A DISASSEMBLY LINE

As mentioned before, balancing a disassembly line is vital for it to work efficiently. This necessitates the development of fast and dynamic disassembly line balancing methods. However, balancing a disassembly line is difficult because it is an NP-hard problem (see Tovey, 2002 for an explanation of such problems). While exhaustive search or exact methods could hypothetically provide the problem's optimal solution, its exponential time complexity quickly reduces the practicality of such methods. Therefore, heuristic and metaheuristic methods are often employed to solve such problems (see Silver, 2004 for an overview of such problems). Heuristic methods aim to find optimal solutions but cannot guarantee them. Instead, these methods often use shortcuts and educated guesses to provide satisfactory results much faster and at low costs. A metaheuristic provides a general framework for heuristics in solving hard problems that can be applied to a wide set of different problems (see Greenberg, 2004). It is for this reason that almost all practical algorithms to solve the assembly or the disassembly line balancing problems are based on heuristic and metaheuristic methods.

As was mentioned in chapter 1, disassembly as a production process was performed in the slaughterhouses more than a century ago. However, the first scientifically oriented paper that was inspired by slaughterhouse lines was published relatively recently (Donnan and Makan, 1983). Although Donnan and Makan titled the paper *disassembly line balancing*, it was more like an assembly line balancing problem, as it had none of the challenges discussed in section 9.2. The first introduction to the disassembly line-balancing problem (DLBP) as it relates to the end-of-life complex products is due to Gungor and Gupta (1999ab, 2001, 2002). The basic DLBP can be stated as the assignment of disassembly tasks to an ordered sequence of workstations such that precedence relationships are satisfied and some measures of effectiveness are optimized.

The basic objective of the DLBP is to utilize the resources of the disassembly line as efficiently as possible. Efficient utilization of resources consists of finding the

**TABLE 9.1**
**Comparison of Operational and Technical Considerations of Assembly and Disassembly Lines**

| Line Considerations | Assembly Line | Disassembly Line |
|---|---|---|
| Demand | Dependent | Dependent |
| Demand sources | Single | Multiple |
| Demanded entity | End product | Individual components/ subassemblies |
| Precedence relationships | Yes | Yes |
| Complexity related to precedence relationships | High (physical and functional precedence constraints) | Moderate (mostly physical constraints) |
| Uncertainty related to quality of components | Low | High |
| Uncertainty related to quantity of components | Low | High |
| Uncertainty related to workstations and the material handling system | Low to Moderate | High |
| Reliability of the workstations and the material handling system | High | Low |
| Multiple products | Yes | Yes |
| Flow process | Convergent | Divergent |
| Line flexibility | Low to Moderate | High |
| Layout alternatives | Multiple | Multiple |
| Complexity of performance measures | Moderate | High |
| Known performance measures | Numerous | N/A |
| Disappearing Work-pieces Phenomena (DWP) | N/A | Yes |
| Exploding Work-pieces Phenomena (EWP) | N/A | Yes |
| Required line robustness[a] | Moderate | High |
| Complexity of "between workstation inventory" handling | Moderate | High |
| Known techniques for line optimization | Numerous | None |
| Problem complexity | NP-hard | NP-hard |

[a] For example, resistance to dirt encountered during the disassembly process.

minimum number of disassembly workstations required and optimally assigning the disassembly tasks to the workstations (such that precedence relations are satisfied) to provide for a given cycle time. A *cycle time* is the amount of time (same for each workstation) allocated to workstations to complete their assigned tasks. The duration of the cycle time depends on factors such as the demand for the components removed from the products and the time required to complete individual disassembly tasks.

## 9.4 DESCRIPTION OF THE DLBP

### 9.4.1 Introduction

We consider the following simple disassembly line to demonstrate the intricacies of the disassembly line balancing procedures. A paced disassembly line is utilized to sequentially disassemble one type of product into its constituent components and subassemblies. We assume that there is a sufficient supply of products. The configuration of each product received is identical which means that the exact quantity of the components in each product received is known. For simplicity, the disassembly times are assumed to be deterministic and known. The precedence relationships among the components are known and must not be violated. Every component in the product may have an associated demand, i.e., complete disassembly is targeted. The demand parameters are deterministic and known. The components disassembled are accepted by the demand source "as is."

The particular disassembly line investigated here seeks to fulfill the following four objectives:

1. Minimize the number of disassembly workstations and hence, minimize the total idle time.
2. Balance the disassembly line (i.e., ensure the idle times at each workstation are similar).
3. Remove hazardous components early in the disassembly sequence.
4. Remove high demand components before low demand components in the case of equal component removal times.

This is an integer, nonlinear, deterministic, multiple criteria decision-making problem with an exponentially growing search space. Testing a given solution against the precedence constraints fulfills the major constraint of precedence preservation.

### 9.4.2 The First Objective

Minimizing the sum of the workstation idle times, which will also minimize the total number of workstations, addresses the first objective. This objective is represented as

$$\text{Minimize } Z_1 = \sum_{j=1}^{NWS} (CT - WS_j) \tag{9.1}$$

where $Z_1$ is the first objective, $j$ is the workstation number, $NWS$ is the number of workstations, $CT$ is cycle time (maximum time available at each workstation), and $WS_j$ is the elapsed time at workstation $j$.

### 9.4.3 The Second Objective

A line balancing procedure seeks to achieve *perfect balance* (all idle times equal to zero). When this is not achievable, either the line efficiency (LE) or the smoothness index (SI) is often used as a performance measure (Elsayed and Boucher, 1994). We use a measure of balance that combines the two and is easier to calculate. SI rewards similar idle times at each workstation, but at the expense of allowing for a large (suboptimal) number of workstations. This is because SI compares workstation elapsed times to the largest $WS_j$ instead of the *CT*, which rewards the minimum number of workstations, but allows unlimited variance in idle times between workstations because no comparison is made between $WS_j$s. Here we make use of the balancing function suggested by McGovern and Gupta (2003b). The McGovern-Gupta balancing function simultaneously minimizes the number of workstations while aggressively ensuring that idle times at all workstations are similar, though at the expense of the generation of a nonlinear objective function. The function is computed based on the minimum number of workstations required as well as the sum of the square of the idle times for all the workstations. This penalizes solutions where, even though the number of workstations may be minimized, one (or more) workstation has an exorbitant amount of idle time when compared to the other workstations. It provides for leveling the workload between different workstations on the disassembly line. Therefore, a resulting minimum performance value is the more desirable solution indicating both a minimum number of workstations and similar idle times across all workstations. The McGovern-Gupta balancing function is represented as

$$F = \sum_{j=1}^{NWS} (CT - WS_j)^2 \tag{9.2}$$

with the second objective (DLBP balancing objective) $Z_2$ represented as

$$\text{Minimize } Z_2 = \sum_{j=1}^{NWS} (CT - WS_j)^2 \tag{9.3}$$

A perfect balance is indicated by

$$Z_2 = 0 \tag{9.4}$$

Note that, mathematically, objective (9.3) effectively makes objective (9.1) redundant due to the fact that it concurrently minimizes the number of workstations.

In addition, we find

$$NWS_{\text{max}} = n \tag{9.5}$$

$$NWS_{\text{min}} = \left\lceil \frac{\sum_{k=1}^{n} PRT_k}{CT} \right\rceil \tag{9.6}$$

$$I = \sum_{j=1}^{NWS} (CT - WS_j) \tag{9.7}$$

where $NWS_{max}$ and $NWS_{min}$ is the maximum and minimum possible number of workstations, respectively, $n$ is the number of components (parts), $PRT_k$ is the component removal time for the $k$th component, $\lceil \alpha \rceil$ is the ceil operation that returns the smallest integer larger than or equal to $\alpha$, and $I$ is the total idle time for a given solution sequence.

### 9.4.4 The Third Objective

A hazard measure was developed to quantify each solution sequence performance, with a lower calculated value being more desirable (McGovern and Gupta, 2003a). This measure is based on binary flags $h_k$ that indicate whether a component is considered to contain hazardous material or not (the binary flag $h_k$ is equal to one if component $k$ is hazardous, else zero) and its position in the disassembly sequence. A given disassembly sequence's hazard measure $H$ is defined as the sum of multiplication of the hazard binary flag values and their components' positions in the disassembly sequence. This measure is represented as

$$H = \sum_{k=1}^{n} (k \cdot h_{PS_k}) \qquad h_{PS_k} = \begin{cases} 1, \text{hazardous} \\ 0, \text{otherwise} \end{cases} \tag{9.8}$$

where $PS_k$ is the $k$th component in a disassembly sequence (e.g., for the disassembly sequence "3, 1, 2," $PS_2 = 1$) with the third objective (DLBP hazardous component objective) $Z_3$ represented as

$$\text{Minimize } Z_3 = \sum_{k=1}^{n} (k \cdot h_{PS_k}) \tag{9.9}$$

The best-case hazard measure $H_{opt}$ is given by

$$H_{opt} = \sum_{p=1}^{NHP} p \tag{9.10}$$

where $NHP$ is the number of hazardous components in the product and $p$ here refers to the components' position in the disassembly sequence. For example, three hazardous components would give a best-case value of $1 + 2 + 3 = 6$. The worst-case hazard measure $H_{\text{nom}}$ is given by

$$H_{\text{nom}} = \sum_{p=n-NHP+1}^{n} p \tag{9.11}$$

or

$$H_{\text{nom}} = (n \cdot NHP) - NHP \tag{9.12}$$

For example, three hazardous components in a total of 20 would give an $H_{\text{nom}}$ value of 18 + 19 + 20 = 57 or equivalently, $H_{\text{nom}} = (20 \times 3) - 3 = 60 - 3 = 57$.

### 9.4.5 The Fourth Objective

A demand measure was developed to quantify each disassembly sequence performance, with a lower calculated value being more desirable (McGovern and Gupta, 2003a). This measure is based on positive integer values that indicate the quantity desired of this component (demand) $d_k$ and its position in the disassembly sequence. A given disassembly sequence's demand measure $D$ is defined as the sum of multiplication of the quantities desired and their components' positions in the disassembly sequence ($PS_k$). This measure is represented as

$$D = \sum_{k=1}^{n} (k \cdot d_{PS_k}) \qquad d_{PS_k} \in \mathrm{N}, \forall PS_k \tag{9.13}$$

where $N$ is the set of natural numbers, with the fourth objective (DLBP demand component objective) $Z_4$ represented as

$$\text{Minimize } Z_4 = \sum_{k=1}^{n} (k \cdot d_{PS_k}) \tag{9.14}$$

The best-case demand measure $D_{opt}$ is given by Expression (9.13) where

$$d_{PS_1} \geq d_{PS_2} \geq \cdots \geq d_{PS_n}$$

For example, three components with demands of 4, 5, and 6, respectively would give a best case value of $(1 \times 6) + (2 \times 5) + (3 \times 4) = 28$. The worst-case demand measure $D_{\text{nom}}$ is given by Expression (9.13) where

$$d_{PS_1} \leq d_{PS_2} \leq \cdots \leq d_{PS_n}$$

For example, three components with demands of 4, 5, and 6, respectively would give a worst case value of $(1 \times 4) + (2 \times 5) + (3 \times 6) = 32$.

## 9.5 HEURISTIC AND METAHEURISTIC METHODS FOR SOLVING THE DLBP

### 9.5.1 Introduction

As was mentioned in section 9.3, balancing a disassembly line is a complex problem that is challenging to solve. It was, therefore, argued that heuristic and metaheuristic techniques seem to be the only practical way to solve such problems. In this section, we discuss a sampling of heuristic and metaheuristic approaches to provide fast, good, or near-optimal solutions to the multi-objective, nonlinear DLBP described in the previous section.

The first approach rapidly provides a feasible solution to the DLBP and a minimum or near minimum number of workstations using a greedy strategy based on the first-fit decreasing (FFD) algorithm (see subsection 9.5.2). However, it does not always address the workstations balancing objective well. The second approach (the 2-Opt algorithm, see subsection 9.5.3) can be implemented, after the greedy approach, to compensate for its inability to balance the workstations. This local search quickly provides near-optimal balance sequence. The final approach, based on the ant colony optimization metaheuristic (ACO) provides a very fast, near-optimal solution to the multiple objective DLBP (see subsection 9.5.4). The DLBP ACO rapidly provides a feasible solution to the DLBP with a minimum or near-minimum number of workstations and a reasonably balanced sequence.

### 9.5.2 Greedy Model Description and the Algorithm

A greedy strategy always makes the choice that appears best at that moment. That is, it makes a locally optimal choice in the hope that it would lead to a globally optimal solution. Greedy algorithms do not always yield optimal solutions but for many problems, they do (Cormen et al., 2001). The DLBP greedy algorithm described here was built around the FFD algorithm (McGovern and Gupta, 2003a). The FFD algorithm looks at each element in a list, from largest to smallest component removal time (*PRT*) in the DLBP, and puts that element into the first workstation in which it fits. When all of the work elements have been assigned to a workstation, the process is complete. The greedy FFD is further modified with priority rules to meet multiple objectives. The most hazardous components are prioritized to the earliest workstations, and are greedy ranked, large removal time to small. The remaining nonhazardous components are greedy ranked next, large removal times to small. In addition, selecting the component with the larger demand ahead of those with lesser demands breaks any ties for components with equal component removal times. This is done to prevent damage to these more desirable components. Once the components are sorted in this multi-criteria manner, the components are placed in workstations in FFD greedy order while preserving precedence. Each component in the sorted list is examined from first to last. If the component had not previously been put into the solution sequence (as indicated by a separately maintained binary value tabu list $ISS_k$), the component is put into the current workstation if enough idle time remains to accommodate it. $ISS_k$ is defined as

$$ISS_k = \begin{cases} 1, & \text{if the } k\text{th part is in the solution sequence} \\ 0, & \text{otherwise} \end{cases} \tag{9.17}$$

If, at a given time in the search, a workstation cannot accommodate the component due to precedence constraints, the component is maintained on the sorted list (i.e., its $ISS_k$ value remains 0) and the next component (not yet selected) on the sorted list is considered. If all components have been examined for insertion into the current workstation on the greedy solution list, a new workstation is created and the process is repeated. The DLBP greedy procedure in pseudo-code format is as follows:

*Procedure DLBP greedy {*

1. $\forall PRT_k \mid 1 \leq k \leq n$, generate sorted component sequence, *PSS*, based upon: hazardous components (i.e., $h_k = 1$), large $PRT_k$ to small, then $h_k = 0$ components, large $PRT_k$ to small; if more than one component has both the same hazard rating and the same $PRT_k$, then sort by demand, large $d_k$ to small
2. $\forall p \in PSS \mid p = PSS_k,\ 1 \leq k \leq n$, generate greedy component removal sequence, *PSG*, by
   IF: $ISS_p = 1$ (i.e., component $p$ already included in *PSG*) $\vee$
   precedence constraints are not met $\vee$
   $I_j < PRT_p$ (i.e., component removal time exceeds idle time available at $WS_j$)
   THEN: IF: $p = n$ (i.e., at the last component in sequence)
   THEN: Increment $j$ (i.e., start a new workstation)
   Start again at $p = 1$
   ELSE: Try next $p$
   ELSE: Assign $p$ to $WS_j$ and set $ISS_p = 1$
3. Using *PSG*:
   Calculate *F*, *H*, and *D*

*}*

The DLBP greedy algorithm provides minimum or near-minimum number of workstations; the more the constraints, the more likely it is that the minimum number of workstations will be found. The level of performance tends to improve with the increase in the number of precedence constraints.

### Example 1

We will demonstrate the application of the DLBP greedy algorithm to provide a solution to a disassembly line balancing problem adapted from Gungor and Gupta (2002), where the objective is to completely disassemble a given product consisting of eight components ($n = 8$) on a disassembly line operating at a speed which allows 40 sec ($CT = 40$) for each workstation to perform its required disassembly tasks. This practical and relevant example consists of the data for the disassembly of a personal computer (PC) as shown in Table 9.2. The eight subassemblies have component removal times $PRT_k = \{14, 10, 12, 18, 23, 16, 20, 36\}$ seconds, binary hazard values $h_k = \{0, 0, 0, 0, 0, 0, 1, 0\}$, and component demands $d_k = \{360, 500, 620, 480, 540, 750, 295, 720\}$ units. The precedence relationships between the various

**TABLE 9.2**
**Input Data for the PC Example**

| Component | Component Description | Time | Hazardous | Demand |
|---|---|---|---|---|
| 1 | PC top cover | 14 | No | 360 |
| 2 | Floppy drive | 10 | No | 500 |
| 3 | Hard drive | 12 | No | 620 |
| 4 | Back plane | 18 | No | 480 |
| 5 | PCI cards | 23 | No | 540 |
| 6 | RAM modules (2) | 16 | No | 750 |
| 7 | Power supply | 20 | Yes | 295 |
| 8 | Motherboard | 36 | No | 720 |

components are captured in the following disassembly precedence matrix (see chapter 6 for a discussion):

$$DPM = \begin{array}{c} \\ 1 \\ 2 \\ 3 \\ 4 \\ 5 \\ 6 \\ 7 \\ 8 \end{array} \begin{array}{c} \begin{matrix} 1 & 2 & 3 & 4 & 5 & 6 & 7 & 8 \end{matrix} \\ \begin{bmatrix} 0 & 1 & 1 & 1 & 1 & 1 & 1 & 1 \\ 0 & 0 & 0 & 0 & 0 & x & 0 & 1 \\ 0 & 0 & 0 & 0 & 0 & -x & 0 & 1 \\ 0 & 0 & 0 & 0 & 0 & 0 & 0 & 0 \\ 0 & 0 & 0 & 1 & 0 & 0 & 0 & 1 \\ 0 & 0 & 0 & 0 & 0 & 0 & 0 & 1 \\ 0 & 0 & 0 & 1 & 0 & 0 & 0 & 0 \\ 0 & 0 & 0 & 0 & 0 & 0 & 1 & 0 \end{bmatrix} \end{array} \tag{9.18}$$

Thus the following precedence relationships hold: $1 \rightarrow 2$, $1 \rightarrow 3$, $1 \rightarrow 4$, $1 \rightarrow 5$, $1 \rightarrow 6$, $1 \rightarrow 7$, $1 \rightarrow 8$, $2 \rightarrow 8$, $3 \rightarrow 8$, (2 OR 3) $\rightarrow 6$, $5 \rightarrow 4$, $5 \rightarrow 8$, $6 \rightarrow 8$, $7 \rightarrow 4$, $8 \rightarrow 7$.

The DLBP greedy algorithm obtained an optimal solution for this problem (Table 9.3) and did so very quickly in spite of a large search space (potentially as large as 8! or 40,320). The speed for the C++ implemented program on this problem was less than 1/100th of a second on a 2.5 GHz P4 ×86 family computer. The DLBP greedy algorithm was able to generate a disassembly sequence that fulfilled the first objective, i.e., minimized the number of workstations. The remaining objectives (viz., balance the disassembly line, remove hazardous components early in the disassembly sequence, and remove high demand components before low demand components) were preempted by rigid enforcement of the precedence constraints.

While being very fast and generally very efficient, the DLBP greedy algorithm often lends itself to filling the earlier workstations as much as possible, whereas the downstream workstations end up with progressively greater idle times as can be seen from the following example.

**TABLE 9.3**
**DLBP Greedy Solution for the PC Example**

| Component Removal Sequence | Workstation 1 | 2 | 3 | 4 | |
|---|---|---|---|---|---|
| 1 | 14 | | | | Time to Remove Component (in seconds) |
| 5 | 23 | | | | |
| 3 | | 12 | | | |
| 6 | | 16 | | | |
| 2 | | 10 | | | |
| 8 | | | 36 | | |
| 7 | | | | 20 | |
| 4 | | | | 18 | |
| **Total Time** | 37 | 38 | 36 | 38 | |
| **Idle Time** | 3 | 2 | 4 | 2 | |

### Example 2

Consider a product that has to be completely disassembled on a disassembly line. The relevant data for this product is given in Table 9.4. In addition to the precedence relationships (which are given in the form of immediate successors), the product includes 10 subassemblies with component removal times of $PRT_k = \{14, 10, 12, 17, 23, 14, 19, 36, 14, 10\}$, binary hazard values of $h_k = \{0, 0, 0, 0, 0, 0, 1, 0, 0, 0\}$, and demands of $d_k = \{0, 500, 0, 0, 0, 750, 295, 0, 360, 0\}$. The disassembly line is operated at a speed that allows 40 sec for each workstation (i.e., $CT = 40$).

The precedence relationships, given in Table 9.4, can be interpreted as follows: (1 AND 8 AND 9 AND 10) $\rightarrow$2, (1 AND 8 AND 9 AND 10)$\rightarrow$ 3, (5 AND 6) $\rightarrow$7, (4 AND 7) $\rightarrow$8.

**TABLE 9.4**
**Input Data for Example 2**

| Component | Immediate Successor Component(s) | Time | Hazardous | Demand |
|---|---|---|---|---|
| 1 | 2,3 | 14 | No | None |
| 2 | — | 10 | No | 500 |
| 3 | — | 12 | No | None |
| 4 | 8 | 17 | No | None |
| 5 | 7 | 23 | No | None |
| 6 | 7 | 14 | No | 750 |
| 7 | 8 | 19 | Yes | 295 |
| 8 | 2,3 | 36 | No | None |
| 9 | 2,3 | 14 | No | 360 |
| 10 | 2,3 | 10 | No | None |

**TABLE 9.5**
**DLBP Greedy Solution for Example 2**

| Component Removal Sequence | Workstation 1 | 2 | 3 | 4 | 5 | |
|---|---|---|---|---|---|---|
| 5 | 23 | | | | | Time to Remove Component (in seconds) |
| 4 | 17 | | | | | |
| 6 | | 14 | | | | |
| 7 | | 19 | | | | |
| 8 | | | 36 | | | |
| 9 | | | | 14 | | |
| 1 | | | | 14 | | |
| 10 | | | | 10 | | |
| 3 | | | | | 12 | |
| 2 | | | | | 10 | |
| **Total Time** | 40 | 33 | 36 | 38 | 22 | |
| **Idle Time** | 0 | 7 | 4 | 2 | 18 | |

The greedy algorithm was able to successfully find a feasible solution (see Table 9.5) having the minimum number of workstations while placing the hazardous component (component number 7) and high demand components (components 6, 7, and 9) relatively early in the removal sequence (the exception being component 2, primarily due to precedence constraints and a smaller component removal time). However, the line is not balanced because the greedy algorithm has a tendency to fill the earlier workstations as much as possible. Using the follow on method presented in the next subsection one can mitigate the DLBP greedy algorithm's inability to address the balancing problem.

### 9.5.3 2-Opt Description and the Algorithm

A second phase for the multi-objective DLBP is implemented to compensate for the DLBP greedy algorithms' inability to balance the workstation assignments. This second phase quickly provides a near-optimal and feasible balance sequence using a tour improvement procedure. The best-known tour improvement procedures are the edge exchange procedures. In an $r$-opt algorithm, all exchanges of $r$ edges are tested until there is no feasible exchange that improves the current solution; this solution is said to be $r$-optimal. Since the number of iterations increases rapidly with increases in $r$, $r = 2$ and $r = 3$ are most commonly used. In addition, a greedy solution is generally considered the best starting point for $r$-opt (Lawler et al., 1985). This section describes a 2-Opt local search algorithm for the DLBP. In a 2-Opt algorithm, two tours are neighbors if one can be obtained from the other by deleting two edges, reversing one of the resulting two paths, and reconnecting. In the DLBP, the 2-Opt neighbors of the tour are pair-wise component exchanges within the current solution sequence. For example, with $n = 4$ and an original solution sequence of {1, 2, 3, 4}, the six resulting pair-wise exchanges are {2, 1, 3, 4}, {3, 2, 1, 4}, {4, 2, 3, 1}, {1, 3, 2, 4}, {1, 4, 3, 2}, and {1, 2, 4, 3}. Exhaustive search visits all $n!$ tours, while the total number of tours visited by an iteration of DLBP 2-Opt, including the original data set, is given by the following expression

$$Tours = 1 + \sum_{p=1}^{n-1} (n - p) \tag{9.19}$$

Thus, for $n = 4$, this results in 1 + 3 + 2 + 1 = 7 tours.

Typically, a random starting tour is generated for the initial application of the 2-Opt algorithm. In this section, the starting point is rapidly attained by the previously discussed DLBP greedy algorithm, which provides a minimum *NWS* (number of workstations), feasible solution. This near-optimal starting point requires the DLBP 2-Opt only to improve the balance measure. The DLBP 2-Opt tries switching the position of any two components in the current solution sequence to get the best tour in this neighborhood. It then repeats this process using this new tour (the best previous solution sequence) until no better solution can be found. The DLBP 2-Opt is designed to swap every task with each task in every subsequent workstation in search of improved balance. It does this while preserving precedence and not exceeding *CT* in any workstation. It also seeks to a) improve hazard performance, or b) maintain hazard performance while improving demand performance. In addition, each DLBP 2-Opt neighbor under consideration recalculates the minimum number of workstations required for a given problem set, enabling the determination of more efficient component removal sequences that may exist in the 2-Opt neighborhood, which did not exist in the DLBP greedy solution. The DLBP 2-Opt algorithm seeks to improve the balance but not at the expense of the hazard measure or a precedence violation. As long as the balance is at least maintained, the DLBP 2-Opt then seeks improvements in hazard measure and demand measure but never at the expense of precedence constraints. The heuristic is given below in pseudo-code format.

*Procedure DLBP 2-Opt {*

1. Initialize:
   Set $PST = PSG$, set permanent sequence equal to the solution sequence from greedy algorithm
   Set $TMP = PSG$, set temporary sequence equal to the solution sequence from greedy algorithm
   Set $BST = PSG$, set permanent best sequence equal to the solution sequence from greedy algorithm
2. $\forall p \in PST \mid p = PST_k$, $1 \le k \le n$, attempt to generate better balanced 2-Opt component removal sequence *PST* by:
   $\forall q \in TMP \mid p + 1 \le q \le n$, perform 2-Opt component swap (i.e., swap $TMP_p$ with every $TMP_q$)
   Calculate new *F*, *H*, and *D*
   IF: (new *F* is better than current best *F* $\wedge$
   new *H* is better than or equal to current best *H*)
   $\vee$

(new $F$ = current best $F\wedge$
new $H$ is better than or equal to current best $H\wedge$
new $D$ is better than or equal to current best $D$)
$\wedge$
precedence constraints are met
THEN: Set $BST = TMP$

3. Save results:
   Set $PST = BST$
   Set $TMP = BST$
4. Repeat STEP 2 until no more improvements in $F$, $H$, or $D$ can be made

}

The DLBP greedy algorithm is run once to determine a solution. 2-Opt, howver, like many combinatorial optimization techniques, is continuously run on subsequent solutions for as long as is deemed appropriate or acceptable by the user or until it is no longer possible to improve, at which point it is assumed that the (local) optima has been reached. Repeating the DLBP 2-Opt method in this way provides improved balance over time.

### Example 3

Consider the product that was examined in Example 2. The relevant data for the product was given in Table 9.4 and the DLBP greedy solution in Table 9.5. Using the solution in Table 9.5 as the starting point, the solution obtained using the DLBP 2-Opt method is given in Table 9.6.

The DLBP 2-Opt was able to dramatically improve the overall balance of the line by 46.3% and the demand measure by 8.1% while maintaining the hazard measure (see Table 9.6). The speed for the C++ implemented program on this problem was less

**TABLE 9.6**
**DLBP 2-Opt Solution for Example 3**

| Component Removal Sequence | Workstation 1 | 2 | 3 | 4 | 5 | |
|---|---|---|---|---|---|---|
| 10 | 10 | | | | | Time to Remove Component (in seconds) |
| 5 | 23 | | | | | |
| 6 | | 14 | | | | |
| 7 | | 19 | | | | |
| 9 | | | 14 | | | |
| 4 | | | 17 | | | |
| 8 | | | | 36 | | |
| 1 | | | | | 14 | |
| 2 | | | | | 10 | |
| 3 | | | | | 12 | |
| **Total Time** | 33 | 33 | 31 | 36 | 36 | |
| **Idle Time** | 7 | 7 | 9 | 4 | 4 | |

than 1/100th of a second on a 2.5GHz P4 ×86 family workstation while an equivalent size DLBP optimally solved by an exhaustive search on the same workstation took almost 10 sec (almost a thousand times as long!)

The first two objectives (minimize the number of workstations and balance the disassembly line) were achieved by the DLBP Greedy/2-Opt. The next objective (remove hazardous components early in the disassembly sequence) was obtained to the greatest extent possible by the DLBP Greedy alone while the DLBP 2-Opt improved upon the last objective (remove high demand components before low demand components) without adversely affecting the hazard measure. The DLBP Greedy/2-Opt technique rapidly found near-optimal solution in an exponentially large search space (potentially as large as 10! or 3,628,800).

### 9.5.4 Ant Colony Optimization Metaheuristic Model Description

An ant-cycle model based ant colony optimization metaheuristic (ACO) is a probabilistic evolutionary algorithm based on a distributed autocatalytic process that makes use of agents called ants (due to these agents' similar attributes to insects). Just as a colony of ants can find the shortest distance to a food source, these ACO agents work cooperatively toward an optimal solution (see, for example, Onwubolu, 2002). Multiple agents are placed at multiple starting nodes, such as cities for the traveling salesman problem or components for the DLBP. Each of the $m$ ants is allowed to visit all remaining (unvisited) edges (i.e., locations) as indicated by a tabu type list. Each ant's possible subsequent steps (i.e., from any node $p$ to a node $q$ giving edge $pq$) are evaluated for desirability and each is assigned a proportionate probability as shown in Expression 9.20 below. Based on these probabilities, the next step in the tour is randomly selected for each ant. After completing an entire tour, the ant with the best solution is given the equivalent of additional pheromone (proportionate to how desirable the tour is and referred to as trail $\tau_{pq}$), which is added to each step it has taken. All paths are also decreased in their pheromone strength according to a measure of evaporation $(1-\rho)$. This process is repeated for a maximum designated number of cycles or until stagnation behavior (i.e., where all ants make the same tour) is demonstrated.

The probability of ant $r$ taking an edge $pq$ at time $t$ is given by

$$p_{pq}^{r}(t) = \begin{cases} \dfrac{[\tau_{pq}(t)]^{\alpha} \cdot [\eta_{pq}]^{\beta}}{\sum\limits_{r \in \text{allowed}_r} [\tau_{pq}(t)]^{\alpha} \cdot [\eta_{pr}]^{\beta}} & q \in \text{allowed}_r \\ 0 & \text{otherwise} \end{cases} \tag{9.20}$$

where $p_{pq}^{r}(t)$ is the probability of ant $r$ taking an edge $pq$ at time $t$, $\tau_{pq}(t)$ is the amount of trail on edge $pq$ at time $t$, $\eta_{pq}$ is the ACO visibility value of edge $pq$, $\alpha$ is the weight of existing pheromone (trail) in path selection, and $\beta$ is the weight of a given edge in path selection.

Trail for all edges is calculated using

$$\tau_{pq}(t+n) = \rho \cdot \tau_{pq}(t) + \Delta\tau_{pq} \tag{9.21}$$

where $\rho$ is the coefficient such that $1-\rho$ represents the pheromone evaporation rate and

$$\Delta\tau_{pq} = \sum_{r=1}^{m} \Delta\tau_{pq}^{r} \tag{9.22}$$

and

$$\Delta\tau_{pq}^{r} = \begin{cases} \dfrac{Q}{L_r} & \text{edge}(p,q)\text{used} \\ 0 & \text{otherwise} \end{cases} \tag{9.23}$$

where $Q$ is the amount of pheromone added if a path is selected and $L_r$ is the ACO delta trail divisor value, here set equal to $F_{nr}$ (balance of ant $r$ sequence at time $n$).

#### 9.5.4.1 DLBP-Specific Qualitative Modifications

The DLBP ACO is a modified ant-cycle algorithm. It is designed around the DLBP problem by accounting for feasibility constraints and provides for multiple objectives (Gupta and McGovern, 2004). We know that, in the DLBP, a solution consists of a sequence of work elements (sometimes also referred to as tasks, components or parts). For example, if a sequence under consideration in a cycle consisted of "5 2 8 1 4 7 6 3," then component 5 would be removed first, followed by component 2, then component 8, and so on. In DLBP ACO, each component is a node on the tour with the number of ants being set equal to the number of components and having one ant uniquely on each component as the starting position. Each ant is allowed to visit all components not already in the solution. Each ant's possible subsequent steps are evaluated for feasibility and the McGovern-Gupta measure of balance then assigned a proportionate probability. Infeasible steps receive a probability of zero and those ants are effectively ignored for the remainder of the cycle. (In the sequence example above, at time $t = 3$, component $p$ would represent component 8 while component $q \in \{1, 4, 7, 6, 3\}$ and possible partial solution sequences "5 2 8 1," "5 2 8 4," "5 2 8 7," "5 2 8 6" and "5 2 8 3" would be evaluated for feasibility and balance.) Based on these probabilities, the next component is randomly selected for each ant. At tour completion, the ant with the best balanced feasible component removal solution adds additional trail to each step it has taken. All paths are decreased in their pheromone strength by the evaporation value. This best solution found is saved and the process is repeated for a maximum designated number of cycles; it does not terminate for stagnation to allow for potential better solutions (due to the ACOs' stochastic edge selection capabilities). The best solution found is evaluated primarily on its measure of balance, which addresses minimizing the number of workstations and minimizing the variance in idle time between workstations. Other objectives include removing hazardous components and high demand components as soon as possible in the disassembly process. DLBP ACO selects the best performing measure of balance solution as given by Expression 9.2. Equal balance solutions are then evaluated for hazardous component removal positions per Expression 9.8 and equal balance and hazard measure solutions are evaluated for high demand component removal positions as measured by Expression 9.13. The DLBP ACO seeks to preserve precedence while not exceeding $CT$ in any workstation. As long as the balance is at least maintained, the DLBP ACO then seeks improvements in hazard measure and demand measure but never at the expense of precedence constraints. For DLBP ACO, the visibility $\eta_{pq}$ was defined as

$$\eta_{pq} = \frac{1}{F_{tr}} \tag{9.24}$$

where $F_{tr}$ is the balance of ant $r$ at time $t$ (i.e., ant $r$'s balance thus far in its incomplete solution sequence generation). The divisor for the change in trail is defined for DLBP ACO as

$$L_r = F_{nr} \tag{9.25}$$

where a small final value for ant $r$'s measure of balance at time $n$ (i.e., at the end of the tour) provides a large measure of trail added to each edge. Though $L_r$ and $\eta_{pq}$ are related in this application, this is not unusual for ACO applications in general; for example, in the traveling salesman problem, $L_r$ is the tour length while $\eta_{pq}$ is the reciprocal of the distance between cities $p$ and $q$. However, this method of selecting $\eta_{pq}$ (effectively a short-term greedy choice) may not always translate into the best long-term (i.e., final tour) solution for a complex problem like the DLBP.

#### 9.5.4.2 DLBP-Specific Quantitative Values and the DLBP ACO Algorithm

In DLBP ACO, the maximum number of cycles is normally set at $NC_{\max} = 300$ since even larger problems than those presented in this section have been observed (via experimentation) to reach their best solution by that count. The process was not run until no improvements were shown but, as is the norm with many combinatorial optimization techniques, was run continuously on subsequent solutions until $NC_{\max}$ (Hopgood, 1993). This also enabled the probabilistic component of ACO an opportunity to leave a potential local minimum. Repeating the DLBP ACO method in this way provides improved balance over time. Per the best ACO metaheuristic performance experimentally determined by Dorigo et al. (1996), the weight of pheromone in path selection $\alpha$ was set equal to 1.00; the weight of balance in path selection $\beta$ set equal to 5.00; the evaporation rate $1 - \rho$ was set to 0.50; and the amount of pheromone added if a path is selected $Q$ was set to 100.00. The initial amount of pheromone on all of the paths $c$ was set equal to 1.00. The following is the DLBP ACO procedure.

*Procedure DLBP ACO {*

1. Initialize:
   Set $t = 0$ {$t$ is the time counter}
   Set $NC = 0$ {$NC$ is the cycle counter}
   For every edge $(p, q)$ set an initial value $\tau_{pq}(t) = c$ for trail intensity and $\Delta\tau_{pq} = 0$
   Place the $m$ ants on the $n$ nodes
2. Set $s = 1$ {$s$ is the tabu list index}
   For $r = 1$ to $m$ do
   Place the starting component of the $r$-th ant in $\mathbf{tabu}_r(s)$
3. Repeat until tabu list is full {this step will be repeated $(n - 1)$ times}
   Set $s = s + 1$
   For $r = 1$ to $m$ do
   Choose the component $q$ to move to, with probability $p^r_{pq}(t)$ given by expression (9.20)
   {at time $t$ the $r$-th ant is on component $p = \mathbf{tabu}_r(s - 1)$}
   Move the $r$-th ant to the component $q$
   Insert component $q$ in $\mathbf{tabu}_r(s)$
4. For $r = 1$ to $m$ do
   Move the $r$-th ant from $\mathbf{tabu}_r(n)$ to $\mathbf{tabu}_r(1)$
   Compute $F$, $H$, and $D$ for the sequence described by the $r$-th ant
   Save the best $F$, or for equal $F$, the best $H$, or for equal $F$ and equal $H$, the best $D$ found
   For every edge $(p, q)$
   For $r = 1$ to $m$ do

$$\Delta\tau^r_{pq} = \begin{cases} \dfrac{Q}{L_r} & edge(p,q)used \\ 0 & otherwise \end{cases}$$

$$\Delta\tau_{pq} = \Delta\tau_{pq} + \Delta\tau^r_{pq}$$

5. For every edge $(p,q)$ compute $\tau_{pq}(t+n)$ according to $\tau_{pq}(t+n) = \rho \cdot \tau_{pq}(t) + \Delta\tau_{pq}$
   Set $t = t + n$
   Set $NC = NC + 1$
   For every edge $(p, q)$ set $\Delta\tau_{pq} = 0$
6. If $(NC < NC_{\max})$ and (not stagnation behavior)
   then
   Empty all tabu lists
   Goto step 2
   else
   Print best balance
   Stop

*}*

### Example 4

Consider the PC that was examined in example 1. Most of the relevant data for the disassembly of the PC were given in Table 9.2 and Expression 9.18. The solutions obtained using the DLBP ACO algorithm are given in Table 9.7.

Due to its probabilistic component, over multiple runs, the DLBP ACO obtained *all four optimal solutions* for this problem (Table 9.7abcd) in what approaches an exponentially large search space (potentially as large as 8! or 40,320) and did so very quickly. For the same reason, it was, on rare occasions, seen to select a suboptimally balanced solution. It also did not find the optimal solutions with the same frequency, finding the solution in Table 9.7a most often, followed by 9.7b, then 9.7c, then 9.7d. This can be attributed to Expression 9.24 and the ACOs' requirement to evaluate partial solutions before making a solution element selection decision, as well as the higher demand performance of the solution in Table 9.7a. Note the larger role played by Expression

**TABLE 9.7**
**DLBP ACO Algorithm Solutions for Example 4**

(a)

| Component Removal Sequence | Workstation 1 | 2 | 3 | 4 |
|---|---|---|---|---|
| | Time to Remove Component (in seconds) | | | |
| 1 | 14 | | | |
| 5 | 23 | | | |
| 3 | | 12 | | |
| 6 | | 16 | | |
| 2 | | 10 | | |
| 8 | | | 36 | |
| 7 | | | | 20 |
| 4 | | | | 18 |
| **Total Time** | 37 | 38 | 36 | 38 |
| **Idle Time** | 3 | 2 | 4 | 2 |

(b)

| Component Removal Sequence | Workstation 1 | 2 | 3 | 4 |
|---|---|---|---|---|
| | Time to Remove Component (in seconds) | | | |
| 1 | 14 | | | |
| 5 | 23 | | | |
| 3 | | 12 | | |
| 2 | | 10 | | |
| 6 | | 16 | | |
| 8 | | | 36 | |
| 7 | | | | 20 |
| 4 | | | | 18 |
| **Total Time** | 37 | 38 | 36 | 38 |
| **Idle Time** | 3 | 2 | 4 | 2 |

(c)

| Component Removal Sequence | Workstation 1 | 2 | 3 | 4 | Time to Remove Component (in seconds) |
|---|---|---|---|---|---|
| 1 | 14 | | | | |
| 5 | 23 | | | | |
| 2 | | 10 | | | |
| 6 | | 16 | | | |
| 3 | | 12 | | | |
| 8 | | | 36 | | |
| 7 | | | | 20 | |
| 4 | | | | 18 | |
| **Total Time** | 37 | 38 | 36 | 38 | |
| **Idle Time** | 3 | 2 | 4 | 2 | |

(d)

| Component Removal Sequence | Workstation 1 | 2 | 3 | 4 | Time to Remove Component (in seconds) |
|---|---|---|---|---|---|
| 1 | 14 | | | | |
| 5 | 23 | | | | |
| 2 | | 10 | | | |
| 3 | | 12 | | | |
| 6 | | 16 | | | |
| 8 | | | 36 | | |
| 7 | | | | 20 | |
| 4 | | | | 18 | |
| **Total Time** | 37 | 38 | 36 | 38 | |
| **Idle Time** | 3 | 2 | 4 | 2 | |

9.24. Although the Table 9.7a solution has the highest demand performance (with balance and hazard measures equal to the remaining solutions), the Table 9.7c solution has a slightly better demand performance than the more often found Table 9.7b solution. The speed for the C++ implemented program on this problem was less than 1/10th of a second (averaging 0.04 seconds per run) on a 1.6 GHz PM $\times 86$ family workstation.

## EXAMPLE 5

Consider the product that was examined in example 2. Most of the relevant data for the disassembly of the product were given in Table 9.4. The solutions obtained using the DLBP ACO method are given in Table 9.8.

Unlike the DLBP greedy algorithm, DLBP ACO method typically was able to get the optimal number of workstations, the best balance, and the best hazardous component removal measure. However, the demand performance was not as good as 2-Opt.

**TABLE 9.8**
**DLBP ACO Algorithm Solutions for Example 5**

(a)

| Component Removal Sequence | Workstation 1 | 2 | 3 | 4 | 5 | |
|---|---|---|---|---|---|---|
| 10 | 10 | | | | | Time to Remove Component (in seconds) |
| 5 | 23 | | | | | |
| 4 | | 17 | | | | |
| 6 | | 14 | | | | |
| 7 | | | 19 | | | |
| 9 | | | 14 | | | |
| 8 | | | | 36 | | |
| 1 | | | | | 14 | |
| 3 | | | | | 12 | |
| 2 | | | | | 10 | |
| **Total Time** | 33 | 31 | 33 | 36 | 36 | |
| **Idle Time** | 7 | 9 | 7 | 4 | 4 | |

(b)

| Component Removal Sequence | Workstation 1 | 2 | 3 | 4 | 5 | |
|---|---|---|---|---|---|---|
| 10 | 10 | | | | | Time to Remove Component (in seconds) |
| 5 | 23 | | | | | |
| 4 | | 17 | | | | |
| 6 | | 14 | | | | |
| 7 | | | 19 | | | |
| 1 | | | 14 | | | |
| 8 | | | | 36 | | |
| 9 | | | | | 14 | |
| 3 | | | | | 12 | |
| 2 | | | | | 10 | |
| **Total Time** | 33 | 31 | 33 | 36 | 36 | |
| **Idle Time** | 7 | 9 | 7 | 4 | 4 | |

For example, even though component 2 and 3 components have equal precedence and hazard measures, if they are in the same workstation, which they are, component 2 should be removed before component 3 since component 2 has the higher demand. However, DLBP ACO does not discover this because ACO does not calculate an entire sequence prior to evaluation, but rather it builds an answer and evaluates it step by step. Since component 3 has the higher component removal time (which is a higher priority than demand), ACO always (suboptimally in demand) puts component 3 in the last workstation before component 2.

## 9.6 CONCLUSION

As has been emphasized throughout this book, disassembly is widely accepted as an important element of product recovery because it allows the separation of desired components and materials from the end-of-life products. It is anticipated that in the foreseeable future, as more and more techniques are developed, disassembly will become more cost effective compared to other alternatives in product recovery especially with the changing regulations and higher penalty costs associated with the waste management and the use of virgin material resources. Therefore, designing efficient disassembly lines is important to automate and optimize the product recovery process. Motivated by this thought, in this chapter, we discussed the complications that are likely to arise on a disassembly line. We also discussed the difficulties in balancing disassembly lines and introduced a sampling of heuristic and metaheuristic approaches to provide fast, good, or near-optimal solutions to the multi-objective, nonlinear disassembly line balancing problems.

The subject of disassembly lines and associated issues is still in its infancy. However, research activity has started to surface in this field. Here we mention some recent studies reported by Tang et al. (2001), Altekin et al. (2003) and Udomsawat et al. (2003).

## REFERENCES


Altekin, F. T., Kandiller, L., and Ozdemirel, N. E., 2003, Disassembly line balancing with limited supply and subassembly availability, *Proceedings of the SPIE International Conference on Environmentally Conscious Manufacturing III*, Providence, Rhode Island, 59–70, October 29–30.

Brennan, L., Gupta, S.M., and Taleb, K.N., 1994, Operations planning issues in an assembly/disassembly environment. *International Journal of Operations and Production Management*, **14**(9), 57–67.

Cormen, T., Leiserson, C., Rivest, R. and Stein, C., 2001, *Introduction to Algorithms*. The MIT Press, Cambridge, MA.

Donnan, D.C. and Makan, K., 1983, Disassembly line balancing. *MTM Journal of Methods, Time Measurement*, 10(2), 20–27.

Dorigo, M., Maniezzo, V., and Colorni, A., 1996, The ant system: optimization by a colony of cooperating *agents, IEEE Transactions on Systems, Man, and Cybernetics–Part B*, **26**(1), 1–13.

Elsayed, E.A. and Boucher, T.O., 1994, *Analysis and Control of Production Systems*. Second Edition, Prentice Hall, Upper Saddle River, NJ.

Ghosh, S. and Gagnon, R.J., 1989, A comprehensive literature review and analysis of the design, balancing and scheduling of assembly systems. *International Journal of Production Research*, **27**(4), 637–670.

Greenberg, H.J., 2004, *Mathematical Programming Glossary*. Available on Internet via: *http://www.cudenver.edu/~hgreenbe/glossary/*

Gungor, A. and Gupta, S.M., 1999a, Disassembly line balancing. *Proceedings of the 1999 Annual Meeting of the Northeast Decision Sciences Institute*, 193–195.

Gungor, A. and Gupta, S.M., 1999b, A Systematic solution approach to the disassembly line balancing problem. *Proceedings of the 25th International Conference on Computers and Industrial Engineering*, 70–73.

Gungor, A. and Gupta, S.M., 2001, A solution approach to the disassembly line balancing problem in the presence of task failures. *International Journal of Production Research*, **39**(7), 1427–1467.

Gungor, A. and Gupta, S.M., 2002, Disassembly line in product recovery. *International Journal of Production Research*, **40**(11), 2569–2589.

Gupta, S.M. and McGovern, S. M., 2004, Multi-objective optimization in disassembly sequencing problems, *Proceedings of the POMS-Cancun Meeting*, Cancun, Mexico, (CD-ROM).

Hopgood, A.A., 1993, *Knowledge-Based Systems for Engineers and Scientists*, CRC Press, Boca Raton, FL.

Lawler, E.L., Lenstra, J.K., Rinnooy Kan, A.H.G., and Shmoys, D.B. 1985, *The Traveling Salesman Problem: a Guided Tour of Combinatorial Optimization*, John Wiley and Sons, New York, NY.

McGovern, S.M. and Gupta, S.M., 2003a, 2-Opt heuristic for the disassembly line balancing problem. *Proceedings of the SPIE International Conference on Environmentally Conscious Manufacturing III*, 71–84.

McGovern, S.M. and Gupta, S.M., 2003b, Greedy algorithm for disassembly line scheduling. *Proceedings of the 2003 IEEE International Conference on Systems, Man, and Cybernetics*, October 5–8, Washington, D.C., 1737–1744.

Onwubolu, G.C., 2002, *Emerging Optimization Techniques in Production Planning and Control*, Imperial College Press, London.

Silver, E.A., 2004, An overview of heuristic solution methods. *Journal of Operational Research Society*, 55 (9), 936–956.

Tang, Y., Zhou, M., and Caudill, R. 2001, A systematic approach to disassembly line design. *Proceedings of the IEEE International Symposium on Electronics and the Environment*, Denver, Colorado, May 7–9, 173–178.

Tovey, C.A., 2002, Tutorial on computational complexity. *Interfaces*, **32**(3), 30–61.

Udomsawat, G., Gupta, S.M., and Al-Turki, Y.A.Y., 2003, Multi-kanban model for disassembly line with demand fluctuation. *Proceedings of the SPIE International Conference on Environmentally Conscious Manufacturing III*, Providence, Rhode Island, 85–93, October 29–30.

# Author Index

## D



## E



## F



## G



## H

## X



## Y



## Z

# Subject Index

## C

## D

## E

## P

## Q



## R

## S

## T



## U



## V

## W



## X



## Y



## Z